Applying Bio-Measuremer
in Science Education Rese

Iztok Devetak · Saša Aleksij Glažar
Editors

Applying Bio-Measurements Methodologies in Science Education Research

Springer

Editors
Iztok Devetak
Faculty of Education
University of Ljubljana
Ljubljana, Slovenia

Saša Aleksij Glažar
Faculty of Education
University of Ljubljana
Ljubljana, Slovenia

ISBN 978-3-030-71537-3   ISBN 978-3-030-71535-9 (eBook)
https://doi.org/10.1007/978-3-030-71535-9

© Springer Nature Switzerland AG 2021
This work is subject to copyright. All rights are reserved by the Publisher, whether the whole or part of the material is concerned, specifically the rights of translation, reprinting, reuse of illustrations, recitation, broadcasting, reproduction on microfilms or in any other physical way, and transmission or information storage and retrieval, electronic adaptation, computer software, or by similar or dissimilar methodology now known or hereafter developed.
The use of general descriptive names, registered names, trademarks, service marks, etc. in this publication does not imply, even in the absence of a specific statement, that such names are exempt from the relevant protective laws and regulations and therefore free for general use.
The publisher, the authors and the editors are safe to assume that the advice and information in this book are believed to be true and accurate at the date of publication. Neither the publisher nor the authors or the editors give a warranty, expressed or implied, with respect to the material contained herein or for any errors or omissions that may have been made. The publisher remains neutral with regard to jurisdictional claims in published maps and institutional affiliations.

This Springer imprint is published by the registered company Springer Nature Switzerland AG
The registered company address is: Gewerbestrasse 11, 6330 Cham, Switzerland

Preface

This peer-reviewed monograph includes several chapters presenting the application of specific bio-measurement methods that can be used in education, with the emphasis on eye-tracking technology. By examining students' verbal or written answers, teachers usually do not have insight into how the students solved specific problems and how they arrived at the right or wrong solution; this means that with methods used in physiology, data can be obtained by scanning the central nervous system (brain): for example, functional magnetic resonance imaging (fMRI), which depicts the increased blood flow through the brain during a specific activity and indicates the increased activity of this part of the brain, or electroencephalography (EEG), which measures the electrical activity of the brain during a specific activity. The solving of mental tasks is accompanied by physiological responses. These not only indicate a stress situation but are supposedly related to the cognitive load during the problem-solving process. For this reason, another set of devices collects data from specific parts of the human body's activities, which have their origin in the dynamics of the autonomic nervous system, such as measurements of heart rate, heart rate variability, blood pressure, skin conductance, skin temperature, facial thermoscan, respiratory rate and amplitude, pupil dilatation, and eye movements.

The primary objective of science education research is to understand and improve the learning and teaching of science and to develop the scientific literacy of citizens. In addition, the understanding of learning scientific concepts is an essential issue so that teaching can be tailored to the students' needs and efficient learning can take place. In this context, research can use various methods of collecting learning data, which are relatively more objective than observations, interviews, questionnaires, and other methods. However, careful consideration should be given to the interpretation of the psychophysiological data collected. All contributions in this book attempt to follow this goal. Most chapters use the eye-tracking method, which enables following the focus of the students' attention and drawing conclusions about the strategies they used to solve the problem. Often several eye movement characteristics are used, such as the location of fixations, the total number of fixations, the proportion of the total duration of fixation and the duration of fixation in specific areas of interest, the number of revisits to those areas, the number of blinks, the fixation sequences, and the pupil size. These characteristics indicate the goal of the student's attention, which reflects the amount of cognitive resources devoted to information processing and problem-solving strategies. This book presents studies that measure eye movements while students solve authentic scientific problems. Each chapter also discusses the practical implications for education.

This book consists of a total of fifteen chapters. Authors from eight countries emphasise the same trends despite their cultural and educational differences. Regarding content, the book begins with general chapters describing cognitive processes and how these processes are measured using eye-tracking methods and other psychophysiology parameters and motivation. This is followed by chapters presenting studies in specific scientific fields from chemistry, biology, physics, and geology.

The first chapter, by Anja Podlesek, Manja Veldin, Cirila Peklaj, and Matija Svetina, entitled "Cognitive Processes and Eye-Tracking Methodology", discusses psychological aspects of eye-tracking approaches to cognition research with a focus on cases of science education.

The second chapter, titled "The Interplay of Motivation and Cognition: Challenges for Science Education Research and Practice", by Mojca Juriševič and Tanja Črne, examines two internal variables—students' motivation and visual attention—and how they influence learning processes.

This is followed by the chapter "Predicting Task Difficulty Through Psychophysiology" by Junoš Lukan and Gregor Geršak, who argue that mental task-solving is accompanied by physiological reactions. These not only indicate a stressful situation but are also associated with cognitive load. They measure various physiological parameters, such as heart rate, heart rate variability, respiratory rate, electrodermal activity, and skin temperature, and attempt to explain a considerable proportion of the difficulty variance.

In the following four chapters, research work from the field of chemistry is presented.

In the fourth chapter, entitled "The Role of the Explanatory Key in Solving Tasks Based on Submicroscopic Representations", authors Vesna Ferk Savec and Špela Hrast analyse Slovenian chemistry textbooks. They focused on the submicroscopic representations integrated into the textbooks to illustrate the particle nature of chemical concepts and processes and to facilitate learning chemistry with understanding. To support the students' learning of chemistry in a meaningful way, submicroscopic representations must be properly understood by the students, and the explanatory key accompanying them can play an essential role in this. The role of the explanatory key in the processing of submicroscopic representations in solving chemistry tasks was investigated. Particular emphasis was placed on the use of pictorial and textual explanatory keys accompanying the submicroscopic representations, which was investigated by using eye-tracking and interviews with the students.

Sevil Akaygun and Emine Adadan, in the chapter "Investigating the Role of Conceptual Understanding on How Students Watch an Experimental Video Using Eye-Tracking", explore how eye-tracking technology can be used to explore the role of students' conceptual understanding of how they navigate tasks to be solved while watching an experimental video of a redox reaction. Eye-tracking technology can be used to investigate how different levels of understanding can induce students to follow different aspects in an experimental video.

In the sixth chapter, entitled "Using an Eye-Tracker to Study Students' Attention Allocation when Solving a Context-Based Problem on the Sublimation of Water", Miha Slapničar, Valerija Tompa, Iztok Devetak, Saša Aleksij Glažar, and Jerneja Pavlin present the importance of 3D-dynamic submicroscopic representations for solving specific chemical tasks in context. The students' attention allocation in solving context-based tasks involving macroscopic and submicroscopic levels of sublimation of water representations was investigated. The research objective was to identify differences between successful and unsuccessful students in overall fixation duration, visit count, and average pupil size. The research results provide an insight into the learning process, especially the information processing of 3D-dynamic submicroscopic representations.

In the seventh chapter, entitled "Using an Eye-Tracking Approach to Explain Students' Achievements in Solving a Task about Combustion by Applying the Chemistry Triplet", by Iztok Devetak, the chemistry triplet (macro-, submicro-, and symbolic levels of chemical concepts representations) is examined as an essential part of teaching and learning chemistry. It is crucial to understand how students cognitively move between these representations when solving specific context-based tasks or problems. The chemical reaction is one of the fundamental concepts in chemistry teaching; as one of the specific examples, burning is often used to illustrate chemical changes. The research problem presented in this chapter relates to students' performance in solving chemistry triplet and context-based natural gas combustion exercises. The results show that students mainly use macroscopic explanations of chemical phenomena, but those students who chose the chemical equation correctly were more successful. The total fixation duration (TFDs) and fixation counts (FCs) of the students' gaze on the 3D-SMR were significantly lower for those students who chose the chemical equation correctly than for those who did so incorrectly. The significance of this research is in understanding how important the different levels of the chemistry triplet are to students in solving specific problems and how teachers can predict which levels should be more emphasised in chemistry lessons, depending on the level of chemical knowledge and skills of other students.

In the next two chapters, studies from the fields of biology and medical education are presented.

Tanja Gregorčič and Gregor Torkar, in their chapter entitled "Pre-Service Teachers' Determination of Butterflies with Identification Key: Studying Their Eye Movements", study preservice teachers' ability to determine butterfly species with a simplified dichotomous identification key containing illustrations, photographs, and written description with the eye-tracking method.

The ninth chapter, by Ilona Södervik and Henna Vilppu, entitled "Case Processing in the Development of Expertise in Life Sciences—What Can Eye Movements Reveal?", presents future experts in the life sciences with innovative forms of reasoning, and the ability to use knowledge and skills adaptively in unforeseen and adverse contexts. The authors synthesise two studies using the eye-tracking method, in which routine and non-routine text-based case tasks were used to investigate processing and problem-solving by medical personnel with different levels and types of expertise. Eye-tracking provides interesting insights into knowledge integration and problem-solving through medical case processing.

The next series of chapters includes studies that focus on physics concepts.

The tenth chapter, by Miroslawa Sajka and Roma Rosiek, entitled "Analysis of Aspects of Visual Attention when Solving Multiple-Choice Science Problems", deals with the aspects of visual attention in solving multiple-choice science problems, which include mathematics or physics concepts, and has been analysed using eye-tracking technology. How the problem-solving strategies influenced the students' visual attention was also examined. They emphasised that the methodology used in this research can be useful in identifying cognitive load while solving multiple-choice tasks during the decision-making process.

Jerneja Pavlin and Miha Slapničar, in their chapter "The Impact of Students' Educational Background, Formal Reasoning, Visualisation Abilities, and Perception of Difficulty on Eye-Tracking Measures when Solving a Context-Based Problem with Submicroscopic Representation", emphasise the importance of several independent variables that can influence the solving of context-based exercises at the macroscopic and submicroscopic levels. The exercise covers the phenomenon of gas release when opening a bottle of soda (mineral water) or other carbonated beverage. Their study aimed at determining how the level of education, formal reasoning, and visualisation abilities influence the way students solve these tasks. Eye-tracker measurements were applied while solving the task.

The twelfth chapter, by Pascal Klein, Stefan Küchemann, Ana Susac, Alpay Karabulut, Andreja Bubic, Maja Planinic, Marijan Palmovic, and Jochen Kuhn, entitled "Students' Understanding of Diagrams in Different Contexts: Comparison of Eye Movements Between Physicists and Non-Physicists Using Eye-Tracking", examines the understanding of line charts as a skill necessary for understanding information in science and everyday life. Eye-tracking technology has been used to explore students' competences in solving problems related to the slope and area under the graph.

The next chapter, by Roman Rosiek and Miroslawa Sajka, "Task-Evoked Pupillary Responses in Context of Exact Science Education", also deals with the importance of graphs in physics and mathematics. The results of eye-tracking research, which involves monitoring and analysing changes in pupil diameter size when solving graph-related tasks, are analysed to determine whether there are significant differences in the physiological response of individuals. The analysis of relevant changes in pupil diameter can be an indicator that describes a subjective assessment of the difficulty of the task and an indicator of motivation.

The final chapter in this series of contributions, with the title "An Investigation of Visual and Manual Behaviors Involved in Interactions Between Users and Physics Simulation Interfaces", by Guo-Li Chiou, Chung-Yuan Hsu, and Meng-Jung Tsai, illustrates the importance of computer simulation in physics education. Although computer simulation has been shown to have a positive impact on improving science learning, little is known about how users interact with simulation interfaces. A study of the visual and manual behaviour of user interaction with physics simulation interfaces was conducted by analysing both eye movements and log data.

The final chapter of this book, by Karen S. McNeal, Rachel Atkins, and Elijah T. Johnson is entitled "Visualising Student Navigation of Geologic Block Diagrams" and deals with the 3D visualisation problems in the geosciences. Geology is an important area of science, and not much research is published in general research journals for science education. For this reason, this book concludes with a chapter in this field. Eye-tracking, as an explorative method, has been used to study students' visual navigation approaches to spatial problems, specifically geologic block diagrams used in geoscience. Spatial and temporal information about the students' gaze patterns was collected and analysed, using the different facets of the block diagrams as prominent locations, and the relationship between spatial ability and visual patterns in problem-solving was investigated.

All chapters of this scientific monograph were reviewed by two reviewers before it was submitted to the publisher. Afterwards, a further peer review round took place with the publisher while the manuscript was being edited.

Ljubljana, Slovenia

Iztok Devetak
Saša Aleksij Glažar

Contents

1 **Cognitive Processes and Eye-Tracking Methodology** 1
   Anja Podlesek, Manja Veldin, Cirila Peklaj, and Matija Svetina

2 **The Interplay of Motivation and Cognition: Challenges for Science Education Research and Practice** 33
   Mojca Juriševič and Tanja Černe

3 **Predicting Task Difficulty Through Psychophysiology** 55
   Junoš Lukan and Gregor Geršak

4 **The Role of the Explanatory Key in Solving Tasks Based on Submicroscopic Representations** 71
   Vesna Ferk Savec and Špela Hrast

5 **Investigating the Role of Conceptual Understanding on How Students Watch an Experimental Video Using Eye-Tracking** 93
   Sevil Akaygun and Emine Adadan

6 **Using an Eye-Tracker to Study Students' Attention Allocation When Solving a Context-Based Problem on the Sublimation of Water** 107
   Miha Slapničar, Valerija Tompa, Iztok Devetak, Saša Aleksij Glažar, and Jerneja Pavlin

7 **Using an Eye-Tracking Approach to Explain Students' Achievements in Solving a Task About Combustion by Applying the Chemistry Triplet** 129
   Iztok Devetak

8 **Pre-service Teachers' Determination of Butterflies with Identification Key: Studying Their Eye Movements** 155
   Tanja Gregorčič and Gregor Torkar

9 **Case Processing in the Development of Expertise in Life Sciences-What Can Eye Movements Reveal?** 169
   Ilona Södervik and Henna Vilppu

10	**Analysis of Aspects of Visual Attention When Solving Multiple-Choice Science Problems**	185
	Miroslawa Sajka and Roman Rosiek	
11	**The Impact of Students' Educational Background, Formal Reasoning, Visualisation Abilities, and Perception of Difficulty on Eye-Tracking Measures When Solving a Context-Based Problem with Submicroscopic Representation**	217
	Jerneja Pavlin and Miha Slapničar	
12	**Students' Understanding of Diagrams in Different Contexts: Comparison of Eye Movements Between Physicists and Non-physicists Using Eye-Tracking**	243
	Pascal Klein, Stefan Küchemann, Ana Susac, Alpay Karabulut, Andreja Bubic, Maja Planinic, Marijan Palmovic, and Jochen Kuhn	
13	**Task-Evoked Pupillary Responses in Context of Exact Science Education**	261
	Roman Rosiek and Miroslawa Sajka	
14	**An Investigation of Visual and Manual Behaviors Involved in Interactions Between Users and Physics Simulation Interfaces**	277
	Guo-Li Chiou, Chung-Yuan Hsu, and Meng-Jung Tsai	
15	**Visualizing Student Navigation of Geologic Block Diagrams**	295
	Karen S. McNeal, Rachel Atkins, and Elijah T. Johnson	
Index		309

Editors and Contributors

About the Editors

Iztok Devetak, Ph.D. is a Professor of Chemical Education at the University of Ljubljana, Faculty of Education, Slovenia. His research focuses on how students, from elementary school to university, learn chemistry at macro-, submicro-, and symbolic level, how chemistry in context and active learning approaches stimulate learning, using eye-tracking technology in explaining science learning, aspects of environmental chemistry education, developing teachers' health-managing competences, etc. He has been involved in research projects in the field of science education and he was the national coordinator of PROFILES project (7th Framework Program) for 4.5 years. He co-edited a Springer monograph about active learning approaches in chemistry. He (co)authored chapters in international books (published by Springer, American Chemical Society, Routledge...) and published papers in respected journals (altogether about 400 different publications). He was a Fulbright scholar in 2009. He is a member of ESERA (European Science Education Research Association) and Vice-chair for Eastern Europe of EuChemMS DivCEd (European Association for Chemical and Molecular Sciences Division of Chemical Education). He is a Chair of Chemical Education Division in Slovenian Chemical Society and President of the national Subject Testing Committee for chemistry in lower secondary school. Dr. Devetak is Editor-in-Chief of *CEPS Journal* and editorial board member of respected journals, such as *Chemistry Education Research and Practice*, *International Journal of Environmental and Science Education*, and *Eurasian Journal of Physics and Chemistry Education*.

Saša Aleksij Glažar, Ph.D. is a Professor Emeritus at the University of Ljubljana, Slovenia. His research is focused mainly on the development of methodology for the organisation of chemical data into networks of knowledge and building chemical relational information systems to be applied in the transfer of knowledge in education and industrial development. Saša A. Glažar is involved in defining, classifying, and categorising science concepts in building relational systems for various levels of education, structuring information into knowledge maps, i.e. hierarchically built relational systems in science education. These systems have been tested through evaluation of knowledge of students at various levels, thus identifying misconceptions. The same approach was also applied in designing study programmes for all levels of education. The new research area of S. A. Glažar is aimed at developing teaching methods based on macroscopic observation and linking it to the microscopic and symbolic interpretations of science phenomena, it is also aimed at implementation of new approaches using computer technology to develop motivation for learning and guidance in the science professions. As an expert, he participates in the development and evaluation of science curricula and in international studies (TIMSS, PISA).

Contributors

Emine Adadan Bogazici University, Istanbul, Turkey

Sevil Akaygun Bogazici University, Istanbul, Turkey

Rachel Atkins North Carolina State University, Raleigh, NC, USA

Andreja Bubic Chair for Psychology, Faculty of Humanities and Social Sciences, University of Split, Split, Croatia

Tanja Černe Counseling Center for Children, Adolescents and Parents Ljubljana, Ljubljana, Slovenia

Guo-Li Chiou Program of Learning Sciences, School of Learning Informatics, National Taiwan Normal University, Taipei, Taiwan

Iztok Devetak Faculty of Education, University of Ljubljana, Ljubljana, Slovenia

Vesna Ferk Savec Faculty of Education, University of Ljubljana, Kardeljeva Ploščad 16, 1000 Ljubljana, Slovenia

Gregor Geršak Faculty of Electrical Engineering, University of Ljubljana, Ljubljana, Slovenia

Saša Aleksij Glažar Faculty of Education, University of Ljubljana, Ljubljana, Slovenia

Tanja Gregorčič Faculty of Education, University of Ljubljana, Ljubljana, Slovenia

Špela Hrast Faculty of Education, University of Ljubljana, Ljubljana, Slovenia

Chung-Yuan Hsu Department of Child Care, National Pingtung University of Science and Technology, Neipu, Pingtung, Taiwan

Elijah T. Johnson Department of Geosciences, Auburn University, Auburn, AL, USA

Mojca Juriševič Faculty of Education, The University of Ljubljana, Ljubljana, Slovenia

Alpay Karabulut Department of Physics, Physics Education Research, Technische Universität Kaiserslautern, Kaiserslautern, Germany

Pascal Klein Faculty of Physics, Physics Education Research, University of Göttingen, Göttingen, Germany

Stefan Küchemann Department of Physics, Physics Education Research, Technische Universität Kaiserslautern, Erwin-Schrödinger-Str. 46, 67663 Kaiserslautern, Germany

Jochen Kuhn Department of Physics, Physics Education Research, Technische Universität Kaiserslautern, Kaiserslautern, Germany

Junoš Lukan Jožef Stefan Institute, Ljubljana, Slovenia

Karen S. McNeal Department of Geosciences, Auburn University, Auburn, AL, USA

Marijan Palmovic Laboratory for Psycholinguistic Research, Department of Speech and Language Pathology, University of Zagreb, Zagreb, Croatia

Jerneja Pavlin Faculty of Education, University of Ljubljana, Ljubljana, Slovenia

Cirila Peklaj Department of Psychology, Faculty of Arts, University of Ljubljana, Ljubljana, Slovenia

Maja Planinic Department of Physics, Faculty of Science, University of Zagreb, Zagreb, Croatia

Anja Podlesek Department of Psychology, Faculty of Arts, University of Ljubljana, Ljubljana, Slovenia

Roman Rosiek Institute of Physics, Pedagogical University of Krakow, Kraków, Poland

Miroslawa Sajka Institute of Mathematics, Pedagogical University of Krakow, Krakow, Poland

Miha Slapničar Faculty of Education, University of Ljubljana, Ljubljana, Slovenia

Ilona Södervik Centre for University Teaching and Learning, University of Helsinki, Helsinki, Finland;
Department of Teacher Education, University of Turku, Turku, Finland

Ana Susac Department of Applied Physics, Faculty of Electrical Engineering and Computing, University of Zagreb, Zagreb, Croatia

Matija Svetina Department of Psychology, Faculty of Arts, University of Ljubljana, Ljubljana, Slovenia

Valerija Tompa Faculty of Education, University of Ljubljana, Ljubljana, Slovenia

Gregor Torkar Faculty of Education, University of Ljubljana, Ljubljana, Slovenia

Meng-Jung Tsai Program of Learning Sciences, School of Learning Informatics, National Taiwan Normal University, Taipei, Taiwan

Manja Veldin Evaluation Studies Centre, Educational Research Institute, Ljubljana, Slovenia

Henna Vilppu Department of Teacher Education, University of Turku, Turku, Finland

Chapter 1
Cognitive Processes and Eye-Tracking Methodology

Anja Podlesek, Manja Veldin, Cirila Peklaj, and Matija Svetina

Introduction

Students' understanding of science concepts can be assessed by their performance in a test, i.e. by their answers to different types of questions related to the content learned. By examining the verbal answers, teachers usually have no insight into how the students solved a particular problem and how they arrived at the answer. They cannot know what information the students have used to solve the problem. The eye-tracking method allows us to follow the focus of the students' overt attention and to draw conclusions about the strategies they used to solve the problem. In this chapter, we present a study that has shown that individual differences in eye movements in solving authentic science problems are related to differences in students' cognitive abilities. Students with comparable prior knowledge but different levels of cognitive ability focused their attention on different parts of the task and used different strategies to solve the problem.

Use of Authentic Tasks with Different Levels of Science Concept Representation in Teaching

Science problems faced by pupils at school can either be routine, well defined, contain proposed solutions and require low-level thinking, or they can be similar to real-life

A. Podlesek (✉) · C. Peklaj · M. Svetina
Department of Psychology, Faculty of Arts, University of Ljubljana, Ljubljana, Slovenia
e-mail: anja.podlesek@ff.uni-lj.si

M. Veldin
Evaluation Studies Centre, Educational Research Institute, Ljubljana, Slovenia

problems which are often not well defined, complicated to solve, have many possible solutions and require higher-level thinking skills (Muhamad et al., 2016). Learning that usually focuses on real, complex and interdisciplinary problems and problem-based activities is called authentic learning (Lombardi & Oblinger, 2007). In the case of abstract science concepts that are difficult to understand, the use of problems simulating an authentic environment can lead to a higher interest and motivation of students to learn and increase their science problem-solving competence (Muhamad et al., 2016).

Students' understanding of science concepts can be supported by using representations of the concepts at different levels (macroscopic, submicroscopic and symbolic). The macroscopic level represents the concept through an experimental approach or observation of the phenomenon. The submicroscopic level of concept representation shows the interaction between particles and can therefore contribute to the understanding of the mechanisms underlying the phenomenon or processes. The symbolic level of representation uses physical artefacts and computer models, structural and empirical formulae and physical and chemical equations (Treagust et al., 2003) to represent the concept. Students' understanding of a particular science concept and their ability to solve problems related to that concept depends on integrating the understanding of all three levels of representation (Ferk Savec et al., 2016).

Learning about science can start within a science class, where students are first introduced to the macroscopic level of the selected concepts and the associated submicroscopic representations, which are not yet very specific (e.g. chemical particles are not yet differentiated into atoms, molecules, electrons). Later, when a certain knowledge of various concepts has already accumulated and students enter the formal operational stage of cognitive development and develop the ability to think about abstract concepts and to apply hypothetical and deductive reasoning (from about 12 years of age; Inhelder & Piaget, 1958), the symbolic level of concept representation can be introduced. However, newcomers usually use only one form of representation and have difficulty in understanding the symbolic level of conceptual representation and relating the symbolic level to the other two levels (e.g. they treat chemical equations as mathematical puzzles rather than as representations of dynamic and interactive chemical processes), while experts combine the use of all three levels and have no difficulty in moving between different levels of representation (Kozma & Russell, 1997). The success of newcomers in integrating different levels of conceptual representation depends not only on their knowledge but also on their cognitive abilities.

In order to illustrate how different cognitive abilities can be involved in solving science problems, we focused on an example of an authentic task that includes the macro- and submicroscopic representation of the concept of ice melting. The task used with Slovenian participants in their native language is shown in Fig. 1.1. The macroscopic part of the task consisted of two photos showing the initial and final state of the glacier, and the submicroscopic part was illustrated with an animation of the changes in the particle structure. The students were instructed to compare the two photos of a Slovenian glacier and to name and explain the process that caused the change in its state.

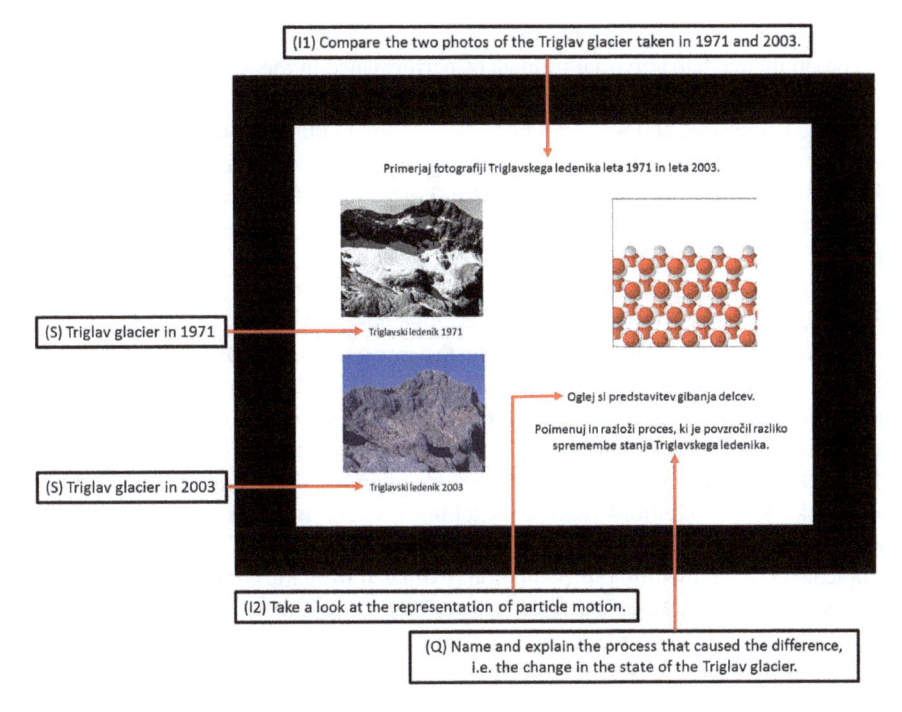

1 Cognitive Processes and Eye-Tracking Methodology

Fig. 1.1 Example of an authentic task in which the pupils had to think about what happens when the glaciers melt. The two photos of the glacier showed the process under consideration on a macroscopic level, and the animation showed it on a submicroscopic level. In the animation, the upper layer of water molecules visibly moved, while the lower layers were almost static. Different parts of the text were divided into: instructions (I1 are instructions relating to the macroscopic level of the representation of melting; I2 are instructions relating to the submicroscopic level of the representation of melting), question (Q) and photo subtitles (S)

Cognitive Factors of Problem-Solving Performance in Science

Various cognitive processes take place during learning and problem-solving in science. During the study of a certain concept, which is presented with different levels of representation, or while solving a problem related to this concept (e.g. a task like the one shown in Fig. 1.1), students need to focus their visual attention on texts or visual external representations of the concept (such as pictures, graphs, diagrams, photos or animations), make internal representations of spoken/written words and visual information, temporarily store multimodal information, make connections between different types of representations and with previous knowledge, make decisions about the problem and actively build new knowledge. It is therefore to be expected that students' abilities in these areas will influence the way in which they deal with a particular problem and the students' performance in solving science problems.

General intelligence. General intelligence, also known as the *g*-factor, represents the broad mental capacity that influences performance in measuring cognitive abilities and includes various cognitive factors such as visuospatial processing, logical thinking, knowledge and working memory. General intelligence appears to facilitate all forms of problem-solving as well as the accuracy and speed of contextualized deductive reasoning that involves if-then inference (Kaufman et al., 2011). It is also closely related to a person's academic success (García et al., 2014; Rosander et al., 2011).

Visuospatial perception and attention. When solving tasks like the one shown in Fig. 1.1, students must use focused or selective attention, i.e. they must focus their attention on the most important parts of the information (certain key concepts in the text, certain details in the photos and animations) while ignoring the superfluous and irrelevant information. In some cases, students may need to divide their attention simultaneously between two different relevant tasks, targets or stimulus streams (Eysenck & Keane, 2015). While divided attention requires parallel processing of both tasks, some situations require switching between two tasks. For example, to solve the task shown in Fig. 1.1, students can switch between observing the macroscopic level of the melting representation (extracted from the comparison of the two photos) and its submicroscopic level (derived from the animation) before connecting the two levels. Inhibition plays an important role in shifting attention and switching between different tasks or stimuli. Miyake and Friedman (2012) identified it as one of the most important executive functions used when a person intentionally overrides dominant and prepotent responses and does not react or respond with the first response that comes to mind. It also helps the individual to resist distractions and successfully complete tasks.

Trexler Holland (1995) pointed out that students also need good visual-spatial perception, visual-spatial memory and logical visual-spatial thinking to solve science problems. In displays with dense visual information, they must be able to carefully observe fine details of objects, find patterns within an image or disembed an image from distracting surroundings. They must also be able to examine an image, an object or a diagram and later retrieve it from visual memory. Last but not least, their performance also depends on their visuospatial and visualization abilities when they need clear three-dimensional internal models of molecular structures and need to manipulate them (mentally rotate visual objects, compare different perspectives of the object to judge whether they are the same object, etc.). Spatial abilities contribute to the prediction of educational and vocational outcomes in the STEM sciences (Berkowitz & Stern, 2018; Kell et al., 2013; Shea et al., 2001).

Working memory. When solving problems, working memory is the most important cognitive process. According to the traditional information processing model of Atkinson and Shiffrin (1968), working memory is a part of the system in which information is encoded, maintained and retrieved. Attkinson and Shifrin proposed three different stores in which the information is processed: sensory register, short-term memory and long-term memory. Information from the environment (visual, auditory, haptic, etc.) is first processed in a sensory register that keeps it for a very short time. If attention is not focused on the incoming information, it is lost. The short-term memory or temporary working memory takes up information from sensory registers or retrieves it from long-term memory. Without active processing of the information, e.g. chunking or rehearsing, it is lost after about 30 s. Rehearsing helps to store this information relatively permanently in the long-term memory and it can be retrieved later in the short-term memory if required.

According to the multicomponent model developed by Baddeley (1996a, 1996b, 2000, 2012), working memory is composed of four subsystems: phonological loop, visuospatial sketchpad, central executive, and episodic buffer. These subsystems can act relatively independently of each other. They have a limited capacity and there is interference between two tasks that are processed in the same subsystem (Schüler et al., 2011). The *phonological loop* processes and stores auditory information for a short time. It consists of two components: a passive phonological store for speech perception and an articulatory control process that provides access to the phonological store and is linked to speech production (Baddeley, 1996b). Spoken words are processed directly in the phonological store, while written words are first converted into articulatory code by a subvocal rehearsal process and then processed further (Schüler et al., 2011). The next working memory subsystem—the *visuospatial sketchpad*—is responsible for the temporary storage and manipulation of visual patterns and spatial movements as well as for image processing (Schüler et al., 2011). It consists of two components: (i) the visual cache, which stores information about colour and shape, and (ii) the inner scribe, which processes spatial and motion information and is also involved in the rehearsal of information (Logie, 1995). The third subsystem of Baddeley's model (2012) is an *episodic buffer* which temporarily stores multimodal information (visual, auditory, kinaesthetic, etc.). It serves as a buffer between other components of working memory and connects it with perception and information stored in long-term memory. It binds information from the phonological loop and the visuospatial sketchpad with information about time and order, thus developing an integrated representation of a particular stimulus or event that can be stored in long-term memory. The fourth subsystem and most important part of Baddeley's model is the *central executive*, a control attention system responsible for manipulating information and controlling other subsystems. Executive processes are defined as "processes that organize and coordinate the functioning of the cognitive system to achieve current goals" (Eysenck & Keane, 2015, p. 220). Numerous executive processes are important in goal-directed activities, such as setting goals, planning, prioritizing, organizing, flexibly shifting between tasks, holding and manipulating information in the working memory and self-monitoring (Meltzer, 2010). Central executive capacity is usually measured by complex tasks, e.g. the Reading Span Task (Daneman & Carpenter, 1980; Engle et al. 1999), Operation Span Task (Conway et al., 2005; Turner & Engle, 1989), Counting Span Task (Case et al., 1982; Conway et al., 2005) or Spatial Span Task (Schüler et al., 2011; Shah & Miyake, 1996). In these complex tasks, the participants are asked to store information in the phonological loop (e.g. memorizing letters) or in the visuospatial sketchpad (e.g. memorizing shapes) and additionally to perform another cognitive activity, e.g. simple arithmetic or mental rotation of the objects. Good performance in such tasks requires the monitoring and coordination of different cognitive processes

(Schüler et al., 2011). According to Baddeley (2012), the central executive comprises three executive functions: (i) focusing attention, (ii) dividing attention between two stimuli or tasks and (iii) switching between tasks. Other researchers use different categorizations of executive functions, e.g. Miyake and Friedman (2012) consider inhibition, shifting and updating the most prominent executive functions; Cornoldi and Giofre (2014) speak of updating in working memory, inhibition, shifting, planning, organizing, categorizing, applying rules and delaying gratification.

The multicomponent working memory model of Baddeley (2012) has some implications for learning and solving science tasks that use textual and pictorial information in concept representations. In science problems such as the one shown in Fig. 1.1, all four subsystems of working memory are active. The two phonological loop components (passive phonological store for speech perception and articulatory control process for speech production) are active when the student works on instructions, descriptions, explanations, questions and tasks. The visuospatial sketchpad is active to store the information about colours, shapes, spatial position and movements in images and animations. The episodic buffer combines all the information in a multidimensional code and enables the development of a holistic and meaningful conceptual representation. The central executive leads the student to the goal (solution of a problem) by directing attention to the most important information and inhibiting irrelevant information, shifting attention between different information and updating the information. According to Schüler et al. (2011), limitations of working memory capacity in different subsystems should have different effects on task solving. Limited phonological loop capacity should have negative effects on the processing of verbal and symbolic information, while constraints in the visuospatial sketchpad should affect the processing of pictorial information at the macroscopic and submicroscopic levels of science concept representations. The integration of verbal and pictorial information should be influenced by limitations in the central executive and episodic buffer.

Reasoning ability. The students' level of cognitive development is another important factor in their performance in solving authentic science problems, including concept representations at the submicroscopic or symbolic level. Learning the so-called formal concepts (Lawson & Renner, 1975), e.g. those represented at the submicroscopic level (such as molecular structure), requires students to go beyond their experiences and draw conclusions based on logic and inference. They need to think abstractly and understand abstract ideas, which is why they need to reach the formal reasoning level (Haidar & Abraham, 1991; Trexler Holland, 1995; Trifone, 1987). According to the structuralist approach, the understanding of different concepts results from general, context-independent frameworks of reasoning, e.g. understanding of proportional relations, probabilities, relations between independent and dependent variables in experimental investigations (i.e. identification of a cause and its effects), combinatorics, etc. (Ma Oliva, 1999). These frameworks represent the common core of different science conceptions and may play an important role when students are confronted with a new problem, i.e. a problem with which they have no experience.

Verbal abilities. Verbal ability is the cognitive ability to understand linguistic information (written or spoken words, sentences, texts, e.g. descriptions of the science concept and the description of the task) and to use language, i.e. to produce words and sentences (e.g. to give a verbal response to the task). Verbal ability comprises a number of components, including vocabulary, language proficiency, comprehension, reading skills, oral and written communication, verbal memory and verbal reasoning. In addition to mathematical/quantitative abilities, verbal reasoning abilities have been shown to be crucial for advanced STEM learning (Berkowitz & Stern, 2018). For solving mathematical problems, Bahar and Maker (2015) found that mathematical knowledge and general intelligence predict closed problem-solving performance, while general creativity and verbal ability predict performance in solving open-ended problems.

Domain knowledge. It has been shown that domain knowledge has a considerable predictive validity for post-secondary academic success and STEM persistence (Ackerman et al., 2013). Among university students, previous knowledge, both general and subject-specific, has been found to be a good predictor of academic success (Binder et al., 2019), as students can more easily build new knowledge on a solid foundation. Domain knowledge also predicts students' problem-solving performance (She et al., 2012). According to Taconis et al. (2001), prior knowledge contains schemas (mental constructs) that enable students to recognize new information as belonging to a previously learned category and to elicit certain responses appropriate to that category. A strong knowledge base therefore supports problem-solving. This is what problem-solvers must have (Taconis et al., 2001): (i) declarative knowledge of facts, principles and laws of the discipline, on the basis of which they interpret the given information, build a representation of the problem situation and draw conclusions, (ii) situational knowledge to identify and classify the problem and select declarative knowledge needed for the solution, (iii) strategic knowledge, in the basis of which they know what combination or series of cognitive activities they must go through to find the solution and (iv) procedural knowledge, on the basis of which they apply declarative knowledge in executing the plan.

Methods for Monitoring How Authentic Science Problems Are Solved

In order to better understand how students solve science problems, it is necessary to observe cognitive processes such as the direction of attention, switching between different information, inhibition of irrelevant information, the process of reasoning and decision-making about answers, etc. Various methods can be used to monitor these processes.

In problem-solving, task difficulty is usually defined in terms of response accuracy, which can also be considered an indirect measure of cognitive effort. Sometimes the time required to solve the problem is also measured, which gives us an additional insight into the problem difficulty. The two indicators are regarded as behavioural measures of the cognitive processes studied. The problem with these measures is that they can only provide an observation of the final results of the cognitive processes and cannot provide data on what happened during the problem-solving process, e.g. how the participants reasoned or what strategies they used to solve the problem.

The "Think-Aloud" method allows us to study the individual's thinking during the execution of a particular task (or the recollection of thoughts immediately after the completion of that task) and could therefore be used to follow the reasoning process of a student, but the method is not very natural and could interfere with the execution of the primary task (Eccles & Arsal, 2017). The thoughts of the participants which are the subject of the study can be influenced by the method itself (Susac et al., 2014), which affects the validity of the method (Lai et al., 2013). Moreover, some participants do not have explicit knowledge that would allow them to report in detail on the strategies used (Green et al., 2007). Verbalizing every thought that comes to their mind and producing verbalized descriptions quickly enough and with enough detail may be too difficult for them. For the correct use of this method they would need to be thoroughly trained in introspection (i.e. they would need to be aware of their reasoning process) and in the rapid verbalization of their thoughts, which can sometimes be difficult because not all thoughts are necessarily expressed in words (e.g. visual patterns can be compared in the absence of verbal analysis).

The framework of the methodology of experience sampling (Dimotakis et al., 2013; Rausch et al., 2019) offers a different way of real-time measurement of the problem-solving process by means of several measurements taken during the process. Specific methods have been developed to follow the process more closely, such as the Warmth Measure method for measuring the development of the Aha! experience (Laukkonen & Tangen, 2018) or the Embedded Experience Sampling method in which participants are briefly interrupted during the problem-solving process and asked to answer short questions about their current experiences, e.g. their interest in the topic (Rausch et al., 2019).

Qualitative interviews with the participants after completion of the task can also provide more in-depth information about the process of problem-solving. However, retrospective self-reports on what happened during the problem-solving process (e.g. description of the strategies used or selection from a list of strategies) have the disadvantage of being subject to memory errors. When subjective rating scales are used, many other errors can also occur (e.g. halo effect, central tendency error, contrast error, etc.), so that the quality of such data can be questionable.

One of the methods that seem to provide more objective data on the problem-solving process than self-reports is the eye-tracking method. Compared to other methods of monitoring cognitive processes presented above, eye-tracking provides more objective, measurable real-time information about cognitive and attention processes involved in problem-solving and does not influence the processes themselves (Susac et al., 2014). Eye movements reveal the focus of our (spatial) attention and can be used to study visuospatial processing and related higher cognitive functions (Hartmann, 2015), the reading process, scene perception and performance in other information processing tasks (Duchowski, 2002), including problem-solving, metacognitive processes, etc. In studies of learning, eye movements can be used to examine patterns of information processing, effects of instructional design, individual differences, effects of learning strategies, patterns of decision-making and conceptual development (Lai et al., 2013).

Studies on eye movements have been conducted for more than 100 years (see e.g. Javal, 1878). Today, eye-tracking is based on non-invasive methods that use light reflected from the cornea (Dodge & Cline, 1901). Most eye trackers measure the spatial (horizontal and vertical coordinates) and temporal characteristics of eye movements (Susac et al., 2014), with a typical sampling frequency between 50 and 500 Hz (Jacob & Karn, 2003). The change in the eye position or speed of eye movement is used to extract fixations and saccades from the gaze location data (Jacob & Karn, 2003). Fixations represent a relatively stable holding of the gaze at a certain point. Saccades are fast and rapid movements of the eyes from one location to another (Susac et al., 2014). Six measures of eye movement are most commonly used in research (Jacob & Karn, 2003): the number of fixations, the proportion of time the gaze is directed to different areas of interest, the average duration of fixation, the number of fixations per individual area of interest, the average viewing time of each area of interest and the fixation rate (number of fixations per second). Temporal measures are used most frequently, followed by count and spatial measures (Lai et al., 2013). For specific purposes, such as measuring affect or cognitive load, pupil dilation may also be an interesting measure (Beatty, 1982).

Eye-Tracking Features as Indicators for Cognitive Processes

Eye movements are not automatically directed at spatial locations, but typically result from the intentions and inner goals of the individual (König et al., 2016) and reflect the dynamic updating of internal, multimodal memory representations (Scholz et al. 2015), which is why they can be used to study intentional processes, such as strategies for problem-solving. By following and comparing the steps of problem-solving by experts and non-experts, the results can be used to improve educational methods (Susac et al., 2014). Due to their superiority in information processing, selective attentional division and extension of the visual span, experts have longer saccades and spend less time until the first fixation on the relevant information (Gegenfurtner et al., 2011). Research shows that experts have fewer fixations than non-experts. Sometimes they have multiple fixations in areas relevant to the task (Susac et al., 2014). A meta-analysis of eye-tracking research also showed that experts have shorter fixations than non-experts and more fixations on areas relevant to the task and less fixations on task-redundant areas (Gegenfurtner et al., 2011). This suggests that experts focus their attention on the areas relevant to the problem (Susac et al., 2014). The correlation between the number of fixations and the efficiency of problem-solving suggests that more effective participants also developed more adequate strategies and "knew where to look" (Susac et al., 2014). Similarly, Madsen et al. (2012) examined individual differences in visual attention in problem-solving (Madsen et al., 2012) and showed that people who answered correctly spent more time in relevant areas than those who answered incorrectly.

The choice of which eye movement measures to use is often motivated by a specific research question (Lai et al., 2013). For example, researchers are often interested in certain parts of the visual field, the so-called areas of interest and the transitions between different areas. The number of fixations can provide information about the participant's attention flow during a task (Susac et al., 2014). Visualizations of fixations, such as heat maps or hot zone images, help us to determine where visual attention was allocated during problem-solving. Such visualizations can be performed either for each participant individually or for selected groups of participants (Lai et al., 2013). In the density diagram, the (relative) cumulative time spent on a particular area of interest is represented by warmer colours, while colder (green and blue) colours represent areas that received less attention and a shorter fixation time (Majooni et al., 2016).

When interested in saccadic movements, especially in reading, researchers analyse progressive (forward) and regressive (backward) saccades (Vitu & McConkie, 2000). During reading, progressive saccades occur more frequently, but whenever participants return to the text they had already read (the regressive saccades), it is assumed that they are reprocessing certain parts of the problem (Moeller et al., 2011).

Eye-tracking can be used as an indirect measure of cognitive load and can indicate the difficulty of the problem for a person (Susac et al., 2014). Cognitive load has been found to be associated with saccade amplitude, blink rate and pupil dilation. Longer saccade amplitudes and longer scan paths (i.e. the spatial arrangement of the sequence of fixations and saccades; Jacob & Karn, 2003) suggest greater task complexity and relative search efficiency (Goldberg et al., 2002). Brookings et al. (1996) found that the blink rate and blink duration decline with increasing mental workload. Some other studies, however, found that the blink rate increases with increasing cognitive load in arithmetic tasks, conversations and mental rehearsal activities (Chen, 2014). Chen and Epps (2014) attempted to explain these inconsistencies by separating perceptual from cognitive load and found that at low perceptual load, the blink rate increased with increasing cognitive load of the limited working memory (e.g. with increasing difficulty of arithmetic tasks), but at high perceptual load (e.g. when there was a requirement to search for more objects), the blink rate did not vary with cognitive load. The blink rate also differed between reading (3–7 blinks per minute) and non-reading periods (15–30 blinks per minute) and correlated positively with the time spent on a task and fatigue (Stern et al., 1994). The results are more consistent regarding the relation between pupil dilation and cognitive load. A larger pupil size indicates attention (Hoeks & Levelt, 1993) and is associated with higher levels of cognitive load (Beatty, 1982; Majooni et al., 2016; Pomplun & Sunkara, 2003) as well as arousal and stress (Baltaci & Gokcay, 2016). Cognitive load does not necessarily correlate with pupil size throughout the problem-solving process (Lin et al., 2008), as it can vary from subtask to subtask, and thus may lead to variable pupil size during the problem-solving process (Iqbal et al., 2005). In order to use pupil size as an indicator of psychological processes, it is also necessary to take into account the fact that the pupil reacts to light with a rapid constriction and to darkness with a slow dilation. For estimating cognitive load, the use of an index which takes into account the changes in light may be recommended instead of directly analysing pupil size (Vogels et al., 2018). Recent studies (Biswas et al., 2016; Krejtz et al., 2018) show that a certain type of microsaccadic movement (i.e. saccadic intrusion) can be an even better indicator of cognitive load than changes in pupil size and blinking.

Issues in Eye-Tracking Experiments

The use of eye-tracking for research purposes has some disadvantages. The precise preparation of equipment and participants is time-consuming. A long use of head or chin rests to prevent head movements and data loss can lead to a lack of comfort for the participants. Some eye trackers work from a distance but are not as accurate. The eye movements of some participants (10–20%) cannot be reliably tracked due to specific characteristics of their corneas, vision irregularities, or the need to use glasses or lenses (Jacob & Karn, 2003). Careful calibration of the instrument is necessary before starting the measurements and sometimes during the experiment.

The analysis of eye-tracking studies can be identified as "top-down", i.e. based on a theory, or as "bottom-up", i.e. based solely on observational data, without a predetermined theory (Goldberg et al., 2002). In the first case, researchers usually define specific areas of interest and investigate the (relative) total fixation time spent in different areas of interest. In the second case, visualizations such as heat maps and scan paths can be analysed.

When interpreting eye-tracking data, it is important to choose an appropriate measure. Most eye measures are not straightforward to interpret (Jacob & Karn, 2003). For example, the total fixation time in a particular area of interest may consist of one long single fixation in that area or many short fixations with intermediate jumps to other areas of interest. Measures of time (e.g. time spent in a particular area of interest) and measures related to the number of occurrences of a particular event (number of blinks, number of fixations) correlate with the time taken to solve the problem. For example, the longer the participant solves the task, the higher the number of fixations and the longer the total time spent on a particular area of interest. For this reason, relative measures often need to be used instead of (or in addition to) absolute measures (such as the average duration of a fixation, the proportion of time spent on a particular area of interest).

The use of eye-tracking data does not always give the right insight into what is going on during problem-solving. Problem-solving can be compromised by shifting the covert attention in the absence of eye movement (Thomas & Lleras, 2009) or by directing the covert attention to the areas where the gaze is not directed. In addition, sometimes when participants solve cognitively challenging problems, their gaze wanders or they look at a particular point in the visual field without actually focusing on and processing the information presented there. In the optics of the visual system, this leads to accommodation lag and lens-induced blur (Kruger, 1980; Mihelčič & Podlesek, 2017). Eye trackers cannot extract this kind of information.

Due to the limitations of the eye-tracking method, multiple data sources (e.g. subjective, physical, physiological and eye-tracking data) should be used to assess the cognitive processes behind problem-solving (Lin et al., 2008). In problem-solving studies, eye movement data can be combined with other performance and self-reporting measures (on task difficulty, qualitative reports on task-solving strategy, etc.).

A Case Study

In our case study we investigated individual differences in eye movements of 7th grade primary school children when solving authentic science problems. Since the understanding of science concepts depends on the previous knowledge of the pupils (Haidar & Abraham, 1991), we selected two pupils for this case study who had comparable previous knowledge in the selected science domain but different cognitive abilities—one of them had low and the other high cognitive abilities. Based on the studies presented above, we assumed that differences in their cognitive abilities would be reflected in their eye movements when solving tasks. The primary goal of this case study was to demonstrate the methodology and observations that such a case study can provide.

The following cognitive abilities have been identified as the most relevant for solving authentic science problems, such as the one shown in Fig. 1.1[1]: (i) visual-spatial abilities (the participants had to compare details on different photographs and observe the patterns and their movement in the animation), (ii) working memory (the participants had to retain and integrate different pieces of information in their short-term memory), (iii) switching abilities (the participants had to switch between and integrate different types of information), (iv) verbal abilities (the participants had to understand the instructions and give competent verbal answers), (v) logical reasoning (the participants had to connect two different states of matter and draw conclusions about what exactly happened during the transition from the first to the second state) and (vi) general intelligence, which was considered to be an agglomeration of different cognitive abilities.

---

[1] In the Slovenian Research Agency project J5-6814, Explaining Effective and Efficient Problem Solving of the Triplet Relationship in Science Concepts Representations, the students were confronted with 10 authentic science problems. Only one of the problems was selected to be presented in our case study.

Method

Sample

The two pupils selected for the case study were chosen from the sample of 31 seventh graders (12- to 13-year olds; 19 men, 12 women) of a primary school in Slovenia. The data for the cognitive tasks were collected in the school premises, while the data for eye-tracking were collected in the university laboratory. None of the pupils included in the study reported visual impairment.

Instruments

The visual-spatial ability, working memory and switching ability of the participants were assessed with computer-based tests from the PEBL version 2.0 battery (Mueller & Piper, 2014) and individually administered on a laptop computer. The participants used a computer keyboard or a mouse to respond to the stimuli. The verbal ability test was conducted individually and the responses of the participants were recorded by a researcher. The tests for logical and abstract perceptual reasoning are paper-and-pencil tests and were conducted in class.

The PEBL Pattern Comparison Test (Mueller, 2012), based on the Visual Pattern Test (Della Sala et al., 1999), was used to assess visual-spatial perception and visual-spatial memory. The participants had to compare two 4 × 4 grids of circles presented simultaneously side by side on the screen and had to indicate as quickly and accurately as possible by pressing the right or left shift key whether the patterns were the same or not. The reaction accuracy and average reaction time were calculated.

The PEBL Digit Span task (Mueller, 2012) was used to assess working memory. This task consisted of sets of digits, whereby the length of the set increased from 3 to 10 elements. The digits were displayed on the screen one by one, one per second, with sound recordings switched off. The participants had to repeat the sequence of digits (by typing them in) in the presented order. Two sets of equal length were presented. If the participant repeated at least one of them correctly, the length of the set was increased for the next trial. The second task was to repeat the presented digits in the reverse order (by typing). While the forward recall task was used to assess the capacity of the short-term memory, the backward recall task was used to assess the working memory (the operation of reversal had to be applied to the stored digits). The maximum length of the sequence of digits with at least one set of the two correctly retrieved was recorded.

The switching capacity was measured using PEBL ptrails (Mueller, 2012), a version of the trail making test (Reitan, 1958). Participants were shown 22 numbers or letters positioned pseudorandomly on the screen, and their task was to click on them as quickly as possible. In the first task, the participants had to click on numbers in their numerical order (1-2-3-4, etc.). In the second task, they had to click on the letters in alphabetical order (A-B-C-D, etc.). In the third task, half of the stimuli contained numbers and the other half letters. The participants had to click on them in alternate order (1-A-2-B, etc.). The number of errors and the time taken to connect the stimuli were recorded. The average time required to connect the stimuli under pure number and letter conditions was calculated to assess the visual search speed. The average time taken to connect the stimuli in the alternating condition was used as an indicator of cognitive flexibility (Arnett & Labovitz, 1995). The switching cost was calculated as the difference between the second and the first measure.

For the measurement of verbal abilities, we used a letter fluency task. Letter fluency is a type of verbal fluency. While another type of fluency, i.e. semantic fluency, refers to the production of words within a certain semantic category (e.g. different vegetables), letter fluency refers to the ease of production of different words beginning with a certain letter. In the task we used, the participants had to produce as many words as possible beginning with a certain letter in 60 s. Following an earlier study on the frequency of words beginning with certain letters in the Slovenian language (Remšak, 2013), the high-frequency letters P, S, D and L were selected. The first two were used as representatives of the letters with the highest production rate and the last two were used as letters with a low production rate. The letters M and V were used in the task in which the participants had to produce words by using both letters interchangeably. The total number of words produced in the five tasks was recorded and the number of errors (repetitions and inaccurate productions, such as names or words with the same root word) was subtracted. In the letter fluency task, the participants have to focus on the task, access their mental lexicon and retrieve words, select words that meet certain conditions and avoid repetition. In this way, the fluency tasks can be used as an effective screening tool for general verbal functioning (Shao et al., 2014). Letter fluency has been found to be related to the organization of thought, basic linguistic knowledge, the ability to initiate a search and retrieve data from the lexicon, and executive function, including attention (Hurks et al., 2006) and updating ability (Shao et al., 2014). The test results in the Slovenian version of the letter fluency task showed a high internal consistency (Cronbach's alpha = .91) and correlated strongly with the score in the semantic fluency test ($r = .66$) and the foreign words test, which measures vocabulary range ($r = .50$), supporting its validity for assessing verbal fluency (Remšak, 2013) and, indirectly, verbal abilities.

Test of Logical Thinking—TOLT (Tobin & Capie, 1984) was used to measure the reasoning ability. This paper-and-pencil test consists of 10 tasks composed of two questions. For the first question, several solutions are proposed and participants choose the correct alternative. The second question requires a justification of the choice. Participants choose between several justifications. The first two items measure the proportion reasoning ability, which is necessary for the interpretation of ratios and equations; the second two items assess the participants' ability to identify and control variables, which is crucial when planning experimental investigations; the third two items assess probabilistic reasoning, i.e. the ability to solve problems using probabilities; the next two items measure correlational reasoning, i.e. the ability to identify and verify the relationships between variables; and the last two items refer to combinatorial reasoning, i.e. the ability to solve problems using combinations of

1 Cognitive Processes and Eye-Tracking Methodology

different elements (Trifone, 1987). The test is suitable for use in groups from grade 6 (age 11 or 12) to university. It has a one-dimensional structure (Tobin & Capie, 1981). Its reliability is high (Cronbach's alpha = .85). The test score is related to academic performance, especially in the natural sciences (Trifone, 1987), and to performance in clinical interviews (Tobin & Capie, 1981).

Set I of Raven's Advanced Progressive Matrices—APM (Raven et al., 1998) was administered in groups to roughly assess the general (fluid) intelligence. APM is a standardized non-verbal test for measuring abstract perceptual reasoning. These items are suitable for adults and adolescents with above-average intelligence. Test set I consists of 12 relatively easy items and was designed either as a short test to place adults in the lower 10, middle 80 or upper 10 per cent of abilities. Each item consists of a 3×3 matrix with geometric patterns drawn with black ink on a white background, with the pattern in the lower right-hand corner missing. The task is to choose one of the eight options presented, which correctly completes the matrix. In a sample of first-year university students, the results of Set I correlated moderately with the results of Set II consisting of 36 items ($r = .53$), with Set I and II reliability of .62 and .84, respectively (Bors & Stokes, 1998). In a sample of German 15-year olds the test-retest reliability of Set I was .73 (Raven et al., 1999).

The pupils' prior knowledge was assessed with a self-constructed prior knowledge performance test consisting of questions on states of matter and chemical and physical change (see also Devetak, this volume), which were developed in the research project. Four items presented different everyday situations (e.g. "It is hot in the summer and you want to refresh yourself with cold juice. You take ice from the fridge and put it in the juice") and asked the pupils what would happen. Four items contained images of particles representing states of matter at the submicroscopic level or transitions between different states, and the pupils had to interpret which states were shown in the images. In the ninth item, the pupils had to complete sentences. In the tenth item, the pupils had to name a scientific concept. The construct validity of the instrument was confirmed by three independent experts in science and chemical education. Members of the research project team evaluated the responses and the test results were converted into the per cent correct score.

Selection of Participants for the Case Study

To select the pupils for the case study, we generated a score for cognitive abilities and assigned this score to each pupil in the sample. The cognitive abilities score was calculated as a linear set[2] of values collected by the cognitive measures used in this study. Three pupils with the lowest and three with the highest scores were

---

[2]The score was calculated as a weighted average (the weights are given in brackets) of the standardized—within the sample—scores on the Digit Span—forward test (1), Digit Span—backward test (1), the visual search score in the PEBL ptrails test (1), the switching cost in the PEBL ptrails test (1), the letter fluency test (1), the APM test (1) and the TOLT (3).

Table 1.1 Raw scores (and within the sample standardized scores in parentheses) on different cognitive tests for the two selected pupils and their performance on science tests

Studied measures	Low-performing pupil (LP)	High-performing pupil (HP)
Visual-spatial abilities (PEBL Pattern Comparison)		
Median RT in the 'same' condition [ms]	1329 (0.18)	1116 (0.71)
Median RT in the 'different' condition [ms]	1431 (−0.31)	1301 (0.17)
Working memory (PEBL Digit Span)		
Forward	4 (−1.60)	7 (0.93)
Backward	3 (−1.58)	5 (−0.05)
Switching ability (PEBL ptrails)		
Median RT in the visual search conditions [ms]	947 (−1.60)	632 (0.84)
Switching cost [ms]	497 (−2.42)	76 (0.74)
Letter fluency	35 (−0.61)	34 (−0.68)
Logical reasoning (TOLT)	2 (−0.44)	10 (2.36)
Fluid intelligence (APM Set I)	10 (0.37)	12 (1.37)
Cognitive abilities score	−0.97	+1.14
Performance in science		
Score on prior knowledge test	88	87
Grade in science class (possible range from 1–5)	4	5
Score on authentic science problems test	21	28

selected to maximize the differences between them in cognitive abilities. From these six pupils, we selected two pupils, one from each group, who had comparable scores on the prior knowledge test, because we wanted to keep this factor under control. The selected high-performing pupil in the cognitive ability tests (HP) was a 12-year-old girl with 87 per cent correct results in the prior knowledge test, while the low-performing pupil in these tests (LP) was a 13-year-old boy with 88 per cent correct results in the prior knowledge test (see also Table 1.1).

Results and Discussion

The results are divided into four parts. In the first part, we describe the differences between the two case study participants in terms of their answers and explanations. In the second part, we compare their heat maps, which show the fixation density. In the third part, we focus on sequences of their saccades to gain additional insights into their problem-solving strategies. In the fourth part, we try to draw conclusions about the cognitive load imposed on them based on the blink rate and pupil dilation.

Participants' answers and explanations. Both pupils called the process shown in Fig. 1.1 correct ("melting"), but the explanation of the process which HP provided was on a higher level than that of LP. HP referred to the submicroscopic level in her explanation and her response was partly correct ("The atmosphere has warmed and the atoms that are in the upper layers go into a liquid state and the water runs down the hill"), while LP referred to the macroscopic level and his answer was wrong ("The ice has melted. It has changed from a solid to a gaseous state"). The time HP and LP spent on this task was similar—73 s and 61 s, respectively.

Fixation densities. Next, we examined the differences in the eye movements of the participants in solving the task. First, we focused on their fixations on different parts of the text, such as the instructions related to the macroscopic level of the science concept representation (I1), instructions related to the submicroscopic level of concept representation (I2), question (Q) and subtitles (S) (see Fig. 1.1), and compared LP and HP in terms of their attention focus. Instructions are an important part of a science problem as they direct the attention, the coding, the information search and the problem-solving processes. We assumed that the participants' cognitive abilities would be reflected in their reading patterns. Heat maps (see Figs. 1.2 and 1.3) show the density plots of fixation duration, i.e. the percentage of time spent on the screen. The heat maps indicate that LP paid more attention to the instructional sections than HP. LP paid more attention to the instructional section at the macroscopic (I1) and submicroscopic (I2) levels of concept representation. LP stopped his gaze at the end of the sentence in I1 as if he needed a little more time to process what he had been instructed to do. He paid more attention to the words "presentation" and "particle" (Fig. 1.3), compared to HP who walked quickly through I2 and spent most of the time on the word "motion" (Fig. 1.2). In comparison to LP, HP focused primarily on the question (Q; to read: "Name and explain the process that caused the difference, i.e. the change of the state of the Triglav glacier".), especially on the part of the question containing the key words, i.e. "the change of the state" (in Slovenian: "spremembe stanja") (Fig. 1.2). LP on the other hand concentrated on other parts of the question, i.e. on the part "that caused the difference" (in Slovenian: "ki je povzročil razliko") (Fig. 1.3). This observation might indicate that LP spent a lot of time trying to find out what the problem was about, while HP read through the instructions quickly and focused more on the key part of the question, as if she had already assumed that the task required the consideration of the transition between the states of matter.

Table 1.2 shows different measures of fixation for both participants in selected areas of interest—the two photos and the animation. The dwell time is the cumulative time spent on a particular area of interest until the participant indicated that the problem was solved. LP spent most of the time (38%) observing the upper glacier photo (see also Fig 1.3), while HP spent most of the time (40%) examining details of the animation. LP had many short fixations on the upper photo and a large number of regressions to the photo. In contrast, HP had a lower number and longer

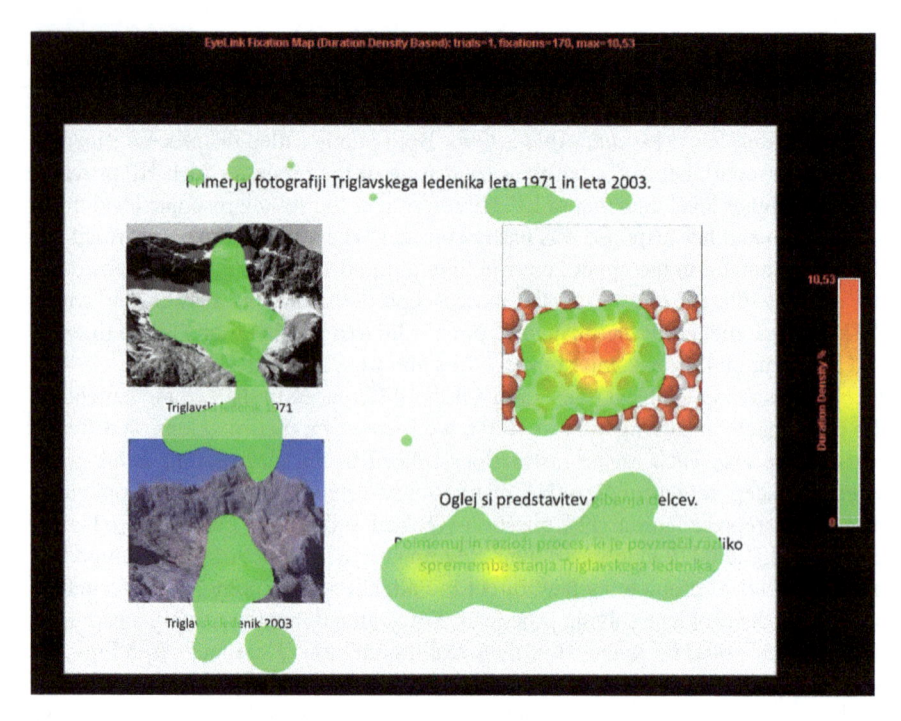

Fig. 1.2 HP's fixation density heat map during problem-solving, showing areas with highest total fixation time (red represents the highest density, followed by yellow and green)

duration of fixations on the animation and fewer regressions on the animation. This could indicate that LP's eye movements resembled frequent and recurring fleeting glances at the photo, while HP watched the animation longer and with fewer eye movements.

The comparison of the time LP and HP spent on viewing the photos and the time spent on animation showed that during the attempt to find relevant information to solve the problem, LP spent more time looking at the photos (Fig. 1.3, Table 1.2), especially the top one, compared to HP who scanned the photos and their subtitles quickly and concentrated mainly on animation (Fig. 1.2, Table 1.2). We assume that focusing on the photo indicated the processing of the macroscopic level of the concept representation, while focusing on the animation promoted a submicroscopic level of the representation of melting.

The difference in the focus of the participants—LP concentrated more on the macroscopic information and HP focused more on the information that directed her reasoning to the submicroscopic level of explanation—was also reflected in their explanations. Their answers were consistent with their fixation density in the areas with macroscopic and submicroscopic representations of melting. HP tried to explain the melting process with a stronger reference to a physical process at the particle level that leads to the phase transition of a substance from a solid to a liquid. Her explanation also included the consequences of the phase transition in the real world: "[…] the

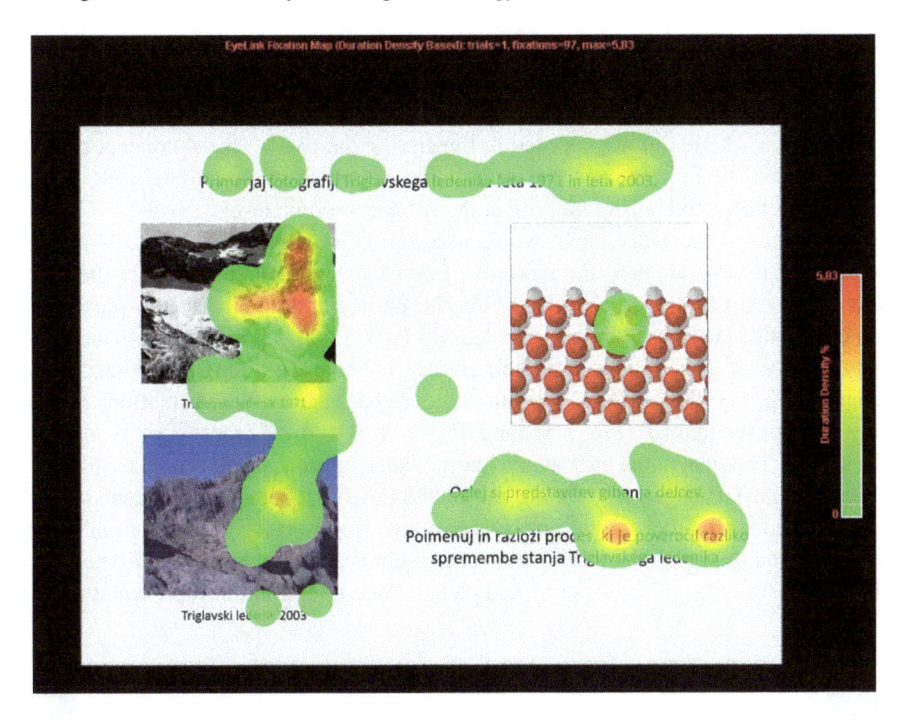

1 Cognitive Processes and Eye-Tracking Methodology

Fig. 1.3 LP´s fixation density heat map during problem-solving, showing areas with highest total fixation time (red represents the highest density, followed by yellow and green)

Table 1.2 Comparison of HP's and LP's measures of fixations on different areas of interest

	Photo 1		Photo 2		Animation	
Fixation measures	LP	HP	LP	HP	LP	HR
Dwell time (ms)	19282	5262	3572	2942	3444	26718
Dwell time (%)	38	8	7	4	7	40
Fixation count	44	17	12	8	9	22
Fixation (%)	27	10	7	5	6	13
Regression count	25	9	9	5	8	4
Average fixation duration (ms)	438	310	298	368	383	1214

atoms that are in the upper layers transition to a liquid state and the water runs down the hill". She seems to have succeeded in integrating different levels of representation of melting—the macroscopic information, i.e. what she observed about the melting process in the two photos ("the atmosphere has warmed up […] the water is running down the hill"), and the submicroscopic information, i.e. what she deduced from observing the animation ("the atoms in the upper layers transition to a liquid state"). She used both levels to explain the melting process. In contrast, LP mostly considered the macroscopic level ("The ice has melted".) and only superficially referred to the submicroscopic level without studying and understanding the animation thoroughly, which led to a "learned" description of the transition from one state of matter to another and to a wrong answer at the end ("It has transitioned from a solid to a gaseous state".). He did not succeed in integrating the two levels of concept representation, and he failed to use the information presented in the animation to think about what happened during melting at the submicroscopic level.

Sequences of saccades. Next, we focused on the scan paths of the participants. In particular, we compared the patterns of saccades in the first 30 s after the task was displayed on the screen to see how the participants looked at the presented materials and started to solve the problem. In Figs. 1.4 and 1.5, the arrows indicate the direction of the saccades, and their colours indicate the timing of the saccades, changing from yellow to grey, with the colour changing every ten fixations in the row. We can see that LP (Fig. 1.5) and HP (Fig. 1.4) took different ways to observe the screen. The scan paths indicate the participants' approaches to problem-solving. Both LP and HP began by reading the instructions that refer to the macroscopic level of the melting representation (I1), continued by observing the photos and their subtitles, and then read the instructions at the submicroscopic level (I2). Remember that this instruction (I2) was short and clear, "Take a look at the representation of

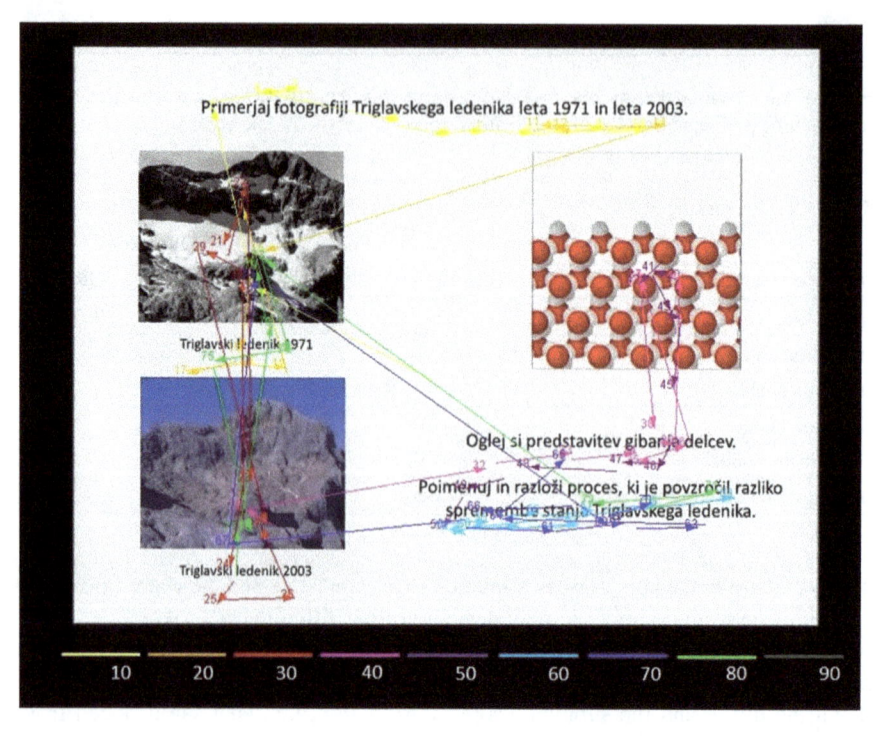

Fig. 1.4 Visualization of HP's saccades in the first 30 s of solving the problem. The colours indicate the gaze sequence (a different colour is used for every 10 fixations). The first 10 saccades are displayed in yellow, the second 10 in orange, the third 10 in red, and so on

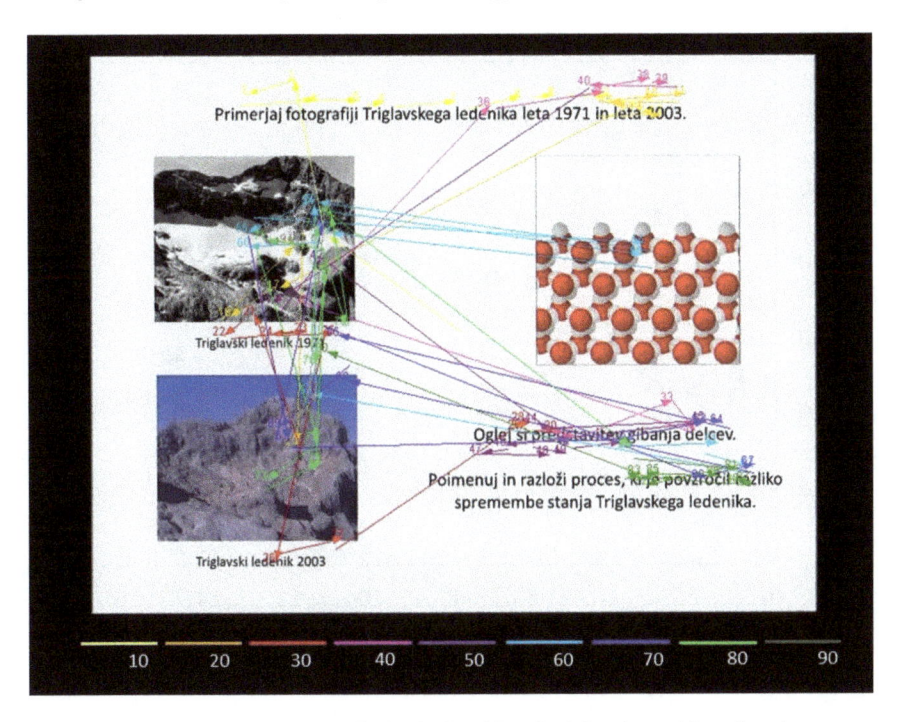

1 Cognitive Processes and Eye-Tracking Methodology

Fig. 1.5 Visualizations of LP's saccades in the first 30 s of solving the problem. See also notes to Fig. 1.4

particle motion" and pointed the participant to the animation. Up to this point, the way LP and HP scanned the screen was the same, but after reading instruction I2, their scanning paths diverged. HP immediately focused her attention on the animation that showed the melting process at the particle level, showing the movement of the increasing number of upper layers against the more static layers below. She observed the animation, then read the question and returned to the photos (Fig. 1.4). Instead of focusing on the animation, LP returned to the top photo, returned to instruction I1, turned again to the top photo, read instruction I2 for the second time, returned to the top photo, then glanced at the animation, looked at the photos again, read instruction I2 and finally read the question (Fig. 1.5). Note that all these sequences took place within the 30-second interval after the screen was displayed. LP did not follow instruction I2, and largely ignored the animation. While HP scanned the display in a predicted order, LP rushed through the displayed information in a seemingly chaotic order. During the next 30 s, which are not shown here, the participants began to return to the previously inspected areas. LP returned to the photos and instruction I2. HP returned to the animation and question Q.

Figure 1.6 shows how often the two participants blinked while solving the problem. Both blinked for the first time after about 20 s. At this point, LP blinked several times in a row. LP also blinked several times in a row after about 40 s of problem-solving. In total, HP blinked four times (twice in the first 30 s) while

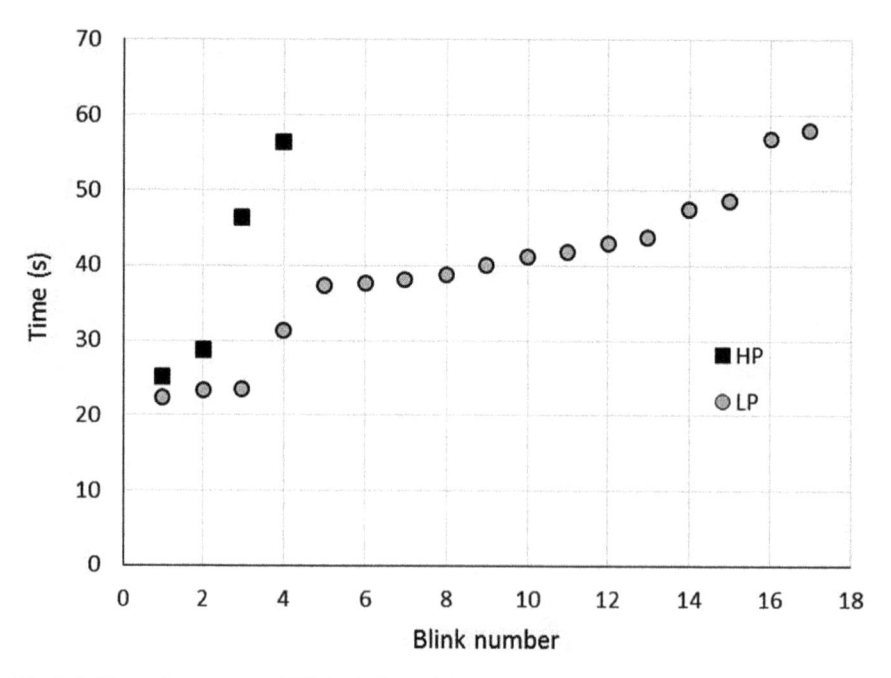

Fig. 1.6 Time of occurence of blinks during solving the problem

solving the problem. LP blinked 17 times (three times in the first 30 s). Since the perceptual load was the same for both participants, the higher blinking rate could indicate a higher cognitive load in LP compared to HP. Figure 1.7 shows the pupil dilation for both participants in the first 80 fixations after the problem was displayed on the screen. A relative increase in pupil size was larger for LP than for HP. Both the blink rate and the relative pupil dilation indicate that the cognitive load was higher for LP than for HP.

Compared to LP, HP seems to have been faster in the visual search, in addition to showing a larger capacity of working memory, better switching ability, and a higher degree of logical reasoning. Although both participants had comparable prior knowledge of the state of matter, HP seemed to have been able to integrate the macroscopic and submicroscopic information presented in the task and develop a better understanding of the transition between different states of matter. The scan paths (Figs. 1.4 and 1.5) reflected differences in the thinking and problem-solving process of the two participants. It appears that HP who had better reasoning abilities and executive functions was able to complete the task more effectively by selectively focusing on the most relevant parts and inhibiting less relevant information, keeping information in working memory and integrating different representations of melting into a meaningful explanation of the process. LP on the other hand, despite having similar prior knowledge of the states of matter (including a similar understanding of the pictorial representations at the submicroscopic level, i.e. at the level of particles), was not able to systematically observe different parts of the display and was not able

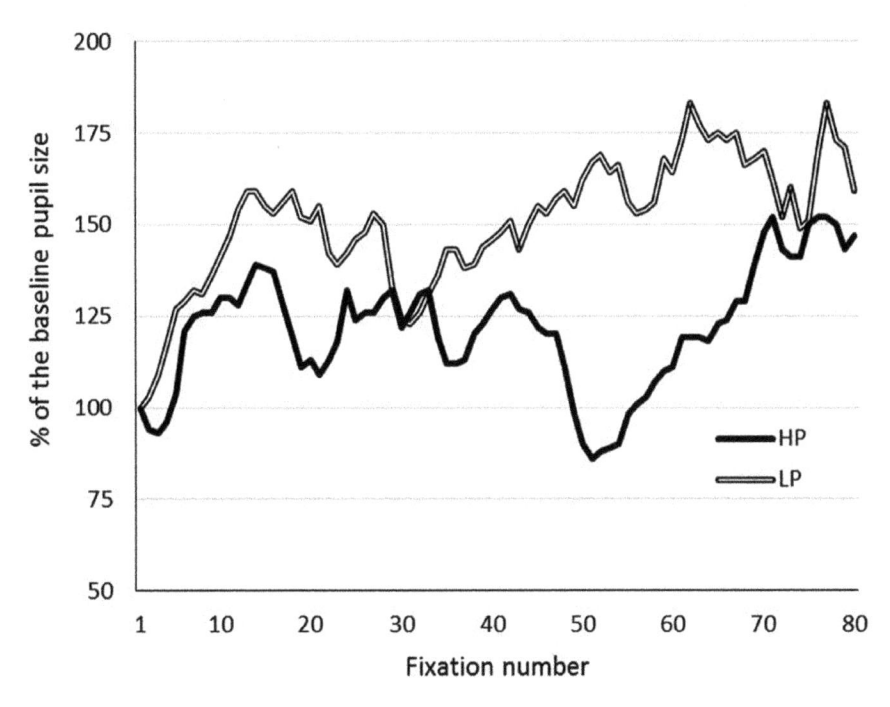

Fig. 1.7 Relative pupil dilation in the first 80 fixations during problem-solving in both participants according to their baseline pupil size

to integrate different levels of representations of melting into a correct understanding of the melting process. Based on various eye movement measures, we conclude that the task was more difficult for him and that he had more difficulties in understanding and following the instructions. Therefore, it seems that cognitive abilities such as those measured in this study could play an important role in the participants' ability to understand the requirements of the task and to integrate different levels of science concept representation into a holistic understanding of the concept. If this is indeed the case, the question of which cognitive abilities are the most important for solving authentic science problems remains a subject of further investigation.

Our results are consistent with the findings of Chanijani et al. (2016), who studied eye movements when students solved physics problems. They found that students with different skill levels preferred different representations for problem-solving and switched between representations with different frequency. Low performers spent more time on questions, while experts focused more on the representations (especially the most informative) and went through the problem more systematically. In our study, HP focused on the most important parts of the instructions and examined the most important representation (i.e. the animation) to find the answer, highlighting the elements of the process that are characteristic for the expert level of problem-solving. In contrast, LP's processing of the information was similar to the problem-solving by beginners.

The aim of the current case study was to show what kind of observations the eye-tracking method can deliver in terms of science problem-solving and the cognitive processes behind it. The cognitive processes can only be derived from their visible and measurable consequences, such as problem-solving behaviour. Eye movements, as manifested by saccade sequences and fixation density heat maps, seem to be a useful tool to gain additional insights into these processes. In our study, the scan paths complemented the information on relative time spent in different display areas as visualized by the heat maps. The classic eye-tracking tool, fixation density heat maps, provide the information about the target and duration of the participants' gaze and indicate the main points of interest in solving the task. In addition to these, as shown in this chapter, the scan paths can provide further insights into the coherence and systematics of the problem-solving processes. While heat maps can provide insightful information about the information that is likely to play an important role in problem-solving, the analysis of the scan paths gives us additional information about how easily this information could be combined. Although the analysis must be performed "by hand" and individually (i.e. with data from a single participant; group analyses are not possible), the temporal dynamics of eye movements provides rich additional information about the reasoning and problem-solving processes and allows a better understanding of these processes.

Conclusions and Implications for Education

The purpose of this chapter was to show that the current level of students' academic knowledge is not the only factor influencing the way students solve science problems. It is likely that different cognitive abilities also play a role. Eye movement measures, such as total fixation duration in specific areas of interest and scan paths used in our case study, may be useful tools in further studies investigating the relationships between interindividual differences in cognitive abilities and science problem-solving. However, the information obtained with these tools also has some limitations. For example, fixation density (total duration of fixation on a particular area of interest and relative measures derived from this) only provides information about the total time spent in a particular area, and cannot tell us whether this total time includes a single or several visits to a particular area, nor can it provide reliable information about how difficult it is for individuals to understand what they are observing, since students sometimes spend long periods of time on an element without much cognitive load being imposed on them (Majooni et al., 2016). Moreover, fixation density does not provide information about the temporal dynamics of eye movements, which can be quite useful when tracking the sequence of information processing and trying to draw conclusions about the person's thinking process. As shown in this chapter, it is therefore recommended to combine fixation density measurements with the analysis of scan paths. In addition, the number of blinks and pupil size can also provide valuable information about the subjective task difficulty and the cognitive load behind problem-solving.

Eye movement measurements provide only indirect information about the cognitive processes involved in the task under study. While eye movement data can be used to infer with some validity where the person's overt attention was directed to during solving the task, it is not possible to be certain how much or how effectively the information from the observed part of the screen was processed, how exactly the processing was done and what exact cognitive processes were used in processing this information. We believe that triangulation of different research approaches could increase the validity of the conclusions drawn. Eye-tracking could be combined with post-test interviews in which students would describe what they were thinking about at a certain point in the problem-solving process and what strategies they used. Stimulated recall could be performed while observing the recording of one's own gaze and the sequence of fixations during the problem-solving process.

These findings have several practical implications. An understanding of the strategies used by both successful and unsuccessful students to solve a specific science problem (e.g. their information processing sequence and fixation density, which indicate the main focus of attention), by comparing their eye movements and cognitive abilities, would lead to a deeper understanding of both the learning processes and the factors that prevent unsuccessful students from solving problems efficiently. This type of insight would enable teachers to help students develop useful and more successful problem-solving strategies, such as focusing on the most important elements of the problem and ways of combining the macro- and submicroscopic levels of concept representation. In addition, by analysing eye movement data, teachers could also gain some understanding of what types of instruction are likely to work best and what types of external representations of science problems are effective in helping students understand science concepts. Additional information on students' cognitive abilities could further clarify interindividual differences in problem-solving and support teaching practices tailored to the predispositions and needs of individual students.

Acknowledgements The authors acknowledge the financial support from the Slovenian Research Agency (research project No. J5-6814, Explaining Effective and Efficient Problem Solving of the Triplet Relationship in Science Concepts Representations, and research core funding No. P5-0110).

References

Ackerman, P. L., Kanfer, R., & Beier, M. E. (2013). Trait complex, cognitive ability, and domain knowledge predictors of baccalaureate success, STEM persistence, and gender differences. *Journal of Educational Psychology, 105*(3), 911.

Arnett, J. A., & Labovitz, S. S. (1995). Effect of physical layout in performance of the Trail Making Test. *Psychological Assessment, 7*(2), 220–221.

Atkinson, R. C., & Shiffrin, R. M. (1968). Human memory: A proposed system and its control processes. In *Psychology of learning and motivation* (Vol. 2, pp. 89–195). Academic Press.

Baddeley, A. (1996a). Exploring the central executive. *The Quarterly Journal of Experimental Psychology Section A, 49*(1), 5–28.

Baddeley, A. (1996b). The fractionation of working memory. *Proceedings of the National Academy of Sciences, 93*(24), 13468–13472.

Baddeley, A. (2000). The episodic buffer: A new component of working memory? *Trends in cognitive sciences, 4*(11), 417–423.

Baddeley, A. (2012). Working memory: Theories, models, and controversies. *Annual Review of Psychology, 63,* 1–29.

Bahar, A., & Maker, C. J. (2015). Cognitive backgrounds of problem solving: A comparison of open-ended vs. closed mathematics problems. *Eurasia Journal of Mathematics, Science and Technology Education, 11*(6), 1531–1546.

Beatty, J. (1982). Task-evoked pupillary responses, processing load, and the structure of processing resources. *Psychological Bulletin, 91*(2), 276.

Berkowitz, M., & Stern, E. (2018). Which cognitive abilities make the difference? Predicting academic achievements in advanced STEM studies. *Journal of Intelligence, 6*(4), 48. https://doi.org/10.3390/jintelligence6040048.

Binder, T., Sandmann, A., Sures, B., Friege, G., Theyssen, H., & Schmiemann, P. (2019). Assessing prior knowledge types as predictors of academic achievement in the introductory phase of biology and physics study programmes using logistic regression. *International Journal of STEM Education, 6*(1), 33.

Biswas, P., Dutt, V., & Langdon, P. (2016). Comparing ocular parameters for cognitive load measurement in eye-gaze-controlled interfaces for automotive and desktop computing environments. *International Journal of Human-Computer Interaction, 32*(1), 23–38.

Bors, D. A., & Stokes, T. L. (1998). Raven's Advanced Progressive Matrices: Norms for first-year university students and the development of a short form. *Educational and Psychological Measurement, 58*(3), 382–398.

Brookings, J. B., Wilson, G. F., & Swain, C. R. (1996). Psychophysiological responses to changes in workload during simulated air traffic control. *Biological Psychology, 42*(3), 361–377.

Case, R., Kurland, M. D., & Goldberg, J. (1982). Operational efficiency and the growth of short-term memory span. *Journal of Experimental Child Psychology, 33,* 386–404.

Chanijani, S. S. M., Klein, P., Al-Naser, M., Bukhari, S. S., Kuhn, J., & Dengel, A. (2016, September). A study on representational competence in physics using mobile eye tracking systems. In *Proceedings of the 18th International Conference on Human-Computer Interaction with Mobile Devices and Services Adjunct* (pp. 1029–1032).

Chen, S. (2014). *Cognitive load measurement from eye activity: acquisition, efficacy, and real-time system design.* Ph.D. thesis, University of New South Wales. http://unsworks.unsw.edu.au/fapi/datastream/unsworks:12261/SOURCE02?view=true.

Chen, S., & Epps, J. (2014). Using task-induced pupil diameter and blink rate to infer cognitive load. *Human-Computer Interaction, 29*(4), 390–413.

Conway, A. R., Kane, M. J., Bunting, M. F., Hambrick, D. Z., Wilhelm, O., & Engle, R. W. (2005). Working memory span tasks: A methodological review and user's guide. *Psychonomic Bulletin & Review, 12*(5), 769–786.

Cornoldi, C., & Giofrè, D. (2014). The crucial role of working memory in intellectual functioning. *European Psychologist, 19*(4), 260–268.

Daneman, M., & Carpenter, P. A. (1980). Individual differences in working memory and reading. *Journal of Verbal Learning and Verbal Behavior, 19,* 450–466.

Della Sala, S., Gray, C., Baddeley, A., Allamano, N., & Wilson, L. (1999). Pattern span: a tool for unwelding visuo–spatial memory. *Neuropsychologia, 37*(10), 1189–1199.

Dimotakis, N., Ilies, R., & Judge, T. A. (2013). Experience sampling methodology. In J. M. Cortina & R. S. Landis (Eds.), *Modern research methods for the study of behavior in organizations* (pp. 345–374). Routledge.

Dodge, R., & Cline, T. S. (1901). The angle velocity of eye movements. *Psychological Review, 8,* 145–157.

Duchowski, A. T. (2002). A breadth-first survey of eye-tracking applications. *Behavior Research Methods, Instruments, & Computers, 34*(4), 455–470.

Eccles, D. W., & Arsal, G. (2017). The think aloud method: what is it and how do I use it? *Qualitative Research in Sport, Exercise and Health, 9*(4), 514–531.

Engle, R. W., Tuholski, S. W., Laughlin, J. E., & Conway, A. R. A. (1999). Working memory, short-term memory and general fluid intelligence: A latent variable approach. *Journal of Experimental Psychology: General, 128,* 309–331.

Eysenck, M. W., & Keane, M. T. (2015). *Cognitive psychology: A student's handbook.* Psychology Press.

Ferk Savec, V., Hrast, Š., Devetak, I., & Torkar, G. (2016). Beyond the use of an explanatory key accompanying submicroscopic representations. *Acta Chimica Slovenica, 63*(4), 864–873.

García, O., López, F., Icaran, E., & Burgos, S. (2014). Relationship between general intelligence, competences and academic achievement among university students. *Personality and Individual Differences, 60,* S67.

Gegenfurtner, A., Lehtinen, E., & Säljö, R. (2011). Expertise differences in the comprehension of visualizations: A meta-analysis of eye-tracking research in professional domains. *Educational Psychology Review, 23,* 523–552. https://doi.org/10.1007/s10648-011-9174-7.

Goldberg, J. H., Stimson, M. J., Lewenstein, M., Scott, N. & Wichansky, A. M. (2002). Eye tracking in Web search tasks: Design implications. *Proceedings of the Eye Tracking Research & Application Symposium,* 51–58. https://doi.org/10.1145/507072.507082.

Green, H. J., Lemaire, P., & Dufau, S. (2007). Eye movement correlates of younger and older adults' strategies for complex addition. *Acta Psychologica, 125,* 257–278.

Haidar, A. H., & Abraham, M. R. (1991). A comparison of applied and theoretical knowledge of concepts based on the particulate nature of matter. *Journal of Research in Science Teaching, 28*(10), 919–938.

Hartmann, M. (2015). Numbers in the eye of the beholder: What do eye movements reveal about numerical cognition? *Cognitive Processing, 16*(1), 245–248. https://doi.org/10.1007/s10339-015-0716-7.

Hoeks, B., & Levelt, W. J. (1993). Pupillary dilation as a measure of attention: A quantitative system analysis. *Behavior Research Methods, Instruments, & Computers, 25*(1), 16–26.

Hurks, P. P. M., Vles, J. S. H., Hendriksen, J. G. M., Kalff, A. C., Feron, F. J. M., Kroes, M., ... & Jolles, J. (2006). Semantic category fluency versus initial letter fluency over 60 seconds as a measure of automatic and controlled processing in healthy school-aged children. *Journal of Clinical and Experimental Neuropsychology, 28*(5), 684–695.

Inhelder, B., & Piaget, J. (1958). *The growth of logical thinking from childhood to adolescence: An essay on the construction of formal operational structures.* New York: Basic Books.

Iqbal, S. T., Adamczyk, P. D., Zheng, X. S., & Bailey, B. P. (2005, April). Towards an index of opportunity: understanding changes in mental workload during task execution. In W. Kellogg & S. Zhai (Eds.), *Proceedings of the SIGCHI conference on Human factors in computing systems* (pp. 311–320). Association for Computing Machinery.

Jacob, R. J. K., & Karn, K. S. (2003). Eye tracking in human–computer interaction and usability research: Ready to deliver the promises. *The Mind's Eye: Cognitive and Applied Aspects of Eye Movement Research,* 573–605.

Javal, E. (1878). Essai sur la physiologie de la lecture. *Annales d'Ocilistique, 80,* 97–117.

Kaufman, S. B., DeYoung, C. G., Reis, D. L., & Gray, J. R. (2011). General intelligence predicts reasoning ability even for evolutionarily familiar content. *Intelligence, 39*(5), 311–322.

Kell, H. J., Lubinski, D., Benbow, C. P., & Steiger, J. H. (2013). Creativity and technical innovation: Spatial ability's unique role. *Psychological Science, 24*(9), 1831–1836.

König, P., Wilming, N., Kietzmann, T. C., Ossandón, J. P., Onat, S., Ehinger, B. V., Gameiro, R. R., & Kaspar, K. (2016). Eye movements as a window to cognitive processes. *Journal of Eye Movement Research, 9*(5), 1–16.

Kozma, R. B., & Russell, J. (1997). Multimedia and understanding: Expert and novice responses to different representations of chemical phenomena. *Journal of Research in Science Teaching: The Official Journal of the National Association for Research in Science Teaching, 34*(9), 949–968.

Krejtz, K., Duchowski, A. T., Niedzielska, A., Biele, C., & Krejtz, I. (2018). Eye tracking cognitive load using pupil diameter and microsaccades with fixed gaze. *PloS one, 13*(9).

Kruger, P. B. (1980). The effect of cognitive demand on accommodation. *American Journal of Optometry and Physiological Optics, 57*(7), 440–445.

Lai, M. L., Tsai, M. J., Yang, F. Y., Hsu, C. Y., Liu, T. C., Lee, S. W. Y., Lee, M. H., Chiou, G. L., Liang, J. C., & Tsai, C. C. (2013). A review of using eye-tracking technology in exploring learning from 2000 to 2012. *Educational Research Review, 10*, 90–115.

Laukkonen, R. E., & Tangen, J. M. (2018). How to detect insight moments in problem solving experiments. *Frontiers in psychology, 9*, 282.

Lawson, A. E., & Renner, J. W. (1975). Relationships of science subject matter and developmental levels of learners. *Journal of Research in Science Teaching, 12*(4), 347–358.

Lin, T., Imamiya, A., & Mao, X. (2008). Using multiple data sources to get closer insights into user cost and task performance. *Interacting with Computers, 20*(3), 364–374.

Logie, R. H. (1995). *Visuo-spatial working memory*. Hove: Erlbaum.

Lombardi, B. M. M., & Oblinger, D. G. (2007). Authentic learning for the 21st century: An overview. *Learning, 1,* 1–7.

Madsen, A.M., Larson, A.M., Loschky, L.C., & Rebello, N. S. (2012). Differences in visual attention between those who correctly and incorrectly answer physics problems. *Physical Review Special Topics—Physics Education Research, 8*(010122), 1–13. https://doi.org/10.1103/physrevstper.8.010122.

Majooni, A., Masood, M., & Akhavan, A. (2016). An eye tracking experiment on strategies to minimize the redundancy and split attention effects in scientific graphs and diagrams. In *Advances in design for inclusion* (pp. 529–540). Cham: Springer.

Ma Oliva, J. (1999). Structural patterns in students' conceptions in mechanics. *International Journal of Science Education, 21*(9), 903–920.

Meltzer, L. (2010). *Promoting executive function in the classroom*. Guilford Press.

Mihelčič, M., & Podlesek, A. (2017). The influence of proprioception on reading performance. *Clinical and Experimental Optometry, 100*(2), 138–143.

Miyake, A., & Friedman, N. P. (2012). The nature and organization of individual differences in executive functions: Four general conclusions. *Current Directions in Psychological Science, 21*(1), 8–14. https://doi.org/10.1177/0963721411429458.

Moeller, K., Klein, E., & Nuerk, H. C. (2011). Three processes underlying the carry effect in addition–evidence from eye tracking. *British Journal of Psychology, 102,* 623–645. https://doi.org/10.1111/j.2044-8295.2011.02034.x.

Mueller, S. T. (2012). The psychology experiment building language, Version 0.13. Retrieved from http://pebl.sourceforge.net.

Mueller, S. T., & Piper, B. J. (2014). The psychology experiment building language (PEBL) and PEBL test battery. *Journal of Neuroscience Methods, 222,* 250–259.

Muhamad, S., Harun, J., Surif, J., & Abd Halim, N. D. (2016). Authentic chemistry problem solving competency for open-ended problems in learning electrolysis: Preliminary study. *Journal Pendidikan Teknik Dan Vokasional Malaysia, 1*(1), 365–373. https://people.utm.my/noordayana/files/2012/10/fullpaper-suraiya-icehots.pdf.

Pomplun, M., & Sunkara, S. (2003). Pupil dilation as an indicator of cognitive workload in human–computer interaction. *Human-Centred Computing: Cognitive, Social and Ergonomic Aspects, 3,* 542–546.

Rausch, A., Kögler, K., & Seifried, J. (2019). Validation of Embedded Experience Sampling (EES) for measuring non-cognitive facets of problem-solving competence in scenario-based assessments. *Frontiers in psychology, 10*.

Raven, J., Raven, J. C., & Court, J. H. (1998). *Manual for Raven's progressive matrices and vocabulary scales*. Oxford: Oxford Psychologists Press.

Raven, J., Raven, J. C., Court, J. H., & Južnič Sotlar, M. (1999). *Priročnik za Ravnove progresivne matrice in besedne lestvice. Zahtevne progresivne matrice : z normami za odrasle in novimi normami za vrsto nacionalnih skupin [Manual for Raven's Progressive Matrices and vocabulary* scales. *Advanced Progressive Matrices with norms for adults and the new norms for several national groups]*. Ljubljana: Center za psihodiagnostična sredstva.

Reitan, R. M. (1958). Validity of the Trail Making Test as an indicator of organic brain damage. *Perceptual and Motor Skills, 8*(3), 271–276.

Remšak, T. (2013). *Razvoj testa črkovne fluentnosti [Development of a test of letter fluency]*. Unpublished graduate thesis, University of Ljubljana, Slovenia.

Rosander, P., Bäckström, M., & Stenberg, G. (2011). Personality traits and general intelligence as predictors of academic performance: A structural equation modelling approach. *Learning and individual differences, 21*(5), 590–596.

Shah, P., & Miyake, A. (1996). The separability of working memory resources for spatial thinking and language processing: An individual differences approach. *Journal of Experimental Psychology: General, 125,* 4–27.

Shao, Z., Janse, E., Visser, K., & Meyer, A. S. (2014). What do verbal fluency tasks measure? Predictors of verbal fluency performance in older adults. *Frontiers in psychology, 5,* 772. https://doi.org/10.3389/fpsyg.2014.00772.

Schüler, A., Scheiter, K., & van Genuchten, E. (2011). The role of working memory in multimedia instruction: Is working memory working during learning from text and pictures? *Educational Psychology Review, 23*(3), 389–411.

Scholz, A., von Helversen, B., & Rieskamp, J. (2015). Eye movements reveal memory processes during similarity-and rule-based decision making. *Cognition, 136,* 228–246.

She, H. C., Cheng, M. T., Li, T. W., Wang, C. Y., Chiu, H. T., Lee, P. Z., Chou, W. C., & Chuang, M. H. (2012). Web-based undergraduate chemistry problem-solving: The interplay of task performance, domain knowledge and web-searching strategies. *Computers & Education, 59*(2), 750–761.

Shea, D. L., Lubinski, D., & Benbow, C. P. (2001). Importance of assessing spatial ability in intellectually talented young adolescents: A 20-year longitudinal study. *Journal of Educational Psychology, 93*(3), 604–614.

Stern, J. A., Boyer, D., & Schroeder, D. J. (1994). Blink rate as a measure of fatigue: A review. *Human Factor, 36*(2), 285–297.

Susac, A., Bubic, A., Kaponja, J., Planinic, M., & Palmovic, M. (2014). Eye movement reveal students' strategies in simple equation solving. *International Journal of Science and Mathematics Education, 12*(3), 555–577. https://doi.org/10.1007/s10763-014-9514-4.

Taconis, R., Ferguson-Hessler, M. G., & Broekkamp, H. (2001). Teaching science problem solving: An overview of experimental work. *Journal of Research in Science Teaching: The Official Journal of the National Association for Research in Science Teaching, 38*(4), 442–468.

Thomas, L. E., & Lleras, A. (2009). Covert shifts of attention function as an implicit aid to insight. *Cognition, 111*(2), 168–174.

Tobin, K., & Capie, W. (1984). The Test of Logical Thinking. *Journal of Science and Mathematics Education in Southeast Asia, 7*(1), 5–9.

Tobin, K. G., & Capie, W. (1981). The development and validation of a group test of logical thinking. *Educational and Psychological Measurement, 41*(2), 413–423.

Treagust, D., Chittleborough, G., & Mamiala, T. (2003). The role of submicroscopic and symbolic representations in chemical explanations. *International Journal of Science Education, 25*(11), 1353–1368.

Trexler Holland, C. (1995). *The effects of formal reasoning ability, spatial ability, and type of instruction on chemistry achievement*. Unpublished doctoral dissertation, University of Florida. https://archive.org/details/effectsofformalr00holl/mode/2up.

Trifone, J. D. (1987). The Test of Logical Thinking: Applications for teaching and placing science students. *The American Biology Teacher,* 411–416.

Turner, M. L., & Engle, R. W. (1989). Is working memory capacity task dependent? *Journal of Memory and Language, 28,* 127–154.

Vitu, F., & McConkie, G. W. (2000). Regressive saccades and word perception in adult reading. In A. Kennedy, R. Radach, D. Heller, & J. Pynte (Eds.), *Reading as a perceptual process* (pp. 301–326). Elsevier.

Vogels, J., Demberg, V., & Kray, J. (2018). The index of cognitive activity as a measure of cognitive processing load in dual task settings. *Frontiers in Psychology, 9,* 2276.

Anja Podlesek, Ph.D. is an Associate Professor of psychological methodology at the Department of Psychology, University of Ljubljana. Her research includes cognitive psychology (cognitive control, cognitive training, visual perception and attention) and measurement in psychology (development of psychological tests, use of physiological measures). Since 2014 she has been the leader of the research programme "Psychological and neuroscientific aspects of cognition" at the Slovenian Research Agency. As a methodologist she has also participated in studies in health psychology, especially suicidal behaviour, developmental and educational psychology (factors of academic achievements, secondary analyses of PISA and TIMSS data, measurement of the quality of the education system). In 2015–2016 she was the leader of the Norway Grants project "Supervised practice of psychologists: Development of a training programme for mentors and a model of supervised practice - SUPER PSIHOLOG". She has published over 90 scientific and professional papers and about 50 edited books or book chapters. She is a member of Horizons of Psychology and Frontiers Human-Media Interaction Editorial Boards. In the years 2012–2020 she was the chair of Slovenian National Awarding Committee for the European Certificate in Psychology (EuroPsy). In 2017 she was awarded the Ljubljana University Golden Plaque prize. Her involvement in the organisation of international scientific events earned her the title Congress Ambassador of Slovenia 2019.

Manja Veldin holds a B.A. degree in Psychology and works as a researcher at the Centre for Evaluation Studies at Educational Research Institute in Ljubljana, Slovenia. She used to cooperate with the Centre for psychodiagnostic assessment, Chameleon Consulting and participated in the study "Quality of life of patients after kidney transplantation" and in a Slovenian project "Explaining effective and efficient problem solving of the triplet relationship in science concepts representations". She also participated in a number of EU projects (HAND in HAND: Social and Emotional Skills for Tolerant and Non-discriminative Societies (A Whole School Approach); BRAVEdu: Breaking the Poverty Taboo: Roles and Responsibilities of Education; ETTECEC: Early attention for the inclusion of children with autism spectrum disorder in ECEC systems) and led the evaluation of a national project (Innovative learning environments supported by innovative pedagogy 1:1). Currently, she is working on several EU projects (HEAD: Empowering School Principals for Inclusive School Culture; DITEAM: Diverse and Inclusive teams for children under 12; NEMO: NEw MOnitoring guidelines to develop innovative ECEC teachers curricula) and a Slovenian project (Positive Youth Development in Slovenia: Developmental Pathways in the Context of Migration). She is a Ph.D. candidate in Experimental Psychology at the Faculty of Arts, University of Ljubljana (dissertation title: "Theoretical Considerations and Development of a Questionnaire to Measure Empathy"). She regularly publishes the results of her research work in professional and scientific journals and presents her findings at scientific conferences.

Cirila Peklaj, Ph.D. is a Professor of Educational Psychology at the University of Ljubljana, Faculty of Arts, Department of Psychology, Slovenia. Her research focuses on the relationships between different learning factors and achievement. She studied learning in different contexts, subjects (biology, mathematics, music, Slovenian language) and with different samples (primary, secondary, and tertiary students). The studies focused on the learning environment (i.e. e-learning, traditional learning) as well as on teachers and their competencies to promote educational goals at school (i.e. reading performance, knowledge, behaviour). They also focused on learning processes, namely cognitive (cognitive and metacognitive strategies) and affective-motivational aspects (self-efficacy, interest, emotions) of self-regulated learning. She has been involved in various research projects on learning. Currently, she is leading a research project on the effectiveness of different types of scaffolds in self-regulated e-learning. She is the editor of a Shaker Verlag monograph on teacher competencies and educational goals and (co-)author of books, chapters, and publications in prestigious journals (about 250 publications in total). She is a member of the editorial board in European Journal of Psychology of Education and Head of the Centre for Pedagogical Education at the Faculty of Arts.

Matija Svetina, Ph.D. is a Professor of Developmental Psychology at the University of Ljubljana, Slovenia, where he teaches courses in developmental and environmental psychology. His research focuses on cognitive and social processes behind developmental and learning changes, developmental transitions throughout lifespan, and decision-making processes related to environmental issues. He is involved in the development of microgenetic analysis, a theoretical and methodological framework for the analysis of behavioural change. He is/was a postdoctoral fellow, visiting lecturer, and researcher at various institutions, Carnegie Mellon University, Pittsburgh, PA; University of Pittsburgh, PA; Augsburg University, Minneapolis, MS; University of Leipzig, Germany; Alps-Adria University, Klagenfurt, Austria.

Chapter 2
The Interplay of Motivation and Cognition: Challenges for Science Education Research and Practice

Mojca Juriševič and Tanja Černe

Learning science at school is a complex learning task with success depending on several internal (e.g., prior knowledge, ability, personal characteristics) and external factors (e.g., teacher's competences, syllabus, classroom climate). Science education and the perception of science classes as "hard" or "difficult" are discussed by scholars, science teachers, and students for at least three decades. Research evidence shows that students do not always achieve their full potential in science classes when they are not properly motivated (Anderman & Young, 1994; Aschbacher et al., 2010; European Commission, 2007; Gilbert, 2006; Holbrook & Rannikmäe, 2014; Johnstone, 1991; Lyons, 2006; Lyons & Quinn, 2010; Osborne et al., 2003; Potvin & Hasni, 2014; Shirazi, 2017; Vossen et al., 2018). According to current research, the "negative" reputation of science in school has two reasons. First, the multifaceted and abstract nature of science, which requires motivation, effort, and higher order thinking in order to learn it meaningfully and relate it to the immediate sociocultural context. Second, the "negative" aspect is related to the ways in which science contents are transferred or taught to students. Another question is interesting in this respect, namely, do new technologies used for science education (Oliveira et al., 2019) contribute to better education (Lodge & Harrison, 2019), and what is their influence on students' motivation and their cognitive processes during science education. In order to understand these questions, one needs to thoroughly understand the processes of teaching and learning (Mayer, 2010).

This chapter aims to explore how two internal variables are affecting learning processes, i.e., students' motivation and (visual) attention. This study focuses on

---

M. Juriševič (✉)
Faculty of Education, The University of Ljubljana, Ljubljana, Slovenia
e-mail: mojca.jurisevic@pef.uni-lj.si

T. Černe
Counseling Center for Children, Adolescents and Parents Ljubljana, Ljubljana, Slovenia

their interrelated role in teaching and learning science at school, which is a relevant problem for the current theoretical and empirical research (Braver, 2016). Moreover, it implies a responsibility for the future research and practice that are both necessary in science education of the twenty-first century (Osborne, 2013). In this chapter we focused on challenges in learning and teaching science and exposed the examples of potential problems in learning motivation in a group of dyslectic students and their difficulties in visual processing, which affect in less effective attention and motivation orientation in the learning process.

In this respect, the framework of the Partnership for 21st Century Learning (P21) lists science among the nine key subjects, and defines motivation—i.e., initiative and self-direction by means of management of goals and time, independent work, and self-directed learning—as one of the five key life and career skills (Battele for Kids, 2019). While this chapter does not provide an overall evaluation of the topic, it aims to define some relevant topics for the evaluation of factors that need to be considered for a quality science education.

Motivation in Learning

Learning motivation is a type of motivation that is expressed by students in the context of learning through their behaviours. Motivation is a psychological process (Weiner, 1992) which appears in the form of various motivational components—interests, self-concept, goals, values, attributions, challenges, or external stimuli. It energises the learning process by activating it and then guiding it until the final learning task or goal is achieved (Juriševič et al., 2012). Koballa and Glynn (2007) argue that

> the study of motivation by science education researchers attempts to explain why students strive for particular goals when learning science, how intensively they strive, how long they strive, and what feelings and emotions characterize them in this process. (p. 85)

Research shows that motivation is closely related to learning and the accompanying cognitive or metacognitive processes (Crede & Phillips, 2011; Pintrich & Schunk, 2002; Schraw et al., 2006; Tuan et al., 2005). It is related to information storage in long-term memory, its recognition, and its retrieval (Cook et al., 2015; Miendlarzewska et al., 2016; Murayama & Elliot, 2011; Murty & Dickerson, 2017; Schiefele & Rheinberg, 1997). According to Stipek (2002), learning motivation is expressed in students' attitudes towards learning and their diverse approaches to learning. Jarvela and Niemivirta (2001) argue that learning motivation advances higher forms of learning and consequentially results in better knowledge, especially at the conceptual level (see also Bomia et al., 1997; Hung-Chih et al., 2010; Linnenbrink & Pintrich, 2002; Phillips et al., 2008; Ryan & Deci, 2009; Thomas & Oldfather, 1997). Based on these findings, motivation is defined as a mediatory variable of learning performance (Linnenbrink-Garcia & Patall, 2016) that is positively, but weakly correlated to school grades; it explains around 10% of learning success, while intellectual abilities explain around 50 or 60%. Nevertheless, motivation is crucial for learning because learning cannot happen without motivation (Juriševič, 2006; Schiefele & Rheinberg, 1997).

By investigating tasks in science school curricula and considering the characteristics of students and their educational context, Rheinberg et al. (2000; see also Vollmeyer & Rheinberg, 2013) identified three levels on which motivation and its components influence learning: (1) the level of time investment that the student devotes to learning and to learning activities, both in the terms of its extent (duration) as well as its frequency, i.e., active learning time; (2) the level of forms and contents of learning activities by which student's efforts are negotiated in relation to activity's level of difficulty on the one hand, and on the other, learning strategies are implemented in order to motivate the student for learning and for achieving learning goals (e.g., superficial vs. deep learning); (3) the level of functional mood, which refers to the optimal psychological state of the student during learning (e.g., positive emotions, devotion, mental concentration). Rheinberg et al. (2000) further define a student's functional state as a motivational state that acts as a mediator for learning success:

> This variable refers to the learner's physiological and psychological activation and concentration during learning. Last but not least, we think that the learner's *motivational state* during learning mediates the effects that the initial learning motivation (i.e., the motivation that led the person to start learning) has on the learning outcome. As the initial learning motivation may change considerably during a long learning period, it is not tautological to regard motivational state during learning as a mediating variable for the impact of initial motivation on learning outcome. (p. 84)

The Relationship Between Motivation and Attention in Learning

Schiefele and Rheinberg (1997) hypothesise that the student's optimal functional state during learning significantly influences her/his conative functioning. The researchers compare this state with "flow", which is according to Csikszentmihalyi (1988) a state of high concentration during which a student best realises his learning abilities and achieves high learning outcomes. According to Posner and Boies (1971), such states of arousal and alertness are the first phase in the process of attention, controlled by the reticular activating system (RAS): if environmental stimuli do not result in student's arousal and alertness, learning does not occur. The phase of selection follows: this is the phase of selecting the most important stimuli, that is, directing the attention in a way that adequately and effectively eliminates distractors and maintains attention for so long and with such intensity as necessary so that the student can understand and use the learned topic. These processes are controlled by the frontal lobes. Moreover, attention allows students to plan, monitor, and regulate their thinking and learning behaviour (Barkley, 1998; Levine, 2002). The relationship between motivation and attention is therefore important for education. However, it has not yet been fully explained (Braver, 2016). Exploring this relationship can take at least two directions: first, studying the impact of motivation (rewards) on attention (Anderson & Sali, 2016; Rothkirch & Sterzer, 2016), and second, studying the role of orientative attention in motivated behaviour (Suri & Gross, 2015). If a student knows, for example, that she/he is learning for a grade or a reward, she/he will be more attentive during the learning process. It is also possible, however, that this type of instrumental behaviour is expected from a student, but she/he will not perform it despite the fact that it would be beneficial for her/him. Suri and Gross (2015) argue that in such cases the lack of attention, which does not cause arousal, is the main problem, as there is no consequential activation (energisation) present. Orientation of behaviour and motivation depends on whether the student chooses the behaviour of approaching or that of avoiding. In the background of this decision are always evaluation processes that negotiate between what is good and what is not good for the individual. These processes require attention. If a student does not pay attention to the evaluation of certain stimuli, this does not result in a motivated learning behaviour.

Early Development of Motivation

Research shows that children who live in a safe and predictable environment develop a healthy motivation system, which is based on a judgement, whether they should begin with an activity or rather avoid it, if the circumstances are threatening. Those children that live in an unpredictable and chaotic environment develop behaviours of avoidance so that their behaviours are based on fear (National Scientific Council, 2018). In behaviours of avoidance, emotions of fear or disgust are present in the amygdala. When the amygdala is activated, norepinephrine and other stress hormones, which prepare the body for a fight or a quick flight, are released. Consequentially, the heart rate increases, levels of blood sugar rise, oxygenation changes, and metabolic and digestive processes slow down (Schabel et al., 2011). Behaviours of avoidance which the child learns in early childhood also affect other life stages, mainly through problems with planning, achieving long-term goals, and motivational orientation. The likelihood of deep learning and higher forms of learning is reduced (National Scientific Council, 2018). It is important to consider this when discussing science education where well-developed cognitive abilities such as reasoning, logical thinking, remembering, understanding sequences, and different types of data integration are important.

Crone and Dahl (2012) found that children who have been emotionally neglected or abused develop an increased susceptibility to social exclusion, especially during adolescence. Positive feedback from important others, including teachers, positively influences their motivational orientation. If in learning an adolescent exhibits a low motivational orientation towards a particular goal, positive feedback can strengthen her/his feeling that she/he is on the right path, and conversely, if an adolescent exhibits a high motivational orientation, critical feedback is more efficient, because it points towards the difference between where she/he currently stands, and where she/he aims to arrive (Fishback et al., 2010). Both motivational orientations, the one oriented towards the expected reward and the one oriented away from threats or dangers, are important for survival. Regulation of motivation depends on the coordination of molecules (peptides, hormones, neurotransmitters) that integrate different signals in the decision-making function (Simpson & Balsam, 2016). Those processes evolve in a supportive environment. However, when both types of motivational orientations are not balanced, over-rewarding or over-avoiding can produce a number of unwanted results, for example, attention deficit hyperactivity disorder, attention deficit disorder, depression, anxiety, substance abuse, and post-traumatic stress disorder (Kasch et al., 2002; Meyer et al., 1999; Muris et al., 2001) as well as other mental disorders (Simpson & Balsam, 2016).

Neural System of Motivation

In recent decades, researchers of cognitive neuroscience use non-invasive methods to explain the neurological basis of motivation. Understanding classical and modern research findings is important for understanding the school setting, especially when discussing about motivational topics and the importance of rewarding, punishing, or asserting repercussions. Using meta-analysis, Cameron and Pierce (1994) found that an individual's interpretation of praise and reward is important. If it is perceived as control over one's actions, it negatively affects internal motivational orientation. If it is perceived as information about one's ability and is understood as recognition of efforts, it can have a positive effect. In his early research, Deci (1971) argued that under certain conditions rewards are not suitable as they reduce learning motivation. Deci et al. (1999) and Lepper et al. (1973) found that external rewards negatively influence internal motivational orientation, especially when intriguing challenges are concerned. Hence, they suggest that external rewards are not always beneficial in learning because the promise of a reward activates the dopamine system and stimulates the performance of an activity, which seems worthy of learning and interesting in terms of gaining new experiences. Fastrich et al. (2018) and McGillivray et al. (2015) also found that tasks, which promote internal motivational orientation, are already interesting so that rewarding is not necessary or may even be harmful. The short-term effect of rewards influences further recognition of those circumstances that lead to an award (Adcock et al., 2006; Murty et al., 2012, 2013). Murayama et al. (2010) also found that internal motivational orientation undermines the prediction of introducing external rewards. This reduces the activation in the striatum, which is the part of the neural network responsible for the reward. As preliminary research, Murayama (2018) found that striatum is related to personal decisions, risk-taking, and the need for curiosity. This proves that internally conditioned "rewards" play a central role in deciding which activity an individual will choose. Murayama and Kitagami (2014) argued that monetary rewards could improve learning. This activates the hippocampus, which creates a neural network that forms the basis for rewarding (Adcock et al., 2006). Regarding external motivational orientation, Schiller et al. (2009) discovered an increased activity in the posterior cingulate cortex, which is activated in decision-making and in social judgement contexts. Lee et al. (2012) found differences in the activation of brain structures when an individual is internally or externally motivated. With regard to the intrinsic motivational orientation, activity in the insular cortex was identified. Insular cortex is responsible for emotional processing in decision-making contexts (Bechara & Damasio, 2005) and is associated with hedonic emotions. Di Domenico and Ryan (2017) suggested that intrinsic motivational orientation is phylogenetically conditioned and that it includes the dopamine system. They found that intrinsic motivational orientation produces patterns of activity in large neural networks, namely, those neural networks that support detection of recognition, attention control, and self-referential cognition.

Self-Determination in Learning

Motivation is one of the most important research topics due to its importance in various areas of human life. In the field of education, various concepts connected with motivation have been studied (e.g., self-efficacy, academic emotions, self-regulation), and various theoretical models were developed in order to explain the relationship between motivation, learning, and learning performance. Examples of such theories include Attributional Theory, Expectancy-Value Theory, Achievement Goal Theory, and Self-Determination Theory (see Hidi & Renninger, 2019; Wentzel & Miele, 2016 for their review). In what follows, we focus on Self-Determination Theory in more detail, because it explains in relative detail those parts of motivation which empirical research identified as particularly relevant for learning and teaching science subjects in schools.

Self-Determination Theory (SDT). SDT (Deci & Ryan, 1985; Ryan & Deci, 2020) is a macro theory that analyses the extent to which human behaviour is self-determined. This means that it analyses the extent to which an individual independently decides to achieve their goals, and how and to what extent they reflect their thoughts during this process. SDT is hence relevant for researching the role of motivation in learning, the promotion of motivation, the evaluation of education, and encouraging students' confidence in their abilities (i.e., self-concept). SDT focuses on motivational orientations or types rather than just the amount of motivation, paying particular attention to autonomous motivation, controlled motivation, and motivation as a predictor of performance, relational, and well-being outcomes (Deci & Ryan, 2008). A common question of SDT in the framework of the school system, therefore, is the following: how to increase learning motivation in education without external pressures, and how to encourage the development of students' internal motivation?

According to the SDT, motivation can be autonomous or controlled. Students who are autonomously motivated feel an inner desire to perform the activity for which they are motivated (Deci & Ryan, 2008). The autonomy-supportive style of teaching is primarily related to a relaxing classroom atmosphere, which, according to neuropsychological research studies, is crucial for effective learning (Reeve, 2016). Education in contexts that promote autonomy also promotes internal motivational orientation (Deci et al., 1981) and internalisation (Grolnick et al., 1991). Patall et al. (2019) report that in science classes students show more genuine interest in learning if teachers are committed to supporting autonomous behaviours. Moreover, such contexts offer better conditions for conceptual learning (Grolnick et al., 1991) and creativity (Koestner et al., 1984), and further increase autonomous motivation and the perceived competences in science labs (Black & Deci, 2000). Skinner et al. (2017) found that autonomy-supportive classes are related to higher identification of students as scientists. Students who aim to choose their goals independently are more likely to achieve success by means of higher educational outcomes (Boggiano et al., 1993; Deci & Ryan, 2008; Deci et al., 1981; Fink et al., 1992; Guay et al., 2008; Juriševič et al., 2012; Toshalis & Nakkula, 2012). Ryan and Grolnick (1986) found that a student's perception of her/his own autonomy varies within a one-hour lesson. They suggested that the perception of a student's control depends on the perception of their competence, the perception of self-concept, and self-esteem. Students have different experiences in classrooms: from subordination to independence. When motivation is controlled, a student feels pressured to think, feel, or behave in a certain way. Research evidence implies that traditional external motivators in schools, such as setting deadlines for tasks (Amabile et al., 1976), supervision (Lepper & Greene, 1975; Plant & Ryan, 1985), competition (Deci et al., 1981), evaluation (Church et al., 2001), and goal-setting (Mossholder, 1980) might not increase students' internal motivation. Guay et al. (2008) found that students who primarily experience controlling motivation are more distracted, anxious, and achieve lower learning results.

Researchers of self-determination are interested in the relationship between autonomy, structure, and discipline, which is an aspect of external motivation. Eckes et al. (2016) argue that teachers most effectively encourage the development of intrinsic motivation by providing appropriate structure and promoting autonomy. Ryan (1982, 2016) emphasises the importance of feedback in providing structure and determines the appropriate relationship between autonomy and structure, which alongside appropriate social cohesion allows for a sense of security and intrinsic motivation.

Self-Determination Theory's taxonomy of motivation. Ryan and Deci (2009, 2020) described the regulation of motivation on a five-level motivational continuum. The first level is without regulation; it is a-motivation or non-regulation. It is present when a student is not motivated for an activity and it describes the lack of intention and purpose, e.g., "Not interested in science subjects at all". A-motivation is therefore a lack of self-determination and motivation (Deci & Ryan, 2000). The individual does not act with an intention to carry out the activity. This level is often related to learned helplessness because a-motivated students feel incompetent and believe that they have no influence over the outcome (Abramson et al., 1978; Ryan & Deci, 2000). On a continuum between a-motivation and introjected regulation, feelings of failure, weak self-concept, anxiety, internal conflicts, loss of self, and consequent loss of intrinsic motivation for life, may be present (Abramson et al., 1978; Ryan & Deci, 2000). The second level is the level of external regulation, also called controlling or controlled motivation. This motivation comes from the outside, namely, from a teacher or a parent. In order to participate in an activity, the student expects to gain or avoid something in return. Both consequences are completely independent of the activity itself: the student has no influence over the external demand but carries it out because this is expected of her/him, e.g., "I study science because I have to as a student". The third level is introjected regulation, which is the first stage of internalised motivation, but is still a controlled form of motivation. The student performs an activity to maintain his self-concept and avoid feelings of shame or internal pressures, e.g., "I study science to satisfy my parents' expectations". The fourth level is identified regulation, which is the first level of autonomous motivation, meaning that the student autonomously decides whether to perform an activity or not, e.g., "I study science to get good grades". The fifth level is integrated motivation, which is the most self-determined type of external motivation. It is similar to internal motivation, but it still has instrumental value, e.g., "I study science because it is important to understand it as a prospective medical doctor", and "I study science because it is interesting to me", respectively.

The SDT describes intrinsic regulation as a "prototype of self-determination". This means that a person fully interested in an activity also has "full choice, experience of deciding what he wants and what he does not want, without a sense of coercion" (Deci & Ryan, 1991, p. 253) and is connected to herself/himself; it is a form of a completely autonomous motivation.

Challenges in Learning and Teaching Science: Examples of Potential Problems in Learning Motivation

Because learning motivation is the result of an interaction between student's personal characteristics and the educational context, it is important to identify and monitor those factors that negatively affect student's readiness for schoolwork. As mentioned, these potential problems or learning difficulties can arise from students themselves or from the environment, either educational or home environment. Two of them that concern students were chosen for the purpose of this chapter. Both are complex in nature, with a fairly high prevalence in children and youth. The first is "learned helplessness", which is of psychological nature and is learned. The second is "developmental dyslexia", which is neurologically conditioned within the context of specific learning difficulties and can prove to be a problem in visual processing that affects in learning science. Both learning difficulties exhibit an inhibitory effect on learning motivation and therefore pose a serious problem in science education for students, regardless of the level of their learning abilities.

Learned Helplessness

The phenomenon of "learned helplessness" in students (Abramson et al., 1978; Maier & Seligman, 2016) is a learned mental state conditioned by stress experiences in which the student feels that they have no control over their situation, i.e., truly no control over the circumstances or a perception that they have no control. It develops from early childhood on as a vicious circle of continuous aversive situations and can be domain- and/or person-specific. Because of the expectation that nothing can be changed, students begin to behave in a helpless manner, overlooking learning opportunities for relief or change and thus influencing overall mental health (Maier & Seligman, 2016). Koballa and Glynn (2007) describe the syndrome of learned helplessness in science education as the students' belief that academic achievement in science subjects is mostly uncontrollable, i.e.,

> Students who develop a learned helplessness are reluctant to engage in science learning. They believe they will fail, so they do not even try. Because they believe they will fail, these students do not practice and improve their science skills and abilities, so they develop cognitive deficiencies. Students with learned helplessness also have emotional problems such as depression and anxiety. (p. 90)

Peterson et al. (1993) further explain the characteristics of learned helplessness; the authors claim that the phenomenon has three main components. The first is contingency, which deals with the uncontrollability of the aversive situation. The second is cognition, which refers to the attributions or explanatory styles people make in relation to the situation of which they are a part (for details see Weiner, 1992). Finally, behaviour allows people to decide whether to give up or continue with the obstacle in front of them. Defined in this way, learned helplessness affects the behaviour of students in three different ways: motivational as a lack of effort, cognitive as passive attributional explanations of situations, and emotional as doubts about one's own ability to learn and consequently the decrease in their self-concept (for details see Gordon & Gordon, 2006).

Developmental Dyslexia

Among the various learning difficulties that hinder school learning, developmental dyslexia is very common (Győrfi & Smythe, 2010; Lyon et al., 2003). Due to neuropsychological and neurophysiological causes, students with dyslexia have continuing difficulties with recognising and decoding words, reading fluency, reading comprehension, pragmatic use of language, naming symbols, manipulating with abstract concepts, time-management, management of school materials, educational sources, and other learning tools (Raduly-Zorgo et al., 2010). Moreover, they have difficulties with metacognitive awareness (Klassen & Lynch, 2007; Kolić-Vehovec et al., 2014), self-observation, and self-evaluation (Gargiulo, 2015), and consequently with self-regulatory learning. In relation to their peers, they perceive themselves as less competent in reading (Soriano-Ferrer & Monte-Soriano, 2017). It seems that less developed metacognitive functions are a consequence of dyslexia (Roth, 2008), but such students can develop independent compensatory skills (Bogdanowicz et al., 2007). Reiter et al. (2004) found that students with developmental dyslexia have deficits in working memory, problem-solving, inhibition, verbal and visual fluency, that is, in cognitive functions associated with the left prefrontal cortex. Brosnan et al. (2002) found that individuals with developmental dyslexia show deficits in executive function, inhibition of distractors, recognising and determining a sequence of events, and in tasks that require the functioning of the prefrontal cortex. By constantly experiencing failure, they develop feelings of helplessness and other negative emotions; self-perception of poor learning effectiveness further affects self-concept in learning, reduces learning effectiveness, and reduces the chance of success (Mitchell, 2014; Sorrenti et al., 2019; Spafford & Grosser, 1993; Sršen Fras, 2016). Research shows that self-efficacy is significantly lower in students with developmental dyslexia than in that of their peers. They perceive themselves as less successful, incompetent, and unappreciated; because they value themselves and their abilities less, they give up faster on difficult learning tasks, and they learn only easier school topics (Bakracevic Vukman et al., 2013). Despite being successful, they may show low interest and a depressed mood (Lackaye et al., 2006). Since they have no automatisation developed, they experience greater cognitive engagement and fatigue in learning and task solving, also due to the inclusion of additional sources of attention, conscious control of the implementation, and the use of higher order cognitive functions (Magajna, 2015).

Various theories aim to explain the etiology of developmental dyslexia, from theories of phonological deficits to theories of automation disorder and to theories of visual attention deficits. Modern findings suggest that students with more severe reading difficulties have a double deficit, a phonological and a visual (Boets, 2014; Boets et al., 2007). Deficiencies in visual processing in students with developmental dyslexia are explained by two main theories. Visual Deficit Theory or Magnocellular Theory suggests that in students with developmental dyslexia, the part of the visual system, which is responsible for controlling eye movements, is hindered. Students with developmental dyslexia experience difficulties in performing precise eye movements, which are necessary for reading (Stein & Walsh, 1997; Stein, 2001, 2012). They also experience difficulties in the following areas: (1) length and direction of saccades (extremely fast eye movements), (2) verge movements in fixations of letters that form a clear image, and (3) coordination between both eyes. The Visual Attention Deficit Theory or Visual Stress Theory, on the other hand, argues that students with developmental dyslexia have difficulty perceiving individual symbols when the attention is asymmetrically distributed between the visual field and the word. This theory assumes that reading difficulties result from hypersensitivity to certain wavelengths of visible light, which results in an unstable or poor/foggy perception of symbols when reading (Wilkins et al., 2016), as well as the reading distress, which increases with the length of the text. Kriss and Evans (2005) found that students with visual stress experience blurring of text, duplication of letters, movement of symbols or words, changes in text format, or disappearance of symbols and words, which all consequently result in learning difficulties (Loew et al., 2014).

The Development of Visual Attention

From the moment of birth, the visual system is exposed to different visual information that it cannot instantly process. Visual attention is the main function of filtering, differentiating, and selectively processing information (Amso & Scerif, 2015). Hendry et al. (2019) suggest that the development of visual attention is multifaceted and non-linear. In everyday life, main functions of attention such as orientation, selective visual attention, and processing visual information are intertwined and influenced by other cognitive components. During task solving, students differ with regard to experience, motivation, cognition, and physical abilities or deficits, while those mechanisms, which determine the success of visual attention development, may change. Visual attention encompasses several functions and processes. Hendry et al. (2019) defined four functions of attention, which are often intertwined in everyday life and are considered to be particularly important in early development. Those are: orienting towards and away from objects, selective filtering of visual inputs, processing visual information, and maintaining focus on a target. Maintaining attention, which can be associated with new information and arousal, as well as the skill of withdrawing attention, can be a mediator in the effective processing of visual information in infants up to six months of age. An increased ability of infants to control their attention means that factors of motivation can influence the recognition of a stimulus and the maintaining of attention. Although skills of executive attention continue to develop during adolescence and later, a child with neurotypical development by the end of the second year of life is able to use control mechanisms to identify and orient objects in space. With this, fundamental aspects of visual attention are established (Hendry et al., 2019).

We will present individual deficiencies in the field of perceptual skills, which significantly affect the less effective learning of science.

Types of visual processing disorder. Visual processing skills are part of a student's cognitive functioning and play an important role in interpreting and understanding visually conveyed information (Brown & Stephen, 2011). Visual processing disorders are not related to myopia, but to deficiencies in understanding visually conveyed information such as object movement, spatial relationships, and recognising the shape of symbols or their orientation. The following types of visual processing disorder were identified: (1) disorders of visual decoding, inference, and identification; (2) disorders of understanding spatial relationships and visual orientation; (3) disorders of visual discrimination; (4) visual memory disorders; (5) visual tracking disorders.

Visual decoding and identification word disorders are manifested in students with developmental dyslexia such as problems with recognising objects and symbols when they are not visible in their entirety, but only in parts.

Disorders in space orientation are manifested in students with developmental dyslexia such as shifting of symbol order in a word and digit order in a number or in a chemical formula. They are directly related to visual sensitivity for spatial relationships and sequences.

Visual discrimination disorders are manifested in students with developmental dyslexia such as mix up letters in a word, digits in a number, and having difficulty reading or scanning visual material.

Visual memory disorders are manifested in students with developmental dyslexia such as retrieving a visual image in the form of letters, numbers, words, or symbols and reading text instructions (Willows, 1998).

Visual sequencing disorders are manifested in students with developmental dyslexia such as harder to follow moving objects because the movement of both eyes is not coordinated (Schulte-Körne, 2010). Their eyes often get tired when reading (Murphy, 2004). Students with visual tracking disorders and disorders of coordinated eyes functioning use a finger or a ruler to focus their visual attention. Schulte-Körne (2010) found that unnecessary and redundant eye movement in students with developmental dyslexia occurs in situations other than reading as well.

Visual processing disorders and visual attention disorders are manifested in students with developmental dyslexia such as various types of attention disorder, such as directing and maintaining attention during specific tasks (Helland & Asbjørnesen, 2000), directing and maintaining attention over a long time period, and following fast and sequential information (Hari et al., 1999). They are more likely to be distracted by insignificant interferences during a task or activity than their peers without developmental dyslexia (Reiter et al., 2004). Compared to their peers without developmental dyslexia, students with developmental dyslexia also exhibit less substantial and quality attention (Segalowitz et al., 1992), which could result in motivational learning difficulties, as such students stop learning earlier and use simpler learning strategies.

Conclusions and Implications

The relationship between motivation and cognition is a complex phenomenon that is not yet fully explained, presenting a challenge for researchers and practitioners in education. This relationship has been the subject of discussion in this chapter, specifically, its relevance for science education and science classes, which are often regarded as highly demanding and thus less motivating in comparison to other school classes. By discussing the dynamics of learning motivation as well as its relationship with cognition, the focus of this chapter was twofold; first, to highlight the importance of autonomy-supportive teaching and learning for the individual student, and second, to present the examples of potential problems in motivation and/or learning outcomes conditioned by both internal and external factors of learning.

The question of the differences between students with regard to their motivational attitudes in science education remains unanswered. It would therefore be beneficial to look more closely at various aspects of teaching in future research, especially the relationship between promoting autonomy, the controlling teaching style, and establishing structure as an aspect of external motivational orientation. Students with learning difficulties should be focused on specifically as they have deficits in visual attention, visual retention of models, images, and illustrations, visual tracking of text and written instructions, visual discrimination in understanding visual information, recognition of objects and symbols due to the inability to integrate and/or synthesise visual information, and spatial relationships and orientations as shifts of symbol in a word order, digits in a number or in chemical formulas. It would be also fruitful to examine the relationship and the impact of visual attention and precise visual processing on learning motivation of students with developmental dyslexia and other types of learning difficulties in different developmental periods, especially in adolescence, which is characterised by the phenomenon of the reward sensitivity in adolescence (Casey & Galván, 2016).

Modern eye-trackers are becoming more sophisticated and transferable so that an expansion of the studied population and the improvement of early detection of developmental deficits are possible. Those deficits are, for example, autism spectrum disorders (Hendry et al., 2019) or hereditary attention deficit hyperactivity disorder (Goodwin et al., 2016). These groups of children are more exposed to the development of cognitive and behavioural problems due to early deficit in the development of attention, and they are often accompanied by developmental dyslexia (see also Sorrenti et al., 2019). Hendry et al. (2019), Putnam et al. (2006, 2008), and Rothbart et al. (2000) point out that the research and assessment of visual processing should use triangulation of sources and methods (EEG, eye-trackers, multiple observation, assessment of visual attention and visual memory, considering data gathered with questionnaires for parents who can provide insight into aspects of attention and motivation outside the laboratory setting, appropriate learning strategies in science education, etc.).

Since students with specific learning difficulties often exhibit comorbid conditions, we suggest individualised interventions for the development of self-regulation in autonomy-supportive learning settings, and with it the development of appropriate motivational skills, motivational strategies, and strategies for the development of visual perception, visual attention, and visual memory used in science to observe similarities, differences, and changes and also structured trainings that improves visual processing skills (Stevenson et al., 2013). When planning the trainings, one should consider that cognitive deficits of memory influence the use of learning self-regulation, which is important in the development of motivational and metacognitive knowledge and skills (Reid et al., 2013). At the same time, it is important to note that problems in learning motivation often arise from the lack of desire to learn and the lack of effort or willpower (Di Domenico & Ryan, 2017; Ryan & Deci, 2000, 2020; Stipek, 1996; Weiner, 1992).

The main question for the future, hence, is how to make science education more approachable to students with motivational and/or learning difficulties, how to encourage them to cope with different obstacles in learning science, how to memorise visual materials and understand the learned topics, and how to attract and retain attention while learning. The use of visual and graphic representations, posters, concept cards, cards with a marked sequence of steps for experiments, aids, schemes, and various animations in the framework of autonomy-supportive classrooms (Reeve, 2016) all affect visual memorisation, procedural knowledge and visual attention retention, and internal representation of mental images (Lodge & Harrison, 2019) and answer to the important question—how to create optimal conditions in the school environment so that students can motivate themselves. When we understand the basic interplay of motivational and cognitive processes that present the inner context for learning and teaching, we can make science education more autonomous and effective.

References

Abramson, L. Y., Seligman, M. E., & Teasdale, J. D. (1978). Learned helplessness in humans: Critique and reformulation. *Journal of Abnormal Psychology, 87*(1), 49–74.

Adcock, R. A., Thangavel, A., Whitfield-Gabrieli, S., Knutson, B., & Gabrieli, J. D. E. (2006). Reward-motivated learning: Mesolimbic activation precedes memory formation. *Neuron, 50*(3), 507–517. https://doi.org/10.1016/j.neuron.2006.03.036.

Amabile, T. M., DeJong, W., & Lepper, M. R. (1976). Effects of externally imposed deadlines on subsequent intrinsic motivation. *Journal of Personality and Social Psychology, 34*(1), 92–98.

Amso, D., & Scerif, G. (2015). The attentive brain: Insights from developmental cognitive neuroscience. *Nature Review Neuroscience, 16,* 606–619. https://doi.org/10.1038/nrn4025.

Anderman, E. M., & Young, A. J. (1994). Motivation and strategy use in science: Individual differences and classroom effects. *Journal of Research in Science Teaching, 31*(8), 811–831. https://doi.org/10.1002/tea.3660310805.

Anderson, B. A., & Sali, A. W. (2016). The impact of reward on attention: Beyond motivation. In T. S. Braver (Ed.), *Motivation and cognitive control* (pp. 50–64). Routledge.

Aschbacher, P. R., Li, E., & Roth, E. J. (2010). Is science me? High school students' identities, participation and aspirations in science, engineering, and medicine. *Journal of Research in Science Teaching, 47*(5), 564–582.

Bakracevic Vukman, K., Funčič Masič, T., & Schmidt, M. (2013). Self-regulation of learning in secondary school students with special educational needs and other students of vocational and technical schools. *The New Educational Review, 33*(3), 295–208. https://doi.org/10.15804/tner.2016.43.1.01.

Barkley, R. A. (1998). *Attention-deficit hyperactivity disorder: A handbook for treatment and diagnosis.* Guilford Press.

Battele for Kids. (2019). *Partnership for 21th century learning.* https://www.battelleforkids.org/networks/p21/frameworks-resources.

Bechara, A., & Damasio, A. R. (2005). The somatic marker hypothesis: A neural theory of economic decision. *Games and Economic Behavior, 52*(2), 336–372. https://doi.org/10.1016/j.geb.2004.06.010.

Black, A. E., & Deci, E. L. (2000). The effects of student self-regulation and instructor autonomy support on learning in a college-level natural science course: A self-determination theory perspective. *Science Education, 84*(6), 740–756. https://doi.org/10.1002/1098-237X(200011)84:6%3C740::AID-SCE4%3E3.0.CO;2-3.

Boets, B. (2014). Dyslexia: Reconciling controversies within an integrative developmental perspective. *Trends in Cognitive Sciences, 18*(10), 501–503. https://doi.org/10.1016/j.tics.2014.06.003.

Boets, B., Ghesquière, P., Van Wieringen, A., & Wouters, J. (2007). Speech perception in preschoolers at family risk for dyslexia: Relations with low-level auditory processing and phonological ability. *Brain Language, 101*(1), 19–30. https://doi.org/10.1016/j.bandl.2006.06.009.

Bogdanowicz, M., Lockiewitz, M., & Bogdanowicz, K. (2007). Life skills. In I. Smithe (Ed.), *Dislexia. A guide for adults, education and culture* (pp. 55–72). Leonardo da Vinci.

Boggiano, A. K., Flink, C., Shields, A., Seelbach, A., & Barrett, M. (1993). Use of techniques promoting students' self-determination: Effects on students' analytic problem-solving skills. *Motivation and Emotion, 17,* 319–336.

Bomia, L., Beluzo, L., Demeester, D., Elander, K., Johnson, M., & Sheldon, B. (1997). *The impact of teaching strategies on intrinsic motivation.* The ERIC Clearinghouse on Elementary and Early Childhood Education.

Braver, T. S. (2016). Motivation and cognitive control: Introduction. In T. S. Braver (Ed.), *Motivation and cognitive control* (pp. 1–20). Routledge.

Brosnan, M., Demetre, H., Hamill, S., Robson, K., Shepherd, H., & Cody, G. (2002). Executive functioning in adults and children with developmental dyslexia. *Neuropsyhologia, 40*(12), 2144–2155. https://doi.org/10.1016/S0028-3932(02)00046-5.

Brown, T., & Stephen, E. (2011). Factor structure of the motor-free visual perception test-3rd edition (MVPT-3). *Canadian Journal of Occupational Therapy, 78*(1), 26–36. https://doi.org/10.2182/cjot.2011.78.1.4.

Cameron, J., & Pierce, W. D. (1994). Reinforcement, reward, and intrinsic motivation: A meta-analysis. *Review of Educational Research, 64*(3), 363–423. https://doi.org/10.2307/1170677.

Casey, B. J., & Galván, A. (2016). The teen brain: "Arrested development" in resisting temptation. In Braver, T. S. (Ed.), *Motivation and cognitive control* (pp. 263–282). Routledge.

Church, M. A., Elliot, J. E., & Gable, S. L. (2001). Perceptions of classroom environment, achievement goals, and achievement outcomes. *Journal of Educational Psychology, 93*(1), 43–54. https://doi.org/10.1037/0022-0663.93.1.43.

Cook, G. I., Rummel, J., & Dummel, S. (2015). Toward an understanding of motivational influences on prospective memory using value-added intentions. *Frontiers of Human Neuroscience, 9,* 278. https://doi.org/10.3389/fnhum.2015.00278.

Crede, M., & Phillips, L. A. (2011). A meta-analytic review of the motivated strategies for learning questionnaire. *Learning and Individual Differences, 21,* 337–346. https://doi.org/10.1016/j.lindif.2011.03.002.

Crone, E. A., & Dahl, R. E. (2012). Understanding adolescence as a period of social-affective engagement and goal flexibility. *Nature Reviews Neuroscience, 13*(9), 636–650. https://www.nature.com/articles/nrn3313.

Csikszentmihalyi, M. (1988). The flow experience and its significance for human psychology. In M. Csikszentmihalyi & I. S. Csikszentmihalyi (Eds.), *Optimal experience: Psychological studies of flow in consciousness* (pp. 15–35). Cambridge University Press.

Deci, E. L. (1971). Effects of externally mediated rewards on intrinsic motivation. *Journal of Personality and Social Psychology, 18*(1), 105–115. https://doi.org/10.1037/h0030644.

Deci, E. L., & Ryan, R. M. (1985). *Intrinsic motivation and self-determination in human behavior.* Plenum.

Deci, E. L., & Ryan, R. M. (1991). A motivational approach to self: Integration in personality. In R. A. Dienstbier (Ed.), *Current theory and research in motivation, Vol. 38. Nebraska Symposium on motivation: Perspectives on motivation* (pp. 237–288). University of Nebraska Press.

Deci, E. L., & Ryan, R. M. (2000). The "what" and "why" of goal pursuits: Human needs and the self-determination of behavior. *Psychological Inquiry, 11*(4), 227–268. https://doi.org/10.1207/S15327965PLI1104_01.

Deci, E. L., & Ryan, R. M. (2008). Self-determination theory: A macrotheory of human motivation, development, and health. *Canadian Psychology, 49*(3), 182–185. https://doi.org/10.1037/a0012801.

Deci, E. L., Koestner, R., & Ryan, R. M. (1999). A meta-analytic review of experiments examining the effects of extrinsic rewards on intrinsic motivation. *Psychological Bulletin, 125*(6), 627–668. https://doi.org/10.1037/0033-2909.125.6.627.

Deci, E. L., Nezlek, J., & Sheinman, L. (1981). Characteristics of the rewarder and intrinsic motivation of the rewardee. *Journal of Personality and Social Psychology, 40*(1), 1–10.

Di Domenico S. I., & Ryan, R. M. (2017). The emerging neuroscience of intrinsic motivation: A new frontier in self-determination research. *Front Human Neuroscience, 11*(145). https://doi.org/10.3389/fnhum.2017.00145.

Eckes, A., Urhahne, D., & Wilde, M. (2016, June 2–5). *Why structure needs to provided autonomy-supportive—The effects of structure and autonomy support on motivation* [Paper presentation]. Self-Determination Conference, Victoria, BC, Canada. http://web.uvic.ca/sdt2016/.

European Commission (2007, June). *Science education now: A renewed pedagogy for the future of Europe* (No. EUR22845). Office for Official Publications of the European Communities. https://www.eesc.europa.eu/sites/default/files/resources/docs/rapportrocardfinal.pdf.

Fastrich, G. M., Kerr, T., Castel, A. D. in Murayama, K. (2018). The role of interest in memory for trivia questions: An investigation with a large-scale database. *Motivation Science, 4*(3), 227–250. https://doi.org/10.1037/mot0000087.

Fink, C., Boggiano, A. K., Main, D. S., Barrett, M., & Katz, P. A. (1992). Children's achievement related behaviors: The role of extrinsic and intrinsic motivational orientations. In A. K. Boggiano & T. S. Pittman (Eds.), *Achievement and motivation: A social-developmental perspective* (pp. 189–214). Cambridge University Press.

Fishback, A., Eyal, T., & Finkelstein, S. R. (2010). How positive and negative feedback motivate goal pursuit. *Social and Personality Psychology Compass, 4*(8), 517–530. https://doi.org/10.1111/j.1751-9004.2010.00285.x.

Gargiulo, R. M. (2015). *Special education in contemporary society: An introduction to exceptionality* (5th ed.). Sage.

Gilbert, J. K. (2006). On the nature of 'context' in chemical education. *International Journal of Science Education, 28*(9), 957–976.

Goodwin, A., Salomone, S., Bolton, P., Charman, T., & Jones, E. J. H. (2016). Attention training for infants at familiar risk of ADHD (INTERSTAARS): Study protocol for randomised controlled trail. *Trials, 17,* 608. https://doi.org/10.1186/s13063-016-1727-0.

Gordon, R., & Gordon, M. (2006). *The turned off child: Learned helplessness and school failure.* Millenial Mind.

Grolnick, W. S., Deci, E. L., & Ryan, R. M. (1991). Inner resources for school achievement: Motivational mediators of children's perceptions of their parents. *Journal of Educational Psychology, 83*(4), 508–517.

Guay, F., Ratelle, C. F., & Chanal, J. (2008). Optimal learning in optimal contexts: The role of self-determination in education. *Canadian Psychology, 49*(3), 233–240. https://doi.org/10.1037/a0012758.

Győrfi, A. in Smythe, I. (2010). *Dyslexia report: Dyslexia in Europe: A pan-european survey.* http://www.doitprofiler.co.za/media/13299/dyslexia_report_2010_final_mep.pdf.

Hari, R., Valta, M., & Uutela, K. (1999). Prolonged attentional dwell time in dyslexic adults. *Neuroscience Letters, 271,* 202–204.

Helland, T., & Asbjørnesen, A. (2000). Executive functions in dyslexia. *Child Neuropsychology, 6*(1), 37–48. https://doi.org/10.1076/0929-7049.

Hendry, A., Johnson, M. H., & Holmboe, K. (2019). Early development of visual attention: Change, stability, and longitudinal associations. *Annual Review of Developmental Psychology, 1,* 251–275. https://doi.org/10.1146/annurev-devpsych-121318-085114.

Hidi, S. E., & Renninger, K. A. (2019). Introduction: Motivation and its relation to learning. In K. A. Renninger & S. E. Hidi (Eds.), *The Cambridge handbook of motivation and learning* (pp. 1–14). Cambridge University Press.

Holbrook, J., & Rannikmäe, M. (2014). The philosophy and approach on which the PROFILES project is based. *CEPS Journal, 4,* 9–29.

Hung-Chih, Y., Hsiao-Lin, T., & Chi-Hung, L. (2010). Investigating the influence of motivation on students' conceptual learning outcomes in web-based vs. classroom-based science teaching contexts. *Research in Science Education, 41*, 211–224. https://doi.org/10.1007/s11165-009-9161-x.

Jarvela, S., & Niemivirta, M. (2001). Motivation in context: Challenges and possibilities in studying the role of motivation in new pedagogical cultures. In S. Volet & S. Jarvela (Eds.), *Motivation in learning context: Theoretical advances and methodological implications* (pp. 105–127). Pergamon.

Johnstone, A. H. (1991). Why is science difficult to learn? Things are seldom what they seem. *Journal of Computer Assisted learning, 7*, 75–83. https://doi.org/10.1111/j.1365-2729.1991.tb00230.x.

Juriševič, M. (2006). *Učna motivacija in razlike med učenci* [Motivation to learn and differences among students]. Univerza v Ljubljani Pedagoška fakulteta.

Juriševič, M., Vrtačnik, M., Kwiatkowski, M., & Gros, N. (2012). The interplay of students' motivational orientations, their chemistry achievements and their perception of learning within the hands-on approach to visible spectrometry. *Chemistry Education Research and Practice, 13*(3), 237–247. https://doi.org/10.1039/C2RP20004J.

Kasch, K. L., Rottenberg, J., Arnow, B. A., & Gotlib, I. H. (2002). Behavioral activation and inhibition systems and the severity and course of depression. *Journal of Abnormal Psychology, 111*(4), 589–597. https://doi.org/10.1037/0021-843X.111.4.589.

Klassen, R. M., & Lynch, S. L. (2007). Self-efficacy from the perspective of adolescents with LD and their specialists' teachers. *Journal of Learning Disabilities, 40*(6), 494–507. https://doi.org/10.1177/00222194070400060201.

Koballa, T. R. Jr., & Glynn, S. M. (2007). Attitudinal and motivational constructs in science education. In S. K. Abell & N. Lederman (Eds.), *Handbook for research in science education* (pp. 75–102). Erlbaum.

Koestner, R., Ryan, R. M., Bernieri, F., & Holt, K. (1984). Setting limits on children's behavior: The differential effects of controlling vs. informational styles on intrinsic motivation and creativity. *Journal of Personality, 52*(3), 233–248.

Kolić-Vehovec, S., Zubković-Rončević, B., & Pahljina-Reinić, R. (2014). Development of metacognitive knowledge of reading strategies and attitudes toward reading in early adolescence: The effect on reading comprehension. *Psychological Topics, 23*(1), 77–98.

Kriss, I., & Evans, B. J. W. (2005). The relation between dyslexia and Meares-Irlen syndrome. *Journal of Research in Reading, 28*(3), 350–364. https://doi.org/10.1111/j.1467-9817.2005.00274.x.

Lackaye, T., Margalit, M., Ziv, O., & Zimen, T. (2006). Comparisons of self-efficacy, mood, effort, and hope between students with learning disabilities and their non-LD-matched peers. *Learning Disabilities Research & Practice, 21*(2), 111–121. https://doi.org/10.1111/j.1540-5826.2006.00211.x.

Lee, E., Reeve, J., Xue, Y., & Xiong, J. (2012). Neural differences between intrinsic reasons for doing extrinsic reasons for doing: An fMRI study. *Neuroscience Research, 73*(1), 68–72. https://doi.org/10.1016/j.neures.2012.02.010.

Lepper, M. R., Greene, D., & Nisbett, R. E. (1973). Undermining children's intrinsic interest with extrinsic reward: A test of the "overjustification" hypothesis. *Journal of Personality and Social Psychology, 28*(1), 129–137.

Lepper, M. R. in Greene, D. (1975). Turning play into work: Effects of adult surveillance and extrinsic rewards on children's intrinsic motivation. *Journal of Personality and Social Psychology, 31*(3), 479–486.

Levine, M. D. (2002). *Educational care: A system for understanding and helping children with learning problems at home and in school* (2nd ed.). Educators Publishing Services.

Linnenbrink, E. A., & Pintrich P. R. (2002). The role of motivational beliefs in conceptual change. In M. Limón & L. Mason (Eds.), *Reconsidering conceptual change: Issues in theory and practice* (pp. 115–135). Springer.

Linnenbrink-Garcia, L., & Patall, E. A. (2016). Motivation. In L. Corno & E. M. Anderman (Eds.), *Handbook of educational psychology* (3rd ed., pp. 91–103). Routledge.

Lodge, J. M., & Harrison, W. J. (2019). The role of attention in learning in the digital age. *Yale Journal of Biology and Medicine, 92*, 21–28.

Loew, S. J., Marsh, N. V., & Watson, K. (2014). Symptoms of Meares-Irlen/visual stress syndome in subjects diagnosed with chronic fatigue syndrome. *International Journal of Clinical and Health Psychology, 14*(2), 87–92. https://doi.org/10.1016/S1697-2600(14)70041-9.

Lyon, G. R., Shaywitz, S. E., & Shaywitz, B. A. (2003). A definition of dyslexia. *Annals of Dyslexia, 53*, 1–14. https://doi.org/10.1007/s11881-003-0001-9.

Lyons, T. (2006). Different countries, same science classes: Students' experiences of school science in their own words. *International Journal of Science Education, 28*(6), 591–613. https://doi.org/10.1080/09500690500339621.

Lyons, T., & Quinn, F. (2010). *Choosing science: Understanding the declines in senior high school science enrolments*. National Centre of Science, ICT and Mathematics Education for Rural and Regional Australia (SiMERR A), Australia.

Magajna, L. (2015). Sodobne raziskave spoznavnih procesov in psiholoških virov pri specifičnih motnjah učenja: izzivi za razvojno delo in prakso [Contemporary research on cognitive processes and psychological resources in specific learning disorders: Challenges for development work and practice]. In Z. Pavlović (Ed.), *60 let podpore pri vzgoji, učenju in odraščanju* [60 years of support in education, learning and growing up] (pp. 141–160). Svetovalni center za otroke, mladostnike in starše Ljubljana.

Mayer, R. E. (2010). Learning with technology. In H. Dumont, D. Istance, & F. Benavides (Eds.), *The nature of learning: Using research to inspire practice* (pp. 179–196). OECD. https://read.oecd-ilibrary.org/education/the-nature-of-learning_9789264086487-en#page1.

Maier, S. F., & Seligman, M. E. (2016). Learned helplessness at fifty: Insights from neuroscience. *Psychological Review, 123*(4), 349–367. https://doi.org/10.1037/rev0000033.

McGillivray, S., Murayama, K., & Castel, A. (2015). Thirst for knowledge: The effects of curiosity and interest on memory in younger and older adults. *Psychology and Aging, 30*(4), 835–841. https://doi.org/10.1037/a0039801.

Meyer, B., Johnson, S. L., & Carver, C. S. (1999). Exploring behavioral activation and inhibition sensitivities among college students at risk for bipolar spectrum symptomatology. *Journal of Psychopathology Behavioral Assessment, 1; 21*(4), 275–292. https://doi.org/10.1023/A:1022119414440.

Miendlarzewska, E. A., Bavelier, D., & Schwartz, S. (2016). Influence of reward motivation on human declarative memory. *Neuroscience and Biobehavioral Reviews, 61*, 156–2176. https://doi.org/10.1016/j.neubiorev.2015.11.015.

Mitchell, D. (2014). *What really works in special education and inclusive education: Using evidence-based teaching strategies* (2nd ed.). Routledge.

Mossholder, K. W. (1980). Effects of externally mediated goal setting on intrinsic motivation: A laboratory experiment. *Journal of Applied Psychology, 65*(2), 202–210.

Murayama, K. (2018). The science of motivation: Multidisciplinary approaches advance research on the nature and effects of motivation. *Psychological Science Agenda, June 2018*. https://www.apa.org/science/about/psa/2018/06/motivation.

Murayama, K., & Elliot, A. J. (2011). Achievement motivation and memory: Achievement goals differentially influence immediate and delayed remember–know recognition memory. *Personality and Social Psychology Bulletin, 37*(10), 1339–1348. https://doi.org/10.1177%2F0146167211410575.

Murayama, K., & Kitagami, S. (2014). Consolidation power of extrinsic rewards: Reward cues enhance long-term memory for irrelevant past events. *Journal of Experimental Psychology: General, 143*(1), 15–20. https://doi.org/10.1037/a0031992.

Murayama, K., Matsumoto, M., Izuma, K., & Matsumoto, K. (2010). Neural basis of the undermining effect of monetary reward on intrinsic motivation. *Proceedings of the National Academy of Sciences of the United States of America, 107*(49), 20911–20916. https://doi.org/10.1073/pnas.1013305107.

Muris, P., Merchelbach, H., Schmidt, H., Gadet B. B., & Bogie, N. (2001). Anxiety and depression as correlates of self-reported behavioural inhibition in normal adolescents. *Behaviour Research and Therapy, 39*(9), 1051–1061. https://www.ncbi.nlm.nih.gov/pubmed/11520011.

Murphy, M. F. (2004). *Dyslexia: An explanation.* Flyleaf Press.

Murty, V. P., & Dickerson, K. C. (2017). Recent developments in neuroscience research on human motivation. In *Advances in motivation and achievement* (Vol. 19, pp. 203–227). Emerald Group Publishing. https://doi.org/10.1108/S0749-742320160000019019.

Murty, V. P., Ballard, I. C., Macduffie, K. E., Krebs, R. M., & Adcock, R. A. (2013). Hippocampal networks habituate as novelty accumulates. *Learning & Memory, 20*(4), 229–235. https://doi.org/10.1016/j.neulet.2018.01.053.

Murty, V. P., Labar, K. S., & Adcock, R. A. (2012). Threat of punishment motivates memory encoding via amygdala, not midbrain, interactions with the medial temporal lobe. *Journal of Neuroscience, 32*(26), 8969–8976. https://doi.org/10.1523/JNEUROSCI.0094-12.2012.

National Scientific Council on the Developing Child. (2018). *Understanding motivation: Building the brain architecture that supports learning, health, and community participation* (Working Paper No. 14). https://developingchild.harvard.edu/.

Oliveira, A., Behnagh, F. R., Ni, L., Mohsinah, A. A., Burgess, K. J., & Guo, L. (2019). Emerging technologies as pedagogical tools for teaching and learning science: A literature review. *Human Behaviour & Emerging Technologies, 1,* 49–160. https://doi.org/10.1002/hbe2.141.

Osborne, J. (2013). The 21st century challenge for science education: Assessing scientific reasoning. *Thinking Skills and Creativity, 10,* 265–279. https://doi.org/10.1016/j.tsc.2013.07.006.

Osborne, J., Simon, S., & Collins, S. (2003). Attitudes towards science: A review of the literature and its implications. *International Journal of Science Education, 25,* 1049–1079. https://doi.org/10.1080/0950069032000032199.

Patall, E. A., Pituch, K. A., Steingut, R. R., Vasquez, A. C., Yates, N., & Kennedy, A. A. (2019). Agency and high school science students' motivation, engagement, and classroom support experiences. *Journal of Applied Developmental Psychology, 62,* 77–92. https://doi.org/10.1016/j.appdev.2019.01.004.

Peterson, C., Maier, S., & Seligman, M. E. P. (1993). *Learned helplessness: A theory for the age of personal control.* Oxford.

Phillips, A. G., Vacca, G., & Ahn, S. (2008). A top-down perspective on dopamine, motivation and memory. *Pharmacology, Biochemistry and Behavior, 90,* 236–249. https://doi.org/10.1016/j.pbb.2007.10.014.

Pintrich, P. R., & Schunk, D. H. (2002). *Motivation in education: Theory, research, and applications* (2nd ed.). Prentice Hall.

Plant, R. W., & Ryan, R. M. (1985). Intrinsic motivation and the effects of self-consciousness, self-awareness, and ego-involvement: An investigation of internally controlling styles. *Journal of Personality, 53*(3), 435–449.

Posner, M. I., & Boies, S. J. (1971). Components of attention. *Psychological Review, 78*(5), 391–408. https://doi.org/10.1037/h0031333.

Potvin, P., & Hasni, A. (2014). Analysis of the decline in interest towards school science and technology from grades 5 through 11. *Journal of Science Education and Technology, 23*(6), 784–802. https://doi.org/10.1007/s10956-014-9512-x.

Putnam S. P., Garstein M. A., & in Rothbart M. K. (2006). Measurement of five-grained aspects of toddler temperament: The early childhood behavior questionaire. *Infant Bahavior & Development, 29,* 386–401. https://doi.org/10.1016/j.infbeh.2006.01.004.

Putnam, S. P., Rothbart, M. K., & Gartstein, M. A. (2008). Homotypic and heterotypic continuity of fine-grained temperament during infancy, toddlerhood, and early childhood. *Infant and Child Development, 17*(4), 387–405. https://doi.org/10.1002/icd.582.

Raduly-Zorgo, E., Smythe, L., & Gyarmathy, E. (2010). *Disleksija – vodnik za tutorje* [Dyslexia—A guide for tutors]. Bravo - društvo za pomoč otrokom in mladostnikom s specifičnimi učnimi težavami.

Reeve J. (2016). Autonomy-supportive teaching: What it is, how to do it. In W. Liu, J. Wang, & R. Ryan (Eds.), *Building autonomous learners* (pp. 129–152). Springer.

Reid, R., Lienemann, T. O., & Hagaman, J. L. (2013). *Strategy instruction for students with learning disabilities*. The Guilford Press.

Reiter, A., Tucha, O., & Lange, K. W. (2004). Executive functions in children with dyslexia. *Dislexia: An International Journal of Research and Practice, 11*, 116–131. https://doi.org/10.1002/dys.289.

Rheinberg, F., Vollmeyer, R., & Rollett, W. (2000). Motivation and action in self-regulated learning. In M. Boekaerts, P. R. Pintrich, & M. Zeidner (Eds.), *Handbook of self-regulation* (pp. 503–531). Academic Press.

Roth, L. S. (2008). *Comprehension, monitoring, cognitive resources and reading disability*. http://search.proquest.com/openview/93dd35a410ddfac44c41f6566cff7fc7/1?pq-origsite=gscholar&cbl=18750&diss=y.

Rothbart, M. K., Derryberry, D., & Hershey K. (2000). Stability of temperament in childhood: Laboratory infant assessment to parent report at seven years. In V. J. Molfese (Ed.), *Temperament and personality development across the life span* (pp. 85–119). Routledge.

Rothkirch, M., & Sterzer, P. (2016). The role of motivation in visual information processing. In T. S. Braver (Ed.), *Motivation and cognitive control* (pp. 23–49). Routledge.

Ryan, R. M. (1982). Control and information in the intrapersonal sphere: An extension of cognitive evaluation theory. *Journal of Personality and Social Psychology, 43*(3), 450–461.

Ryan, R. M. (2016, June 2–5). *Motivation in education: Research and practice using SDT* [Paper presentation]. Self-Determination Conference, Victoria, BC, Canada. http://web.uvic.ca/sdt2016/.

Ryan, R. M., & Deci, E. L. (2000). Intrinsic and extrinsic motivations: Classic definitions and new directions. *Contemporary Educational Psychology, 25*(1), 54–67.

Ryan, R. M., & Deci, E. L. (2009). Promoting self-determined school engagement: Motivation, learning, and well-being. In K. R. Wentzel & A. Wigfield (Eds.), *Handbook of motivation in school* (pp. 171–196). Routledge.

Ryan, R. M., & Deci, E. L. (2020). Intrinsic and extrinsic motivation from a self-determination theory perspective: Definitions, theory, practices, and future directions. *Contemporary Educational Psychology, 25*(1), 54–67. https://doi.org/10.1016/j.cedpsych.2020.101860.

Ryan, R. M., & Grolnick, W. S. (1986). Origins and pawns in the classroom: Self-report and projective assessments of individual differences in children's perceptions. *Journal of Personality and Social Psychology, 50*, 550–558.

Schabel, S. J., Schairer, W., Donahue, R. J., Powell, V., & Janak, P. H. (2011). Similar neural activity during fear and disgust in the rat basolateral amygdala. *PLoS ONE, 6*(12), e27797. https://doi.org/10.1371/journal.pone.0027797.

Schiefele, U., & Rheinberg, F. (1997). Motivation and knowledge acquisition: Searching for mediating processes. In M. L. Maehr & P. R. Pintrich (Eds.), *Advances in motivation and achievement* (Vol. 10, pp. 251–301). JAI Press.

Schiller, D., Freeman, J. B., Mitchell, J. P., Uleman, J. S., & Phelps, E. A. (2009). A neural mechanism of first impressions. *Nature Neuroscience, 12*(4), 508–514. https://doi.org/10.1038/nn.2278.

Schraw, G., Crippen, K. J., & Hartley, K. (2006). Promoting self-regulation in science education: Metacognition as part of a broader perspective on learning. *Research in Science Education, 36*, 111–139.

Schulte-Körne, G. (2010). The prevention, diagnosis, and treatment of dyslexia. *Deutsches Arzteblatt International, 107*(41), 718–726.

Segalowitz, S., Wagner, J. W., & Menna, R. (1992). Lateral versus frontal ERP predictors of reading skill. *Brain and Cognition, 20*, 85–103.

Shirazi, S. (2017). Student experience of school science. *International Journal of Science Education, 39*(14), 1891–1912. https://doi.org/10.1080/09500693.2017.1356943.

Simpson, E. H., & Balsam, P. (2016). The behavioral neuroscience of motivation: An overview of concepts, measures, and translation applications. In E. H. Simpson & P. Balsam (Eds.), *Behavioral neuroscience of motivation*. Current Topics in Behavioral Neurosciences, 27 (pp. 1–12). Springer. https://doi.org/10.1007/7854_2015_402.

Skinner, E., Saxton, E., Currie, C., & Shusterman, G. (2017). A motivational account of the undergraduate experience in science: Brief measures of students' self-system appraisals, engagement in coursework, and identity as a scientist. *International Journal of Science Education, 39*(17), 2433–2459. https://doi.org/10.1080/09500693.2017.1387946.

Sorrenti, L., Spadaro, L., Mafodda, A.V., Scopelliti, G., Orecchio, S., & Filippello, P. (2019). The predicting role of school learned helplessness in internalizing and externalizing problems: An exploratory study in students with specific learning disorder. *Mediterranean Journal of Clinical Psychology, 7*. https://cab.unime.it/journals/index.php/MJCP/article/view/2035/0.

Soriano-Ferrer, M., & Monte-Soriano, M. (2017). Teacher perception of reading motivation in children with developmental dyslexia and average readers. *Procedia – Social and Behavioral Sciences, 237,* 50–56. https://doi.org/10.1016/j.sbspro.2017.02.012.

Spafford, C., & Grosser, G. S. (1993). The social misperception syndrome in children with learning disabilities: Social causes versus neurological variables. *Journal of Learning Disabilities, 26*(3), 178–189.

Sršen Fras, A. (2016). *Afektivni dejavniki samoregulacijskega učenja pri učencih s specifičnimi učnimi težavami* [Affective factors of self-regulatory learning in students with specific learning difficulties]. Unpublished MA thesis, Univerza v Ljubljani Filozofska fakulteta.

Stein, J. F. (2001). *The magnocellular theory of developmental dyslexia*. https://www.researchgate.net/publication/12029637_The_Magnocellular_Theory_of_Developmental_Dyslexia.

Stein, J. F. (2012). *The neurobiological basis of dyslexia: The magnocellular theory*. https://www.researchgate.net/publication/286045035_The_Neurobiological_Basis_of_Dyslexia_The_Magnocellular_Theory.

Stein, J. F., & Walsh, W. (1997). To see but not to read: The magnocellular theory of dyslexia. *Trends in Neuroscience, 20*(4), 147–152. https://doi.org/10.1016/S0166-2236(96)01005-3.

Stevenson, R. A., Wilson, M. M., Powers, A. R., & Wallace, M. T. (2013). The effect of visual training on multisensory temporal processing. *Experimental Brain Research, 225*(4), 479–489. https://doi.org/10.1007/s00221-012-3387-y.

Stipek, D. (1996). Motivation and instruction. In D. C. Berliner & R. C. Calfee (Eds.), *Handbook of educational psychology*. A project of division 15 (pp. 85–113). Macmillan.

Stipek, D. J. (2002). *Motivation to learn: Integrating theory and practice*. Allyn & Bacon.

Suri, G., & Gross, J. J. (2015). The role of attention in motivated behavior. *Journal of Experimental Psychology: General, 144*(4), 864–872. https://doi.org/10.1037/xge0000088.

Thomas, S., & Oldfather, P. (1997). Intrinsic motivations, literacy, and assessment practices: "That's my grade. That's me". *Educational Psychologist, 32,* 107–123. https://psycnet.apa.org/doi/10.1207/s15326985ep3202_5.

Toshalis, E., & Nakkula, M. J. (2012). Motivation, engagement, and student voice. *Education Digest: Essential Readings Condensed for Quick Review, 78*(1), 29–35. https://eric.ed.gov/?id=EJ999430.

Tuan, H. L., Chin, C. C., & Shieh, S. H. (2005). The development of a questionnaire for assessing students' motivation toward science learning. *International Journal of Science Education, 27,* 639–654.

Vollmeyer, R., & Rheinberg, F. (2013). The role of motivation in knowledge acquisition. In R. Azevedo & V. Aleven (Eds.), *International handbook of metacognition and learning technologies* (pp. 697–706). Springer.

Vossen, T. E., Henze, I., Rippe, R. C. A., Van Driel, J. H., & De Vries, M. J. (2018). Attitudes of secondary school students towards doing research and design activities. *International Journal of Science Education, 40*(13), 1629–1652. https://doi.org/10.1080/09500693.2018.1494395.

Weiner, B. (1992). *Human motivation: Metaphors, theories, and research.* Sage.

Wentzel, K. R., & Miele, D. B. (2016). Overview. In K. R. Wentzel & D. B. Miele (Eds.), *Handbook of motivation in school* (2nd ed., pp. 1–8). Routledge.

Wilkins, A., Allen, P. M., Monger, L. J., & Gilchrist, J. M. (2016). Visual stress and dyslexia for the practising optometrist. *Optometry in Practice, 17*(2), 103–112. http://repository.essex.ac.uk/id/eprint/17161.

Willows D. M. (1998). Visual processes learning disabilities. In B. Y. Wong (Ed.), *Learning about learning disabilities* (2nd ed., pp. 203–236). Routledge.

Mojca Juriševič, Ph.D is a Professor of Educational Psychology in the Faculty of Education at the University of Ljubljana, Slovenia. Her areas of expertise include learning and teaching with special focus on motivation for learning, psychological aspects of gifted education, creativity, professional development of teachers, and the translation of psychological research findings into school practice. She is the Head of the Faculty's Center for Research and Promotion of Giftedness. She also serves as a council member of the European Talent Support Network and is a national representative in the World Council for Gifted and Talented Children. She leads the division of psychologists in education at the Slovenian Psychologists' Association and is a national representative in the Standing Committee of Psychology in Education at the European Federation of Psychologists' Associations. She is an associate member of the American Psychological Association.

Tanja Černe, Ph.D is a Special Needs Teacher. She works at the Counselling Center for Children, Adolescents and Parents in Ljubljana, where she conducts diagnostic, corrective, and counselling treatment of children and adolescents with developmental and learning difficulties, in collaboration with their parents and teachers. She is involved in the research project, Protection and promotion of mental health and learning competencies of children and adolescents, where she is a Senior Research Assistant. She also collaborates as a Teaching Assistant of Special Education at the Faculty of Education, University of Ljubljana. Her pedagogical work includes educating professionals in education, educating parents of children with specific learning difficulties, and ADHD. She participated in numerous conferences and seminars and has written several professional and research papers, and a monograph.

Chapter 3
Predicting Task Difficulty Through Psychophysiology

Junoš Lukan and Gregor Geršak

Introduction

Psychophysiology investigates changes in the activity of physiology caused by psychological input (Ravaja, 2004). In principle, psychophysiological measuring devices can be divided into two types, the brain scanning apparatuses for central nervous system and devices for autonomous nervous system dynamics' observation. The latter enable measurement of different physiological parameters, e.g. heart rate, heart rate variability, blood pressure, skin conductance, skin temperature, facial thermal scan, breathing rate and breathing amplitude, pupil dilatation, etc. They have been used with increasing regularity to study different constructs, like mental load or effort, stress, emotions, level of focus, difficulty of a task, etc. (Benedek & Kaernbach, 2010a; Collet et al., 1997, 2009; Fauvel et al., 2000; Kivikangas et al., 2014; Olsson & Phelps, 2007; Storm et al., 2005; Wen et al., 2014)

The relationship between physiological measures and task difficulty has been widely studied. The level of arousal can be linked with the level of challenge, focus, and excitement associated with the difficulty of the task (Cacioppo et al., 2007; Lewis et al., 1993; Mandryk & Atkins, 2007). The reports are sometimes contradictory, but some general conclusions can be drawn. In the majority of related work increase in skin conductance (usually called electrodermal activity, EDA) was found with increased arousal (Boucsein, 2012; Boucsein et al., 2012; Brouwer et al., 2013, 2014; Lisetti & Nasoz, 2004). A number of studies reported correlation between EDA and task engagement. In general, EDA increases with difficulty of the task (Clark et al., 2018; Mandryk & Atkins, 2007; Mehler et al., 2009; Nourbakhsh et al.,

J. Lukan (✉)
Jožef Stefan Institute, Jamova 39, 1000 Ljubljana, Slovenia

G. Geršak
Faculty of Electrical Engineering, University of Ljubljana, Tržaška 25, 1000 Ljubljana, Slovenia

2012; Novak et al., 2014; Pecchinenda, 1996). Heart rate and heart rate variability are subject to effort and workload level associated with task difficulty (Aasman et al., 1987; Veltman & Gaillard, 1998). Similarly, respiration rate increases with mental effort (Butler et al., 2006; Karavidas et al., 2010; Mehler et al., 2009; Veltman & Gaillard, 1998). Skin temperature decreases with arousal due to psychologically induced vasodilatation of peripheral veins (Cacioppo et al., 2007).

In this paper, an attempt of physiology-based prediction of the task difficulty as perceived by the subject is described. Elementary school children were instructed to solve science problems, while their physiology was monitored. Science problems were composed of questions from physics and biology on the state of the matter, microscopic and macroscopic representation of the matter (Slapničar et al., 2017). After each task, they rated the difficulty of the problem. The relationship between its perceived difficulty and physiological parameters was studied.

Methods

Participants

The non-random sample of this pilot study included 10 participants: five were 12 years old (three girls and two boys) and five were 14 years old (three girls and two boys). To ensure anonymity, pupils were assigned a code consisting of a serial number and their age. The subjects were selected from a mixed urban population. They were first familiarised with the purpose and the content of this study and then asked for consent to participate.

Task

The subjects were presented four computer-displayed tasks from the field of science, such as identification of the solid, liquid and gaseous state of water (iceberg, lake, kettle steam), describing the melting of a glacier and opening the gassed beverage bottle, identification of warming of air in a hand pump, estimating water concentration in plants, describing sugar dissolution in water, etc. (Fig. 3.1).

To ensure the least movement possible, subjects were instructed to answer vocally. They could decide from three given choices (Fig. 3.1) and were asked additional questions in case of incomplete or incomprehensible answers. The tasks were designed by three higher education teachers of chemistry and physics and evaluated by six elementary and secondary school teachers (Slapničar et al., 2017). The tasks contained a text presentation (socio-scientific context), visualisations (pictures, schemes, animations) and three-choice answers to be answered by the subjects. After each task, the subjects were asked to grade the properties of the task by a questionnaire of the

Dissolving of a sugar cube in water.

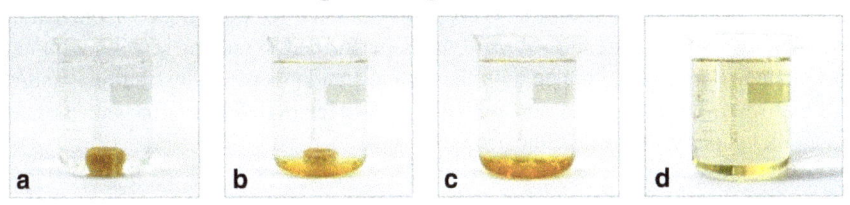

Which representation of particle motion is the most appropriate representation of dissolution of sugar in water? Explain why.

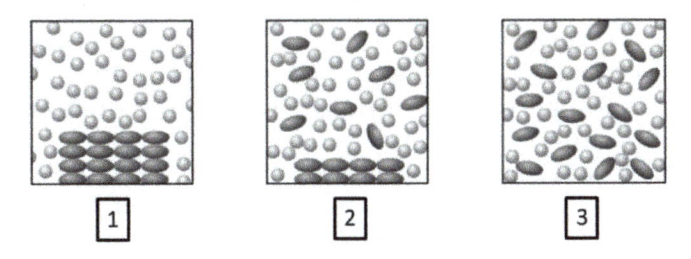

Fig. 3.1 An example of a task from chemistry; dissolving of a sugar cube in water, represented in the lower part using video clips showing movement of the particles

paper-pencil type (five-point Likert-type-scale). The following items were evaluated (1) the difficulty of each task, (2) the confidence of writing the correct answer and (3) whether the task was interesting.

Measuring Protocol

After the initial introduction of the measurement instruments to the test subject, sensors were attached and a rest period was allowed for the initial instructions. This period was set to 5 min to enable the electrolyte gel absorption resulting in an optimal electrical contact (see Ogorevc et al., 2013 for the importance of gel in skin-conductance measurements). Rest period served as a baseline for all the physiological parameters.

The subjects were instructed to sit in a relaxed manner and perform only slow movements if at all. The disturbances from the surroundings (noise from the corridor, activities in the neighbouring rooms) were minimised and were considered negligible. Room temperature and air humidity, the major environmental error sources for skin-conductance measurements, were monitored throughout the experiment. The room's lightning was kept at the same level for all subjects to avoid errors in heart rate estimation by means of photoplethysmography.

This study was approved by the Ethics Committee of the Faculty of Education, University of Ljubljana, resulting in consents obtained for the subjects from school boards, teachers and parents. All procedures performed in studies involving human participants were in accordance with the ethical standards of the institutional and/or national research committee and with the 1964 Helsinki declaration and its later amendments or comparable ethical standards.

Instrumentation

The physiology was acquired by means of a multi-parameter psychophysiology measuring system (MP150 by Biopac, USA) enabling measurements of electrodermal activity, skin temperature, heart rate and respiration with a sampling frequency of 1 kHz. The measuring system was validated in a prior study (Ogorevc et al., 2013). Measuring sites were selected according to Ogorevc et al. (2011) and van Dooren et al. (2012).

Electrodermal activity was monitored by using reusable wet Ag-AgCl electrodes attached to the skin by means of Velcro bands and a Biopac EDA100C amplifier. Measuring sites were the distal phalanges of pointing and middle finger. Skin temperature (ST) was recorded on the ring finger using SKT100C amplifier and a small-size fast response thermistor. Heart rate was calculated from the raw photoplethysmograph signal recorded by a transducer placed on the little finger and connected to amplifier Biopac PPG100C (Fig. 3.2). Respiratory rate was calculated from the

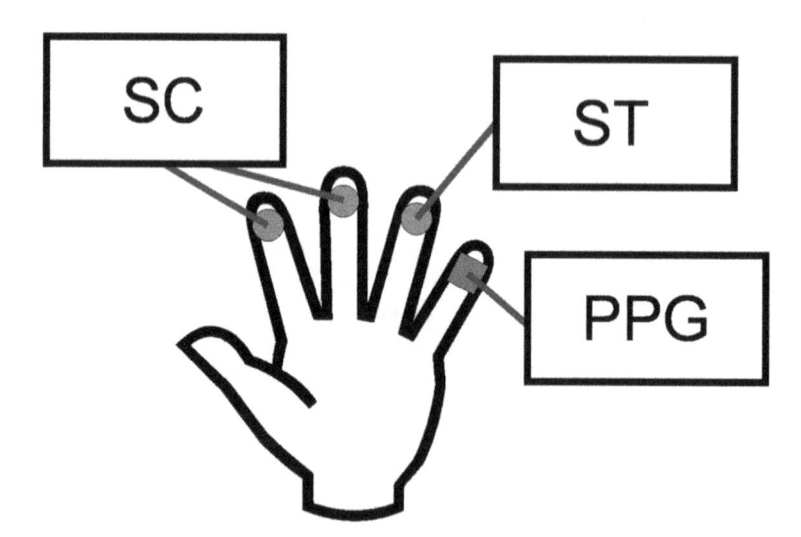

Fig. 3.2 Placement of the physiological sensors on subject's non-dominant hand; SC—skin conductance, ST—skin temperatures, PPG—photoplethysmography for heart rate measurements

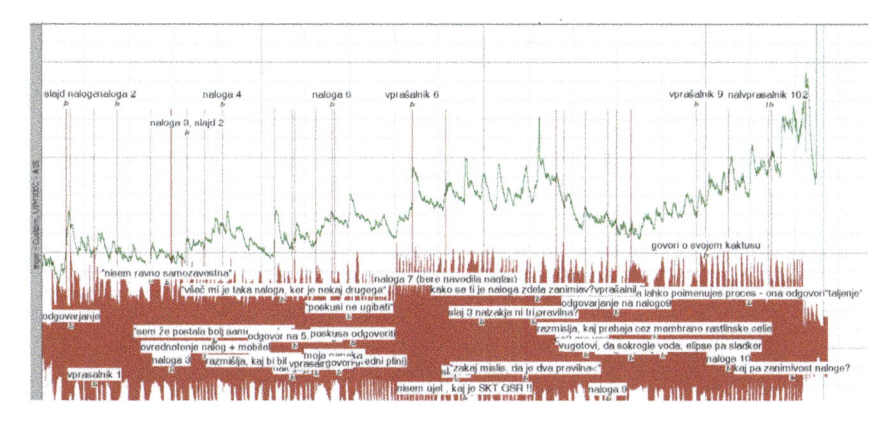

3 Predicting Task Difficulty Through Psychophysiology 59

Fig. 3.3 Raw signal of the skin conductance (upper, green curve) with trigger points (red vertical lines) for marking the time span of the tasks. Manual notations are visible in the lower part of the figure, overlaid over the respiration rate signal

acquired signal of the subject's chest displacement while breathing by means of a chest-belt transducer connected to Biopac RSP100C amplifier.

In addition, an operator used a manual trigger for timestamping the beginning and end of each task and of answering the questionnaires. Manual notations were possible for additional information like unexpected events, observation of subject's behaviour and emotional state, plausible explanation of sudden physiological change, etc. These are visible in the lower part of Fig. 3.3.

Psychophysiological parameters were monitored throughout the task and the subsequent filling out of the questionnaire. The physiological signals were recorded using the software AcqKnowledge 4.1 (Biopac, 2014). The acquired raw signals were pre-processed (filtered, outliers removed) and stored for further signal processing.

Signal Processing

The photoplethysmogram was pre-processed using the AcqKnowledge software. First, a bandpass filter was applied with frequency cut-offs at 0.5 Hz and 3 Hz, using a Hamming window. Next, the AcqKnowledge's in-built cycle detector was employed, where cycles were determined from peaks and a specific threshold configuration was determined for every subject. The peaks' locations were then exported to a CSV file.

Heart rate and heart rate variability were analysed using an R package RHRV (Rodriguez-Linares et al., 2016). Effectively, the raw heart rate signal was first extracted from heartbeat positions previously exported from AcqKnowledge. Next, this signal was further filtered to only include heart rate between 50 bpm and 140 bpm. The signal was then interpolated using spline interpolation. Finally, a Fourier transform was calculated by shifting a 300-second window 0.5 s at a time. The large

Table 3.1 Frequency limits as specified in heart rate variability calculation. These roughly correspond to the limits suggested by European Society of Cardiology (Malik, 1996; see page 360, Table 2) with a wider ULF band

Frequency band	Lower frequency limit (Hz)	Upper frequency limit (Hz)
Ultra low frequency (ULF)	0.00	0.03
Very low frequency (VLF)	0.03	0.05
Low frequency (LF)	0.05	0.15
High frequency (HF)	0.15	0.4

window chosen here places a lower limit of 48 bpm on heart rate due to the Nyquist–Shannon sampling theorem (Clifford, 2002). Thus, a time dependence of heart rate was obtained with a resolution of 0.5 s and heart rate variability was calculated in bands as specified in Table 3.1.

Respiration rate was analysed similarly to heart rate. Since the package RHRV is intended for heart rate analysis and some parameters are set in the source code, the breath time positions were first divided by 3. Using this linear transformation, the respiration rate signal is within pre-specified limits of the heart rate analysis. The time dependence of the respiration rate was finally divided by 3 again to account for the previous transformation. Then the signal's mean, standard deviation and maximum and minimum values were calculated.

Mean, standard deviation, maximum and minimum values, mean slope and the last value of skin temperature were also calculated. The last value was added because skin temperature changes are relatively slow due to skin's slow thermal response to psychological stressors and due to high tissue heat-capacity. The latest temperature value during the task was expected to be more representative of the psychological effect the task had on the subject (Novak et al., 2011).

Skin conductance was analysed using a MATLAB-based program Ledalab (Benedek, 2014). Specifically, its continuous decomposition analysis and the traditional trough-to-peak analysis were chosen (Benedek & Kaernbach, 2010a cf. Benedek & Kaernbach 2010b for (nonnegative) discrete decomposition). The software is intended for analysing a specific time interval after an event. To accommodate our experimental design, one of their scripts, called export_era.m, was modified so that the pre-set time intervals corresponding to experimental conditions could be analysed. Table 3.2 lists the skin-conductance feature calculated using Ledalab. The first one is a classical trough-to-peak method (TTP), which simply counts the peaks as defined by a minimum amplitude criterion and determines their amplitudes by measuring the distance from the peak to the preceding trough. A more complex method, continuous decomposition analysis (CDA), takes a deconvolution of the signal and thus decomposes it into tonic activity (i.e. skin-conductance level, SCL) and phasic activity (which corresponds well to the skin-conductance responses, SCR). Finally, since the number of SCRs and the time integral of the phasic driver are all dependent on the task duration, these three features were divided by the duration of the task.

Table 3.2 Skin-conductance features and their meanings as calculated using Ledalab (Benedek, 2014; Benedek & Kaernbach, 2010b). The prefix of the feature denotes the method used: continuous decomposition analysis (CDA) and the standard trough-to-peak (TTP) method

Feature	Meaning
CDA.nSCR	Number of skin-conductance responses (SCRs) within the task
CDA.Latency	Time to the first SCR within the task
CDA.AmpSum	Sum of amplitudes of SCRs within the task
CDA.SCR	Mean phasic driver, which corresponds to the average of SCRs' amplitudes
CDA.ISCR	Time integral of the phasic driver
CDA.PhasicMax	Maximum of phasic activity, corresponding to the max. SCR amplitude
CDA.Tonic	Mean tonic activity, corresponding to mean SCL
TTP.nSCR	Number of SCRs within the task
TTP.AmpSum	Sum of amplitudes of SCRs within the task
TTP.Latency	Time to the first SCR within the task
Global.Mean	Mean of the skin-conductance (SC) signal
Global.MaxDeflection	Maximum positive deflection in the SC signal

Results

Self-Reports on Perceived Difficulty

Before attempting a psychophysiological analysis, the subjects' self-reports were analysed. Bivariate relationships between the psychological variables were observed in scatter plots and the strongest relationship was found between the subjects' perceived difficulty of the task and the perceived accuracy of their answer. The harder the task was perceived as, the less certain they were that they solved it correctly, the Pearson correlation coefficient was $r = -0.58$ with a $p < 0.001$. The remaining relationships were less apparent with some being non-linear. Such was the relationship between the task order and its perceived difficulty, where later tasks were assessed as more difficult (the Spearman correlation coefficient had a value of $\rho = 0.45$, $p < 0.001$), but not consistently. Other correlations had $|\rho| < 0.4$, albeit some were statistically significant. It is also worth noting that the correlation between the comprehension of the task and its perceived difficulty was low ($\rho = -0.20$, $p = 0.040$) and that the rating of the comprehension was high on average (mean comprehension across all subjects and all tasks was out 4.65 of 5).

Thus it was determined that perceived difficulty of the task was a sufficiently distinct psychological construct, not identical to the solver's comprehension, engagement with the task or conviction about their own answer. The task difficulty was therefore sought to be predicted by physiological parameters.

Choosing Parameter's Best Features

In attempting to answer how to best predict task perceived difficulty based on physiological reactions, a measure of each physiological process was first needed to be chosen. Different physiological parameters (such as skin conductance) were analysed according to different measures (such as the number of skin-conductance responses and the mean skin-conductance level). These were sometimes unrelated or relatively independent features, but could also be the same physical attributes, but calculated according to different methods (cf. skin-conductance measures in Table 3.2). Therefore, a decision was first made as to the best way to assess each of the measured parameters, since feature selection was found to be important in other studies (e.g. Kukolja et al., 2014).

To determine the best predictor, linear regression was employed. The predictors were included in the model step-wise and the correlations between them, their β coefficients and the Akaike information criterion (AIC) were taken into account when deciding which ones to include and which ones to keep in the model. Only simple linear effects and no interactions were considered in regression models in this part of the analysis.

Both, the physiological and psychological variables were linearly transformed before the following analysis. Specifically, they were standardised within each subject, so that the mean of every variable within the subject was 0 and its standard deviation was 1. This was done in order to circumvent the problem of different baselines and to make the distributions closer to the normal distribution.

Skin Temperature

As noted, mean, standard deviation, maximum and minimum values, mean slope and the last value of the temperature were calculated. The predictor that explained the most variance of the difficulty was the mean skin temperature. Compared to the zeroth-order model it lowered the Akaike information criterion (AIC) by the largest amount and was also the most statistically significant predictor in the full model, containing all of the calculated features. The model with only the mean temperature as the predictor was statistically significant ($F(1, 104) = 10.3$, $p = 0.002$) with the beta regression coefficient $\beta = 0.300$ ($t = 3.21$, $p = 0.002$). The mean skin temperature explained 8.1% of the difficulty variance, compared to the adjusted $R^2_{adj} = 0.19$ of the full model. Its correlation was the highest with the minimum ($r = 0.56$) and the final value of skin temperature ($r = 0.55$, both $p < 0.001$). On the other hand, the information about the standard deviation and the slope might be lost with this model since the correlations between them and the chosen feature are $r < 0.1$ and indistinguishable from zero.

Respiration Rate

The respiration rate feature that lowered the AIC the most was the maximum value of the respiration rate during the task. If all of the calculated features were included as predictors in linear regression, the standard deviation of the respiration rate had the highest absolute value of the β-coefficient. This suggests that these two predictors are correlated, which was indeed the case ($r = 0.47$, $p < 0.001$). It should be noted, however, that the predictions took opposite directions. Specifically, a greater maximum value of respiration rate was related to higher task difficulty, while there was more deviation in respiration rate at lower difficulties.

Of these two, the maximum value seems to be the simpler feature. It is also, however, more prone to moving artefacts and thus less reliable. The standard deviation is the simplest indicator of the respiration rate variability. Since the correlations of these two features with task difficulty were in opposite directions, it was decided to keep both as predictors in the subsequent analysis. Together, they explained over 11% of difficulty variance.

Heart Rate

Simple heart rate measures and heart rate variability as calculated using Fourier transform were separated for the purpose of statistical analysis. The simple features of heart rate showed a very similar pattern to the features of the respiration rate. The two predictors that lowered the AIC the most were the maximum value of the heart rate and its (simple) standard deviation. They were also found to be correlated ($r = 0.52$, $p < 0.001$) and their relation to the task difficulty was in opposite directions. The maximum value, however, was the strongest predictor in the full model ($\beta = 0.487$, $t = 4.72$, $p < 0.001$ for maximum value compared to the $\beta = -0.402$, $t = -3.57$, $p < 0.001$ for the standard deviation).

The power in different regions of the power spectrum (see Table 3.1) showed moderately strong linear relationships. It was tested whether the heart rate variability calculated in this fashion could predict task difficulty and replace the simple standard deviation. However, neither the full model containing power in all of the regions in the power spectrum nor the model with only the strongest (by the information criterion) predictor, the power in the very low frequency band (VLF), were statistically significant.

It was thus decided to keep both the heart rate maximum and the simple standard deviation (as a simple measure of heart rate variability) as the predictors for further analysis, since they were only moderately correlated and they explained over 16% of the difficulty variance.

Skin Conductance

Many skin-conductance features were calculated by different methods (see Table 3.2). The predictors were first separated into three blocks, according to the method used. They were then analysed separately and only the best predictors from each method were considered for further analysis. Of the calculated features, both global features (mean and maximum deflection) turned out to be good parameters. The number of SCRs and the sum of their amplitudes as calculated by trough-to-peak method (TTP.nSCR and TTP.AmpSum) were also statistically significant. Among those calculated by the continuous decomposition analysis, four of the predictors were statistically significant: the sum of SCR amplitudes (CDA.nSCR), the time integral of the phasic driver (CDA.ISCR) and its maximum amplitude (CDA.PhasicMax) and the mean tonic activity (CDA.Tonic).

Some of these predictors represent the same physiological processes and this fact was reflected in high correlations. Indeed, the correlations between the number of SCRs as calculated by two different methods, between the sums of SCR amplitudes, and between the mean tonic component and a simple average of the skin-conductance signal were all nearly perfect. It was thus prudent to keep only one of each pair in the final model. The choice was made by determining which of the two lowered the AIC more. It was found that in these three pairs, the features calculated by the continuous decomposition analysis were superior in their predictive power compared to the ones calculated by other methods.

Finally, the best features among those calculated by CDA were chosen. Three predictors were found to be statistically significant: the sum of amplitudes (CDA.AmpSum), the maximum amplitude (CDA.PhasicMax) and the time integral of the SCRs (CDA.ISCR). Naturally, the maximum SCR amplitude and the sum of all of them were correlated ($r = 0.62$, $p < 0.001$), which resulted in a higher variance inflation factor ($VIF \approx 1.7$) for both. Of these two features only the sum of amplitudes was therefore chosen for further analysis, since it is less susceptible to moving artefacts compared to a point feature and also had a higher β-coefficient. The sum of amplitudes and the time integral of the SCRs together explained more than 33% of variability in task difficulty.

Choosing the Best Parameter

In the previous section, features of individual parameters were explored. Specifically, their predictive power pertaining to task difficulty was analysed. After one or two features of each parameter were chosen, the aim was to construct a regression model consisting of any number of physiological parameters.

The predictors included in the full physiological model were:

- mean skin temperature,
- the maximum value and standard deviation of respiration rate,

- standard deviation of heart rate and the maximum heart rate and
- sum of skin-conductance responses amplitudes and their time integral.

Most of these predictors were correlated with the task difficulty, but only moderately correlated between themselves: the highest correlation remained that between the maximum heart rate and its standard deviation ($r = 0.52$, $p < 0.001$). Several possibilities for a regression model were therefore tested, all constructed in a 'backwards' way, eliminating predictors one by one.

In the full regression model, consisting of all physiological predictors listed above, all but the maximum respiration rate ($\beta = 0.106$, $t = 1.16$, $p = 0.250$) were statistically significant. Furthermore, eliminating any predictor (other than the maximum respiration rate mentioned) only increased the Akaike information criterion. The predictors with the highest β-coefficients were both related to skin-conductance responses, with the sum of SCRs amplitudes having $\beta = 0.444$ ($t = 4.45$, $p < 0.001$) and the time integral of the SCRs $\beta = -0.388$ ($t = -4.55$, $p < 0.001$). A model that included all chosen predictors had an $R^2_{adj} = 0.450$ and so it explained almost half of the task difficulty variability.

A model that included only the best two predictors, namely the amplitude and the time integral of the SCRs, fared significantly worse than the full model ($F(97, 92) = 4.17$, $p = 0.002$). However, it still explained over a third of the variance of task difficulty ($R^2_{adj} = 0.361$).

Effect of Task Duration

An attempt was made to explore the role of task duration. Including task duration as a predictor in the full physiological model rendered all other predictors insignificant. Indeed, predicting task difficulty using task duration as a sole predictor in a linear regression model explained more than a half of the task difficulty variance ($R^2_{adj} = 0.506$). Due to collinearity with duration, the variance of other regression coefficients is increased: the most so for the predictors related to SCRs. The *VIF*s for the sum of SCRs amplitudes, their time integral and task duration were 3.49, 2.14 and 3.26, respectively.

To diminish the effect of task duration in the model, another, limited, physiological model was considered. It excluded the features of SCRs and only included mean skin temperature, standard deviation of respiration rate and its maximum and standard deviation of heart rate and its maximum. In this model, all predictors were statistically significant and the model had $R^2_{adj} = 0.296$.

The collinearity of the task duration and the SCR features might be able to illuminate the nature of skin-conductance response, however. The time integral of the SCRs and the sum of their amplitudes are, theoretically, related to the task duration. The first relationship should be monotonously increasing: the longer the task, the more SCRs are generally expected. This means that the sum of their amplitudes, too, is increasing with time. The second relationship is less straightforward. The time integral is broadly expected to increase with task duration. In contrast with discrete (nonnegative) deconvolution method (Benedek & Kaernbach, 2010b), however, the phasic driver representing the SCRs can be negative as well as positive when using continuous decomposition analysis. This means that the time integral of the phasic driver does not increase with time in a simple monotonous manner.

These two different relationships between two features of skin conductivity and task duration are reflected in different correlation coefficients we measured. While the sum of SCRs amplitudes correlated moderately with task duration ($r = 0.45$, $p < 0.001$), the correlation between task duration and the time integral of the phasic driver was not statistically significant ($r = -0.18$, $p = 0.073$).

Despite this, the nature of skin-conductance response cannot be unambiguously inferred from our data. The sum of SCRs amplitudes was higher during the tasks of higher perceived difficulty ($\beta = 0.648$, $t = 7.34$, $p < 0.001$, in a model consisting of only this predictor and the CDA.ISCR). It remains unclear, however, whether this is due to a larger number of SCRs (the time-normalised number of SCRs was not a good predictor in a model consisting of skin conductivity features, $\beta = -0.115$, $t = -1.41$, $p = 0.257$) or to their higher average amplitude. Additionally, the evidence was inconclusive regarding the relationship between the shape of the SCRs and the task difficulty, since the integral of the phasic driver had a negative β coefficient but was not statistically significantly correlated with difficulty when the effect of the sum of amplitudes was not controlled for.

Discussion

In the previous chapter, features of several physiological parameters were considered as predictors of task difficulty. This was done both, by considering the β coefficients of the individual predictors in a comprehensive model and by testing smaller models consisting of only selected features.

Regardless of the physiological parameter under consideration, it was possible to build a statistically significant regression model. Different parameters had different predictive power for task difficulty, however.

In terms of the proportion of explained variance, skin conductivity and its features would seem to be the best physiological parameter for predicting task difficulty, at least judging naively by the R^2_{adj} coefficient of a model composed of two of its features. Taking into account their relationship with the task duration, this conclusion is less convincing.

Task duration was the strongest predictor of task difficulty when comparing it to selected physiological features. It did not render the physiological features redundant, however. Even the features related to skin-conductance responses might still hold valuable information, since their time dependence remains unclear (see "Effect of Task Duration").

Other physiological parameters were good predictors of task difficulty even when skin conductance was not included in the regression model; they explained almost 30% of the variance of the task difficulty. Of these, the best predictors were the maximum heart rate and the mean skin temperature during the task. Indeed, by themselves they explained 11 and 8% of the variance, respectively, when considering regression models consisting of only one predictor.

It would seem then that an attempt to predict how difficult a task is from physiological processes could take several different forms. One way would be to only measure skin conductivity and infer task difficulty from SCRs: specifically, from their amplitude sum and their time integral. One should consider, however, what are the benefits of this set-up compared to simply timing the task and concluding about its difficulty from its duration. More specifically, the relationship between the SCRs features and task duration should be considered before determining whether this would be a fruitful approach.

An alternative approach would be to measure heart and respiration rate and skin temperature, instead, and using appropriate features to predict difficulty. It could be argued that this provides additional information about the task difficulty compared to a simple duration measurement. Furthermore, task duration is a *post festum* measure, while physiological features could in theory be calculated concurrently with solving of the problem of which difficulty is predicted.

There are other factors to consider when choosing a physiological process for task difficulty prediction. The two physiological parameters from which the best predictors were chosen—skin conductivity and heart rate, measured by photoplethysmography—are, arguably, also the most delicate. First, they are both sensitive to moving artefacts and demand careful attachment of the transducers. In addition, skin conductance suffers from errors due to non-stable electrical contact between electrodes and the skin. Secondly, to extract individual heartbeats and skin-conductance responses, significant (pre)processing of the signals is required. This often demands some manual inspection of the signal or results in erroneous detections. Furthermore, doing such analysis on the fly would require considerably more processing power. Finally, while the heart rate measurements are readily accessible via wearable instruments such as smart watches and bracelets, skin conductivity on the other hand is less commonly reported.

There are several limitations to the present study. The small size of the sample made it difficult to assess the distribution of individual physiological features. Assumptions of normality could therefore not be reliably tested. This was partly compensated for by transforming (standardising) the data. In addition, the design of the experiment was such that the sequence of the tasks was not independent of their difficulty. There was a moderate correlation between the task order number and its difficulty (Spearman's $\rho = 0.45$, $p < 0.001$), but the relationship was non-linear and was not accounted for in regression analysis.

In conclusion, the results of our study showed that the perceived difficulty of a task could be predicted by measuring physiological processes and calculating some of their features. Future work could be focused on determining more reliably what the most advantageous processes and features are and to establish the nature of this relationship more definitely.

Acknowledgements The authors acknowledge that the study was part of the project, Explaining Effective and Efficient Problem Solving of the Triplet Relationship in Science Concepts Representations (J5-6814), which was financially supported by the Slovenian Research Agency.

The authors declare that they have no conflict of interest.

References

Aasman, J., Mulder, G., & Mulder, L. J. M. (1987). Operator effort and the measurement of heart-rate variability. *Human Factors, 29*(2), 161–170. https://doi.org/10.1177/001872088702900204.

Benedek, M. (2014). *Ledalab*. Institut für Psychologie, University of Graz, Austria; University of Graz. Retrieved from http://www.ledalab.de/.

Benedek, M., & Kaernbach, C. (2010a). A continuous measure of phasic electrodermal activity. *Journal of Neuroscience Methods, 190*(1), 80–91. https://doi.org/10.1016/j.jneumeth.2010.04.028.

Benedek, M., & Kaernbach, C. (2010b). Decomposition of skin conductance data by means of nonnegative deconvolution. *Psychophysiology, 47*(4), 647–658. https://doi.org/10.1111/j.1469-8986.2009.00972.x.

Biopac. (2014). *MP system hardware guide* (Ver. 21. 1). Goleta, CA: BIOPAC Systems. Retrieved August 10, 2015, from http://www.biopac.com/Manuals/mp_hardware_guide.pdf.

Boucsein, W. (2012). *Electrodermal activity* (2nd ed., p. 618). New York City: Springer Science and Business Media.

Boucsein, W., Fowles, D. C., Grimnes, S., Ben-Shakhar, G., Roth, W. T., Dawson, M. E., & Filion, D. L. (2012). Publication recommendations for electrodermal measurements. *Psychophysiology, 49*, 1017–1034. https://doi.org/10.1111/j.1469-8986.2012.01384.x.

Brouwer, A. M., Hogervorst, M. A., Holewijn, M., & van Erp, J. B. (2014). Evidence for effects of task difficulty but not learning on neurophysiological variables associated with effort. *International Journal of Psychophysiology, 93*(2), 242–252. https://doi.org/10.1016/j.ijpsycho.2014.05.004.

Brouwer, A.-M., van Wouwe, N., Mühl, C., van Erp, J., & Toet, A. (2013). Perceiving blocks of emotional pictures and sounds: Effects on physiological variables. *Frontiers in Human Neuroscience, 7*(June), 1–10. https://doi.org/10.3389/fnhum.2013.00295.

Butler, E. A., Wilhelm, F. H., & Gross, J. J. (2006). Respiratory sinus arrhythmia, emotion, and emotion regulation during social interaction. *Psychophysiology, 43*(6), 612–622. https://doi.org/10.1111/j.1469-8986.2006.00467.x.

Cacioppo, J., Tassinary, L. G., & Berntson, G. G. (Eds.). (2007). *The handbook of psychophysiology* (3rd ed., p. 914). New York: Cambridge University Press. https://doi.org/10.1017/cbo9780511546396.

Clark, D. J., Chatterjee, S. A., McGuirk, T. E., Porges, E. C., Fox, E. J., & Balasubramanian, C. K. (2018). Sympathetic nervous system activity measured by skin conductance quantifies the challenge of walking adaptability tasks after stroke. *Gait and Posture, 60*(August 2017), 148–153. https://doi.org/10.1016/j.gaitpost.2017.11.025.

Clifford, G. D. (2002). *Signal processing methods for heart rate variability analysis*. Doctoral dissertation. University of Oxford. Retrieved from http://www.ibme.ox.ac.uk/research/biomedical-signal-processing-instrumentation/prof-l-tarassenko/publications/pdf/gdcliffordthesis.pdf.

Collet, C., Averty, P., & Dittmar, A. (2009). Autonomic nervous system and subjective ratings of strain in air-traffic control. *Applied Ergonomics, 40*(1), 23–32. https://doi.org/10.1016/j.apergo.2008.01.019.

Collet, C., Vernet-Maury, E., Delhomme, G., & Dittmar, A. (1997). Autonomic nervous system response patterns specificity to basic emotions. *Journal of the Autonomic Nervous System, 62*(1–2), 45–57. Retrieved from http://www.ncbi.nlm.nih.gov/pubmed/9021649.

Fauvel, J. P., Cerutti, C., Quelin, P., Laville, M., Gustin, M. P., Paultre, C. Z., & Ducher, M. (2000). Mental stress-induced increase in blood pressure is not related to baroreflex sensitivity in middle-aged healthy men. *Hypertension, 35*(4), 887–91. Retrieved from http://www.ncbi.nlm.nih.gov/pubmed/10775556.

Karavidas, M. K., Lehrer, P. M., Lu, S.-E., Vaschillo, E., Vaschillo, B., & Cheng, A. (2010). The effects of workload on respiratory variables in simulated flight: A preliminary study. *Biological Psychology, 84*(1), 157–160. https://doi.org/10.1016/J.BIOPSYCHO.2009.12.009.

Kivikangas, J. M., Kätsyri, J., Järvelä, S., & Ravaja, N. (2014). Gender differences in emotional responses to cooperative and competitive game play. *PLoS ONE, 9*(7), https://doi.org/10.1371/journal.pone.0100318.

Kukolja, D., Popović, S., Horvat, M., Kovač, B., & Ćosić, K. (2014). Comparative analysis of emotion estimation methods based on physiological measurements for real-time applications. *International Journal of Human Computer Studies, 72*(10–11), 717–727. https://doi.org/10.1016/j.ijhcs.2014.05.006.

Lewis, M., Haviland-Jones, Barrett, J. M., & Feldman, L. (1993). *Handbook of emotions* (p. 720). New York: The Guilford Press.

Lisetti, C. L., & Nasoz, F. (2004). Using noninvasive wearable computers to recognize human emotions from physiological signals. *EURASIP Journal on Advances in Signal Processing, 2004*(11), 1672–1687. https://doi.org/10.1155/S1110865704406192.

Malik, M., Bigger, J. T., Camm, A. J., Kleiger, R. E., Malliani, A., Moss, A. J., & Schwartz, P. J. (1996). Heart rate variability: Standards of measurement, physiological interpretation, and clinical use. *European Heart Journal, 17*(3), 354–381. https://doi.org/10.1093/oxfordjournals.eurheartj.a014868.

Mandryk, R. L., & Atkins, M. S. (2007). A fuzzy physiological approach for continuously modeling emotion during interaction with play technologies. *International Journal of Human-Computer Studies, 65*(4), 329–347. https://doi.org/10.1016/J.IJHCS.2006.11.011.

Mehler, B., Reimer, B., Coughlin, J., & Dusek, J. (2009). Impact of incremental increases in cognitive workload on physiological arousal and performance in young adult drivers. *Transportation Research Record: Journal of the Transportation Research Board, 2138,* 6–12. https://doi.org/10.3141/2138-02.

Nourbakhsh, N., Wang, Y., Chen, F., & Calvo, R. A. (2012). Using galvanic skin response for cognitive load measurement in arithmetic and reading tasks. In *Australian Computer-Human Interaction Conference* (pp. 420–423).

Novak, D., Beyeler, B., Omlin, X., & Riener, R. (2014). Workload estimation in physical human-robot interaction using physiological measurements. *Interacting with Computers.* https://doi.org/10.1093/iwc/iwu021.

Novak, D., Mihelj, M., & Munih, M. (2011). Psychophysiological responses to different levels of cognitive and physical workload in haptic interaction. *Robotica, 29*(3), 367–374. https://doi.org/10.1017/S0263574710000184.

Ogorevc, J., Geršak, G., Novak, D., & Drnovšek, J. (2013). Metrological evaluation of skin conductance measurements. *Measurement: Journal of the International Measurement Confederation, 46*(9), 2993–3001. https://doi.org/10.1016/j.measurement.2013.06.024.

Ogorevc, J., Podlesek, A., Geršak, G., & Drnovšek, J. (2011). The effect of mental stress on psychophysiological parameters. *IEEE International Symposium on Medical Measurements and Applications, 2011,* 294–299. https://doi.org/10.1109/MeMeA.2011.5966692.

Olsson, A., & Phelps, E. A. (2007). Social learning of fear. *Nature Neuroscience, 10*(9), 1095–1102. https://doi.org/10.1038/nn1968.

Pecchinenda, A. (1996). The affective significance of skin conductance activity during a difficult problem-solving task. *Cognition and Emotion, 10*(5), 481–504. https://doi.org/10.1080/026999396380123.

Ravaja, N. (2004). Contributions of psychophysiology to media research: Review and recommendations. *Media, 6*(2), 193–235. https://doi.org/10.1207/s1532785xmep0602.

Rodriguez-Linares, L., Vila, X., Lado, M. J., Mendez, A., Otero, A., & Garcia, C. A. (2016). RHRV: Heart rate variability analysis of ECG data. Retrieved from https://cran.r-project.org/package=RHRV.

Slapničar, M., Devetak, I., Glažar, A. S., & Pavlin, J. (2017). Identification of the understanding of the states of matter of water and air among slovenian students aged 12, 14 and 16 years through solving authentic tasks. *Journal of Baltic Science Education, 16*(3), 308–323.

Storm, H., Shafiei, M., Myre, K., & Raeder, J. (2005). Palmar skin conductance compared to a developed stress score and to noxious and awakening stimuli on patients in anaesthesia. *Acta Anaesthesiologica Scandinavica, 49*(6), 798–803. https://doi.org/10.1111/j.1399-6576.2005.00665.x.

van Dooren, M., de Vries, J. J. G. G. J., & Janssen, J. H. (2012). Emotional sweating across the body: Comparing 16 different skin conductance measurement locations. *Physiology & Behavior, 106*(2), 298–304. https://doi.org/10.1016/j.physbeh.2012.01.020.

Veltman, J. A., & Gaillard, A. W. (1998). Physiological workload reactions to increasing levels of task difficulty. *Ergonomics, 41*(5), 656–669. https://doi.org/10.1080/001401398186829.

Wen, W., Liu, G., Cheng, N., Wei, J., Shangguan, P., & Huang, W. (2014). Emotion recognition based on multi-variant correlation of physiological signals. *IEEE Transactions on Affective Computing, 5*(2), 126–140.

Junoš Lukan has an M.A. degree in Psychology from the University of Ljubljana and an M.Sc. degree in Physics from the Imperial College London. He is a Ph.D. student at the Jožef Stefan International Postgraduate School and a researcher in the Ambient Intelligence Group at the Department of Intelligent Systems at Jožef Stefan Institute, Ljubljana, Slovenia.

Gregor Geršak, Ph.D. is an Associate Professor lecturing measurements, measuring methods, and instrumentation at the Faculty of Electrical Engineering (FE), University of Ljubljana, Slovenia. Apart from classical metrology, his main research area is psychophysiology used in human–computer interaction, affective computing, education, and cognitive science. He makes use of expertise in metrology and calibration, metrology of biomedical sensors and instrumentation, wearables, signal processing, medical thermography, and virtual reality. He was a visiting scholar at Physikalisch-Technische Bundesanstalt, Braunschweig and University of California, Berkeley.

Chapter 4
The Role of the Explanatory Key in Solving Tasks Based on Submicroscopic Representations

Vesna Ferk Savec and Špela Hrast

Introduction

Since the building blocks of matter—atoms, molecules and ions—cannot be perceived naturally by our senses, the desire to reveal "the world of the invisible" has inspired philosophers and scientists for many centuries. From Plato, or even earlier, to the present day, people have tried to visualise their ideas about the nature of matter by building mental and concrete models (Gregory, 2000). The important role of using models and modelling in science discoveries to visualise concepts and processes at the particle level has been manifested since the nineteenth century by many leading chemists such as Kekulé, Van't Hoff, Pauling, Watson and Crick (Justi & Gilbert, 2002), often related with corresponding Nobel Prizes awards in chemistry, physics and medicine. In contemporary science, new developments related to the use of models and modelling are supported by the application of computer methods and computer graphics. A recent example of this is the Nobel Prize award in physiology or medicine in 2019, which was awarded to Kaelin, Ratcliffe and Semenza. In particular, the researchers identified molecular machinery that regulates the activity of genes in response to varying levels of oxygen, their discoveries are therefore recognised as paving the way for promising new strategies to combat anaemia, cancer and many other diseases (Nobel Assembly at Karolinska Institutet, 2019).

Models have played an important role not only in science research, but also in science education. The pioneering role in the introduction of models in the teaching of chemistry was attributed as early as 1811 to John Dalton, who used wooden spheres connected by metal pins in his lectures (Francoeur, 1997; Hardwicke, 1995).

V. Ferk Savec (✉) · Š. Hrast
Faculty of Education, University of Ljubljana,
Kardeljeva Ploščad 16, 1000 Ljubljana, Slovenia
e-mail: vesna.ferk@pef.uni-lj.si

© Springer Nature Switzerland AG 2021
I. Devetak and S. A. Glažar (eds.), *Applying Bio-Measurements Methodologies in Science Education Research*,
https://doi.org/10.1007/978-3-030-71535-9_4

The "golden age" of molecular models began with the spread of commercial molecular model sets based on Stuart's space-filling models after their invention by Stuart in 1934. The first commercially produced set of Stuart-type models, the so-called Fisher-Herschfelder-Taylor models, had accurate bond angles and adequate mechanical stability (Petersen, 1970). The CPK models are named after the initials of the family names of the chemists Corey, Pauling and Koltun, who created the first concrete, space-filling molecular models of amino acids, peptides and proteins, which were painted in different colours to indicate the respective chemical element, e.g. white for hydrogen, black for carbon, red for oxygen and blue for nitrogen (Corey & Pauling, 1953), and patented their improved version in 1965 (Koltun, 1965). The use of colour conventions in these models is known as the CPK colour scheme, which is still the most commonly used colour scheme today.

Computer versions of CPK models have successfully imitated the appearance of their physical analogues and have enabled additional features, e.g. with respect to the possibilities of simultaneous representation of molecules by different model types (e.g. ball-and-stick, space-filling, wire-frame, valence-shell electron pair repulsion model). The use of computer graphics has also considerably extended the possibilities for labelling respective chemical elements in the models with specific colours. An example of a colour convention for computer models is the CPKnew scheme, which applies to Rasmol v 2.7.3. or later ("Jmol Colors", n.d.). One of the most commonly used software packages for the visualisation of molecules today is called Jmol. Jmol has assigned colours (RGB colour and Hexadecimal-Web-Colour) to almost every element likely to be found in a molecule and even to some common isotope colours. These isotopes include deuterium and tritium of hydrogen, carbon-13 and carbon-14 and nitrogen-15 (Helmenstine, 2019).

However, as Francoeur pointed out, the awareness of the limitations of three-dimensional structural representations in physical or computer models remains in its essence: The "gap" between molecular models and other representations of molecules appears particularly obvious when the latter are based on a quantum mechanical understanding of chemistry. "It is clear that a model for a quantum mechanical system like a molecule itself cannot be quantum mechanical" (Francoeur, 1997, p. 17). As a result, a wealth of literature has been collected over the decades to address students' misconceptions about the nature of models as submicroscopic representations (SMR) of phenomena (Barke et al., 2009; Nakhleh, 1992; Slapničar et al., 2014; Van Driel & Verloop, 1999). Johnstone (1991) was the first to point out that the representation of scientific concepts and processes is based on representations on three levels: macroscopic (observable phenomena), submicroscopic or particulate (various representations of atomic, molecular and particle models) and symbolic (mathematical and chemical symbols). In this respect, it has been shown that the integration of three levels of conceptual representations in the learning process enables students to create mental images of the corresponding phenomena, which supports their better understanding (Al-Balushi & Al-Hajri, 2014; Barke & Wirbs, 2002; Ferk Savec et al., 2009; Gilbert et al., 2008).

Although there are several options for technology-enhanced learning in the digital age, textbook sets continue to play a central role in supporting the effective teaching and learning of science. In science teaching, much attention has been paid to the analysis of textbooks, for example, Devetak et al. (2010) examined explanations of states of matter in Slovenian science textbooks; Laçin-Şimşek (2011) studied female scientists in Turkish science and technology textbooks; Majidi and Mäntylä (2011) examined the knowledge organisation in magnetostatics in Finnish textbooks; Mumba et al. (2007) studied inquiry levels and skills in Zambian chemistry textbooks for high schools. It is often assumed that students understand the SMRs and learn efficiently with them because experienced chemists (e.g. textbook authors) can use them simultaneously as part of a triple representation of chemistry concepts (Johnstone, 1991). However, research (Harrison, 2001; Furió-Más et al., 2005; Gkitzia et al., 2011) indicates that the abundant presence of SMRs in a textbook is not a guarantee of efficient learning. It seems that the integration of SMRs in textbook sets by textbook authors and/or editors has often not been given sufficient attention to support the development of students' representational competence through the curriculum topics from the beginning to the end of the textbook (e.g. through the meaningful integration of explanatory keys), and therefore further studies in this area are needed.

Eye tracker has been used in science education to investigate how students process data, e.g. text data, data diagrams, images, photos, etc. (Havanki & Vanden Plas, 2014; Hinze et al., 2013; Mason et al., 2013; Pavlin et al., 2019; Slykhuis et al., 2005; Torkar et al., 2018), because it enables the monitoring of cognitive processes as a consequence of the links between eye movements and cognition (Rayner, 2009; Yen & Yang, 2016). It seems useful to use the eye tracker also for studying selected examples of SMRs and to collect information on eye movements in order to examine the role of an explanatory key in the processing of SMRs by students in relation to the findings from textbook analysis on their integration into textbooks.

The Context and the Purpose of the Study

The paper focuses on the integration of SMRs in chemistry learning materials in the higher grades of primary school, with emphasis on the accompanying descriptors that support students' recognition of the informational value of SMRs. Based on their own experience with the simultaneous use of SMRs as part of a triple representation of chemical concepts and processes, experienced chemists, such as textbook authors, could assume that the use of SMRs would enable efficient learning by students even without explaining the meaning of the symbols used in these representations. However, understanding the types of information and conclusions provided by the visualisations in the different learning materials requires explicit guidance and practise (Akaygun & Jones, 2014; Ferk Savec et al., 2005; Jones, 2013). To support the effectiveness of chemistry learning through the use of SMRs, it is useful to include in textbooks and teaching materials tools that help students to recognise the informational value of SMRs. Therefore, as indicated in previous studies (Hrast & Ferk

Savec, 2017a, 2017b), SMRs in textbook sets can be accompanied by different types of descriptors (e.g. pictorial, textual, combined, indirect).

In this paper, the considerations on SMRs descriptors in Slovenian chemistry textbooks are elaborated to reconcile their meaning with the general definition of a legend or key. Merriam-Webster Dictionary (n.d.), states that the legend is "an explanatory list of symbols on a map or diagram" and the key is: "something that gives an explanation or identification or provides a solution" (Merriam-Webster Dictionary, n.d.) In this paper we have used the term SMRs with an explanatory key to address the different possibilities of descriptors that accompany SMRs and allow learners to identify their informational value directly and unambiguously. Given the variety of possible representations of SMRs with descriptors that allow particle recognition, there are many useful options for explanatory keys, such as pictorial, textual, integrated structural or other symbolic notations. When SMRs are used without discussing the meaning of the particles, we refer to them in this paper as SMRs without an explanation key.

The main objective of the first part of the paper is to present the integration of SMRs with/without an explanatory key in Slovenian chemistry textbook sets in relation to the topics of the National Chemistry Curriculum for Primary School (Bačnik et al., 2011).

In the second part of the paper we wanted to investigate whether a certain type of explanatory key accompanying the SMRs provides added value for students in solving certain chemistry tasks. For reasons of clarity of the results, only two types of explanatory keys (pictorial, textual) were selected for the study with eye tracker, and in order to gain additional insight into the students' perception of the added value of the explanatory key, the interview with the students was chosen.

The following research questions (RQ) were specified:

RQ1 How (with/without explanatory key) are SMRs integrated in Slovenian chemistry textbook sets in relation to curriculum topics?

RQ2 Does the type of explanatory key (pictorial/textual) that accompanies SMRs affect students in solving simple chemical tasks?

Method

Sample

SMRs in Slovenian chemistry textbook sets for primary schools (related to RQ1)

In the first part of the paper, we focused on the chemistry textbook sets in primary school (8th and 9th grade), which are obligatory in Slovenia based on the objectives of National Chemistry Curriculum and consequently confirmed by the National Commission for Textbook Approval at the Ministry of Education, Science and Sport. The National Chemistry Curriculum for Primary School (Bačnik et al., 2011) includes general objectives of the school subject Chemistry and specific objectives with suggested contents on how to implement the objectives in chemistry teaching for each of the ten listed chemistry topics. Teachers can distribute the curriculum topics in 70 h in grade 8 and 64 h in grade 9. The list of textbook sets whose use in Slovenian schools is currently confirmed is shown in Table 4.1. These textbook sets were the subject of the textbook analysis in the first part of the paper.

Participants of the study of SMRs with eye tracker, also involved in the final interview (related to RQ2)

In the second part of the paper we describe a study with eye tracker, in which 44 students participated, who were selected from the pool of 118 non-chemistry freshmen of the Faculty of Education of the University of Ljubljana on the basis of their performance in a chemistry knowledge pre-test. Four participants were excluded due to their absence from the eye-tracking session and five participants due to poor eye calibration. The final sample consisted of 35 participants with high overall scores on the Chemistry Knowledge Test (the top third of students with the highest scores). The same 35 students also participated in the final interview.

Instruments

Rubrics for the analysis of the explanatory key accompanying SMRs in Slovenian chemistry textbook sets for primary schools (related to RQ1)

For the study presented in the first part of this paper, a rubric for the evaluation of SMRs in textbook sets was developed on the basis of examples from similar studies (Kahveci, 2010; Devetak & Vogrinc, 2013; Hrast & Ferk Savec, 2017a). The rubric is based on the assumption that in order to recognise the structure of SMRs, the learner can use different information in the role of the explanatory key. To ensure the validity of the rubric, 283 pages (10% of all analysed textbook set pages) were analysed by both authors and the criteria for SMRs with an explanatory key and SMRs without an explanatory key were defined. The main criteria were that SMRs with an explanatory key should enable the learner to identify all particles directly and unambiguously, although this can be achieved through different types of explanatory keys, such as pictorial, textual, integrated structural or other symbolic notations used, etc. On the other hand, SMRs without an explanatory key do not enable the learner to recognise particles directly, although it is possible that different types of related information such as the name of the compound, the description of its properties, etc. are provided. In order to reduce bias issues related to the use of the rubric for categorising SMRs with/without explanatory key through discussion and agreement, a 98% inter-rater reliability of the rubric has been established.

Materials and apparatus of the study of SMRs with eye tracker (related to RQ2)

Chemistry knowledge pre-test

With the aim of recruiting participants with a highlevel of prior knowledge for the eye tracker study, 118 students completed the chemistry knowledge paper- and-pencil pre-test developed by Ferk Savec et al. (2016). The pre-test with $\alpha = 0.62$, consists of 30 multiple-choice chemistry questions based on SMRs (M = 12.38; SD

Table 4.1 The list of the analysed textbook sets

Textbook set* title	Author(s)	Publisher	Year of publication (Edition) Textbook/ workbook	Number of Pages Textbook/workbook	Grade/ Learner's age
Kemija danes 1	Gabrič, A., Glažar, S. A., Graunar, M., Slatinek-Žigon, M.	DZS	2014 (1st Ed.)/ 2013 (1st Ed.)	125/106	8/13
Kemija 8, i-učbenik	Sajovic, I., Wissiak Grm, K., Godec, A., Kralj, B., Smrdu, A., Vrtačnik, M., Glažar, S.	Zavod RS za šolstvo	2014	264	8/13
Moja prva kemija	Vrtačnik, M., Wissiak Grm, K. S., Glažar, S. A., Godec, A.	Modrijan	2015 (1st Ed.)/ 2014 (1st Ed.)	240/92, 61	8, 9/13, 14
Peti element 8	Devetak, I., Cvirn Pavlin, T., Jamšek, S.	ROKUS KLETT	2010 (1st Ed.)/2010 (1st Ed.)	103/71	8/13
Pogled v kemijo 8	Kornhauser, A., Frazer, M.	MK	2003 (1st Ed.)/ 2004 (1st Ed.)	140/126	8/13
Od atoma do molekule	Smrdu, A.	JUTRO	2012 (2nd Ed.)/2012 (2nd Ed.)	128/160	8/13
Kemija danes 2	Graunar, M., Podlipnik, M., Mirnik, J. (textbook) Dolenc, D., Graunar, M., Modec, B. (notebook)	DZS	2016(1st Ed.)/ 2016 (1st Ed.)	152/96	9/14
Kemija 9, i-učbenik	Jamšek, S., Sajovic, I., Wissiak Grm, K., Godec, A., Boh, B., Vrtačnik, M., Glažar, S.	Zavod RS za šolstvo	2014	271	9/14

(continued)

4 The Role of the Explanatory Key in Solving Tasks …

Table 4.1 (continued)

Textbook set* title	Author(s)	Publisher	Year of publication (Edition) Textbook/ workbook	Number of Pages Textbook/workbook	Grade/ Learner's age
Peti element 9	Devetak, I.,Cvirn Pavlin, T., Jamšek, S.	ROKUS Klett	2011 (1st Ed.)/2011 (1st Ed.)	77/79	9/14
Pogled v kemijo 9	A. Kornhauser, M. Frazer	MK	2005(1st Ed.)/ 2006 (1st Ed.)	140/115	9/14
Od molekule do makromolekule	Smrdu, A.	Jutro	2013 (2nd Ed.)/2013 (2nd Ed.)	128/152	9/14

The term "textbook set (*)" refers to all materials for students in the written or electronic form

= 4.52). Based on the results of the pre-test, 44 participants with a highlevel of prior knowledge were included in the eye-tracking sub-sample with an average score of over 16.71 points (SD = 2.86).

Eye tracker

For monitoring students' eye movements when solving chemistry tasks based on SMRs with different types of explanatory key, we used the screen-based Tobii Pro X2-30 eye tracker. Gaze data were captured at 30 Hz with an accuracy of 0.4 degrees of visual angle at distances ranging between 40 and 90 cm.

Problem set

The original problem set (Ferk Savec et al., 2016) consisted of 8 tasks based on SMRs, but for the purposes of the present study only 4 tasks were selected to investigate in detail the value of pictorial versus textual explanatory keys accompanying SMRs. Intentionally, all tasks were at the same taxonomy level (application) and the models of simple common compounds were used in all tasks. All tasks are comparable in terms of complexity and type of visual representation. The tasks were displayed one after the other on the computer screen. There was no time limit for solving the task, and the tasks were presented in random order.

The four examined tasks in the problem set can be viewed in full text (including the English translation of the Slovenian instructions) in the results section of this paper through print screens of the heat maps.

Eye-movement measures

In order to determine the visual attention of students for different elements of the tasks they solved, we focused on the total amount of time (total fixation duration, TFD; in some studies also referred to as dwell time) and the number of fixations (fixation count, FC) spent in particular areas of interest (AOI). For this purpose, the tasks displayed on the computer screen were divided into several AOIs (see Fig. 4.1).

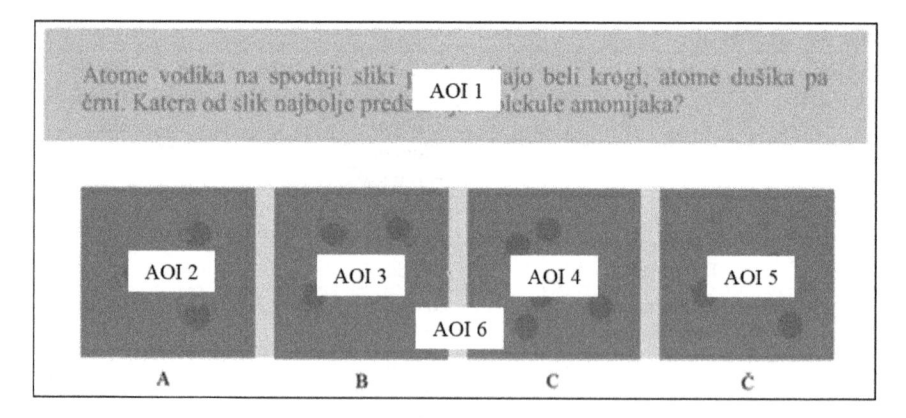

Fig. 4.1 Example of a task divided into AOIs [AOI 1 = Instructions and explanatory key; AOI 2 to AOI 5 = Model 1 (choice A) to Model 4 (choice Č); AOI 6 = All presented models]

A fixation was determined as a process when the participant held his eye for at least 60 ms at a specific AOI.

Since the data from eye movements were collected as part of the information processing during task solving, the total time spent in a given AOI was interpreted as a reflection of the relative amount of attention and consequently reliance paid towards each AOI was given in the service of task solving.

Data Collection and Analysis

Application of the rubrics for the analysis of the explanatory key accompanying SMRs in Slovenian chemistry textbook sets for primary schools (related to RQ1)

The rubric described in the section on instruments was used in the analysis of the chemical representations of the entire sample of chemistry textbook sets, which are presented in Table 4.1. The textbook sets were analysed individually. The SMRs were categorised with respect to the curriculum topics of the National Chemistry Curriculum for Primary School. The core topics in which SMRs were categorised are as follows: (1) Chemistry is a World of Matter (orig. Kemija je svet snovi); (2) Atom and the Periodic System of Elements (orig. Atom in periodni sistem elementov); (3) Compounds and Bonding (orig. Povezovanje delcev/gradnikov); (4) Chemical Reactions (orig. Kemijske reakcije); (5) The Elements in the Periodic Table (orig. Elementi v periodnem sistemu); (6) Acids, Bases and Salts (orig. Kisline, baze in soli); (7) Hydrocarbons and Polymers (orig. Družina ogljikovodikov s polimeri); (8) Organic Compounds Containing Oxygen (orig. Kisikova družina organskih snovi); (9) Organic Compounds Containing Nitrogen (orig. Dušikova družina organskih spojin) and (10) The Mole (orig. Množina snovi). Finally, the number of SMRs in each of the topics were counted and the frequencies calculated.

Collection of data in the SMRs study with eye tracker followed by an interview with students and their analysis (related to RQ2)

The data collection took place in two parts. In the first part of the study, the participants solved the group from the chemistry knowledge test under standard conditions that were the same for all participants. This part lasted up to 45 min. In the second part, in which the cognitive processes of the students were monitored using eye-tracking technology and an interview, the students were selected based on their performance in the chemistry knowledge test, and they participated individually. After calibrating the eye tracker, the students were introduced to a pre-task to avoid any impact due to difficulties in understanding the type of the tasks or the process of recording the answers and moving on to the next task before starting the main testing. Participants were asked to write down one answer for each task on a piece of paper and then press the space bar to move on to the next screen. Afterwards, the students completed eight tasks for the problem set displayed on the computer screen at their own pace, while eye movements were recorded. Each task was presented on the computer screen without time limit and in random order.

After the eye-tracking data were collected, the participants were interviewed and asked to compare tasks with different explanatory keys (textual versus pictorial explanatory key) with the following question: "When you compare these two tasks, was there a difference in difficulty between them? If there was a difference, please explain possible reasons". The oral answers of the participants were transcribed. The collection of eye- tracking and interview data took 30–55 min.

In order to investigate the role of the explanatory key in solving tasks based on submicroscopic representations, the collected eye movement data was first analyzed with Tobii Studio Enterprise. A further analysis was conducted using the Statistical Package for the Social Sciences (SPSS), version 21. The nonparametric test Wilcoxon Ranks Test (Z) was used to evaluate significant differences in the absolute and relative total fixation duration (TFD) and fixation count (FC) within particular areas of interest (AOIs) with respect to the pictorial versus textual explanatory key accompanying the SMRs.

The interview responses of the participants were coded using a coding table. The coding table was derived from a qualitative analysis of 25% ($n = 9$) of the interviews; the reliability of the coding was ensured by independent coding by two researchers (the two authors of this paper). Subsequently, the two evaluations were compared at the points where differences occurred and the more appropriate one was selected after consideration. Overall a reliability of 98% was achieved.

Results and Discussion

The results in paper are presented with regard to the stated research questions.
The integration of SMRs with/without explanatory key in Slovenian chemistry textbook sets with respect to curriculum topics (related to RQ1)

Table 4.2 The proportion of SMRs in the particular topics of the textbook sets with regard to the presence of explanatory key

The topics of the National Chemistry Curriculum for Primary School (8th and 9th Grade)	All SMRs N	f (%)	SMRs with explanatory key N	f (%)	SMRs without explanatory key N	f (%)
1—Chemistry is a World of Matter	179	12.61	50	3.52	129	9.09
2—Atom and the Periodic System of Elements	29	2.04	12	0.85	17	1.20
3—Compounds and Bonding	150	10.57	31	2.18	119	8.39
4—Chemical Reactions	69	4.86	22	1.55	47	3.31
5—The Elements in the Periodic Table	16	1.13	2	0.14	14	0.99
6—Acids, Bases and Salts	160	11.28	49	3.45	111	7.82
7—Hydrocarbons and Polymers	407	28.68	32	2.26	375	26.43
8—Organic Compounds Containing Oxygen	287	20.23	12	0.85	275	19.38
9—Organic Compounds Containing Nitrogen	116	8.17	0	0.00	116	8.17
10—The Mole	6	0.42	0	0.00	6	0.42
SUM	1419	100.00	210	14.80	1209	85.20

The number of SMRs in the Slovenian chemistry textbook for the 8th and 9th grade of primary school varies from one curriculum topic to another. SMRs with an explanatory key represent 14.80% of all SMRs integrated in analysed chemistry textbook sets, which means that the majority (85.20%) of SMRs in all chemistry curriculum topics are not accompanied with an explanatory key (Table 4.2).

Table 4.2 also shows that the SMRs were most frequently used in the Topic 7—Hydrocarbons and Polymers (407 SMRs; 28.68%) and Topic 8—Organic Compounds Containing Oxygen (287 SMRs; 20.23%). The lowest frequency of use of SMRs was found in the Topic 10—The Mole (6 SMRs; 0.42%) and Topic 5—The Elements in the Periodic Table (16 SMRs; 1.13%) and Topic 2—Atom and those Periodic System of Elements (29 SMRs; 2.04%). However, the SMRs with explanatory key were most frequently found in the Topic 1—Chemistry is a World of Matter (50 SMRs; 3.52%) and Topic 6—Acids, Bases and Salts (49 SMRs; 3.45%). In contrast, in the Topic 9—Organic Compounds Containing Nitrogen (0 SMRs; 0.00% of SMRs with explanatory key) and Topic 10—The Mole (0 SMRs; 0.00% of SMRs with explanatory key) no SMRs with an explanation key were found.

It was assumed that the authors systematically plan the integration of SMS into the textbook sets and that the explanatory keys are also meaningfully integrated with SMRs and continuously upgraded through the curriculum topics from the 1st

4 The Role of the Explanatory Key in Solving Tasks … 81

topic towards 10th topic in order to support the development of the representational competence of the learner. From the Fig. 4.1 it can be derived that in order to support the learners' representational competence, the introduction of SMRs accompanied by their explanatory keys into the textbook sets could probably have been more systematic, since the majority of SMRs are not accompanied by the explanatory key and the number of SMRs with explanatory key is zero in some cases (Topic 9—Organic Compounds Containing Nitrogen; Topic 10—The Mole) or very low, e.g. Topic 5—The Elements in the Periodic Table (2 SMRs; 0.14% of the SMRs with an explanatory key).

Figure 4.2 also indicates, that SMRs without an explanatory key are used quite often in the text book sets, especially in the second part of the curriculum topics (topics: 6—Acids, Bases and Salts; 7—Hydrocarbons and Polymers; 8—Organic Compounds Containing Oxygen; 9—Organic Compounds Containing Nitrogen). The high number of SMRs used in these topics could be explained through prepositions in general curriculum objectives, which require students to systematically develop an understanding of the relationship between structure, properties and application of the substances (Bačnik et al., 2011, p. 5), in addition these topics are elaborated with specific curriculum objectives. For example, in the curriculum topic 7—Hydrocarbons and Polymers, which has the largest number of SMSs (out of 407 SMRs: 32 SMRs with explanatory key; 375 SMRs without explanatory key),

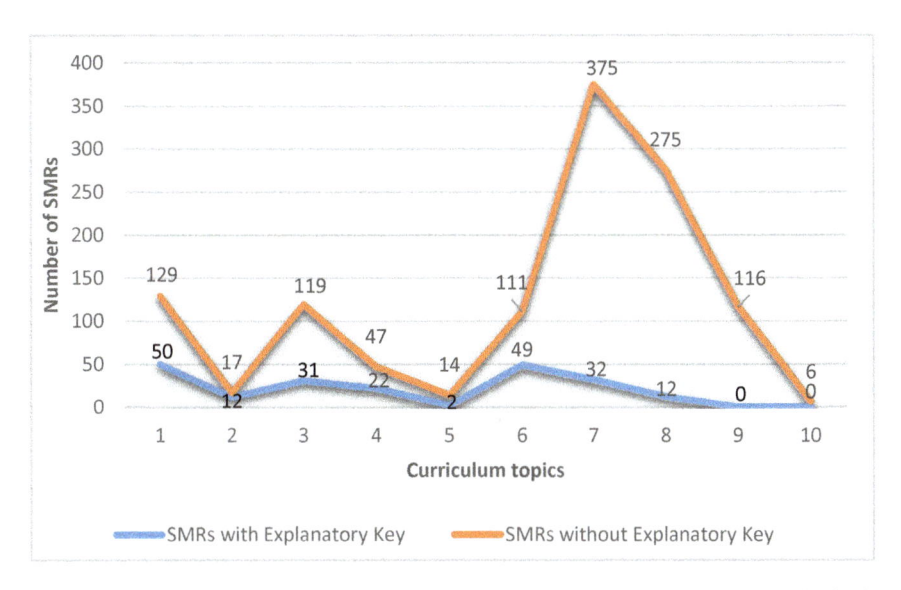

Fig. 4.2 The number of SMRs with/without Explanatory key within the curriculum topics in the textbook sets [Explanation of abbreviations: 1—Chemistry is a World of Matter; 2—Atom and the Periodic System of Elements; 3—Compounds and Bonding; 4—Chemical Reactions; 5—The Elements in the Periodic Table; 6—Acids, Bases and Salts; 7—Hydrocarbons and Polymers; 8—Organic Compounds Containing Oxygen; 9—Organic Compounds Containing Nitrogen; 10—The Mole]

the specific objectives indicate that students should learn that carbon and hydrogen are the fundamental elements of organic compounds, that students should be able to recognise the reasons for the abundance and diversity of organic compounds, the isomerism and that students should know the basic properties of hydrocarbons, correlate them with their use and act accordingly (Bačnik et al., 2011, pp. 11–12). On the other hand, it is not easy to find a reason for a high number of SMRs without an explanatory key, as they do not provide additional information to support the learner's recognition process. It may be that textbook authors assume that learners are already able to recognise the meaning of SMRs without explanation, because their representational competence has been adequately developed in previous topics, or that authors integrate SMRs into textbook sets without considering how correct recognition of SMRs by learners might affect the learning process based on them. The latter assumption is consistent with Johnstone's assertion that experienced chemists make the transition between levels of representation very easily, and they assume that learners can do this as easily as they do themselves (Johnstone, 1991).

The value of the pictorial versus textual explanatory key accompanying SMRs for the learners (related to RQ2)

In order to better understand the potential added value of the explanatory key accompanying SMRs for students, in the second part of the paper we present a study based on four simple cases of SMRs accompanied by pictorial versus textual explanatory keys, which were examined with the help of an eye tracker, followed by a short interview with the students.

First, based on the study with eye tracker, Figs. 4.2 and 4.3 present examples of eye movements using a heat map for the task in which SMRs were accompanied by

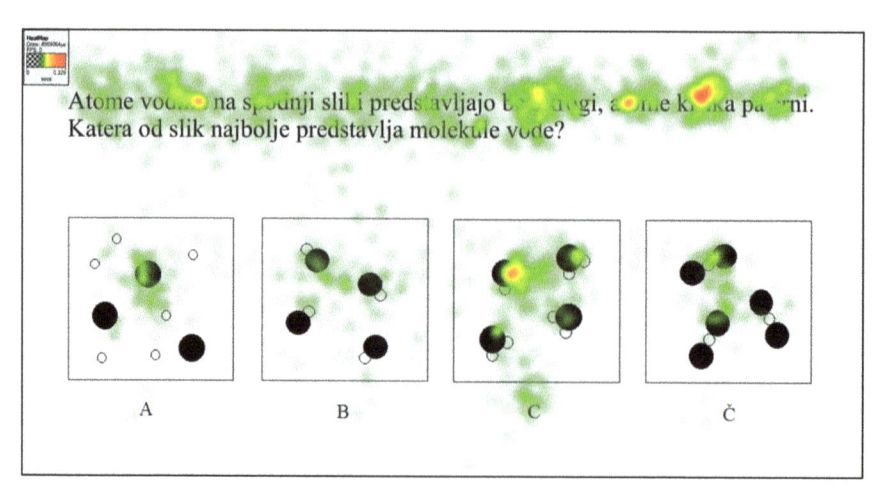

Fig. 4.3 A heat map for Task 1, which includes textual explanatory key, shows the relative density of fixations using a colour gradient [Task 1 Instruction (above): On the pictures bellow, the hydrogen atoms are represented by white circles, oxygen atoms are represented by black circles. Which of pictures best presents the water molecule?]

4 The Role of the Explanatory Key in Solving Tasks ...

textual explanatory key (Task 1 and Task 2). Figures 4.4 and 4.5 show tasks in which SMRs were accompanied by pictorial explanatory key (Task 3 and Task 4). In heat maps, the red colour stands for a high relative density of fixations and green for a low relative density of fixations by students.

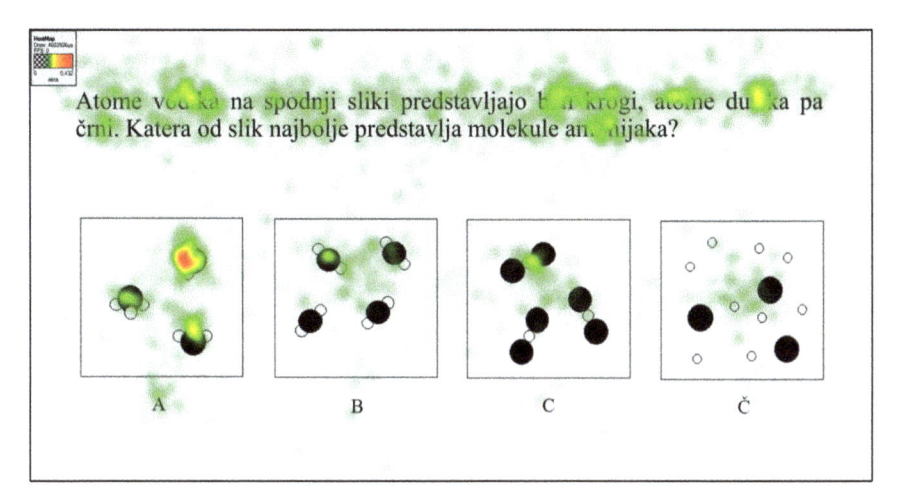

Fig. 4.4 A heat map for Task 2, which includes textual explanatory key, shows the relative density of fixations using a colour gradient [Task 2 Instruction (above): On the pictures bellow, the hydrogen atoms are represented by white circles, nitrogen atoms area represented by black circles. Which of pictures best presents the ammonia molecule?]

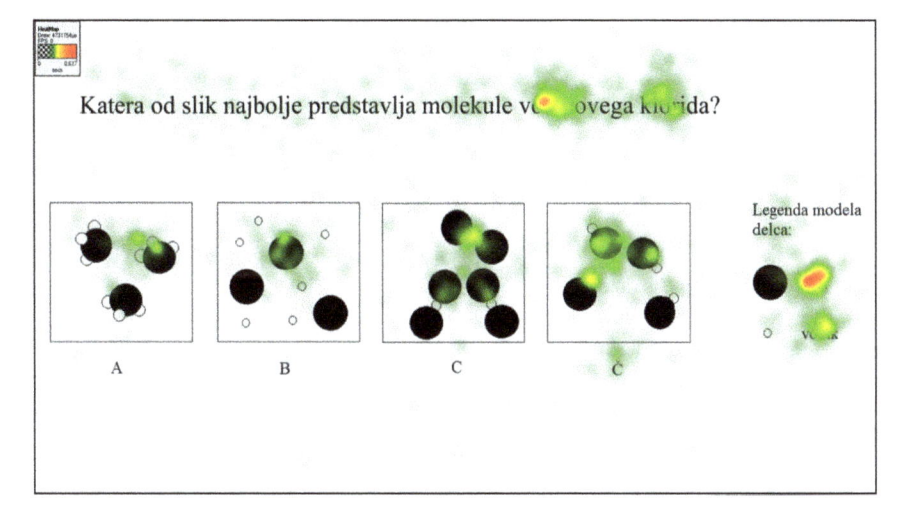

Fig. 4.5 A heat map for Task 3, which includes pictorial explanatory key, shows the relative density of fixations using a colour gradient [Task 3 Instruction (above): Which of pictures best presents the hydrogen chloride molecule? Task 3 Instruction (right): Legend of the particles in the model: bigger black circle—chlorine, smaller white circle—hydrogen]

In addition to the heat maps, Table 4.2 shows the corresponding mean values of the absolute and relative TFD and FC for the areas of interest of particular models. The Spearman correlation coefficients ($r = 0.777 – 0.969, p < 0.001$) indicate that there is a strong correlation between absolute and relative TFD and FC in all tasks (Task 1–Task 4).

From Fig. 4.3 it can be seen that the AOI of choice B and choice C of Task 1 have a higher relative density of fixations compared to other choices, since the students' gaze was more often fixed on these representations when solving a task at particle level. This can also be seen from Table 4.3, which shows the mean values of the absolute and relative TFD and FC are presented. The choice C, which attracted the attention of the majority of students (TFD 13.90%; FC 13,12%), is also the correct answer of the Task 1, where the accuracy of the students' answers is 100%.

Figure 4.4 shows that the AOI of choice A and choice B have a higher relative density of fixations. The same student focus can also be seen from Table 4.3, where the choice A attracted the most student attention (TFD 16.93%; FC 12.18%), as it is the correct answer of Task 2, with a student response accuracy of 97.01%.

From Fig. 4.5 it can be seen that the AOI of choice C and choice Č have a higher relative density of fixations, since the students' gaze was more often fixed on these representations when solving the task. The observation is also evident from Table 4.3,

Table 4.3 Mean values of absolute and relative TFD and FC for tasks 5 to 8 with the focus on AIO for particular possible choices of answers in tasks

Type of explanatory key	Task	Eye-movement measures	Area of interest (AOI)				
			Model 1 (choice A)	Model 2 (choice B)	Model 3 (choice C)	Model 4 (choice Č)	All presented models
Textual	5	TFD [s]	0.51	0.59	1.44	0.56	3.11
		TFD [%]	4.94	5.71	13.90	5.42	29.96
		FC [count]	2.73	3.84	7.14	2.73	16.44
		FC [%]	5.02	7.06	13.12	5.02	30.22
	6	TFD [s]	2.07	0.88	0.82	0.54	4.29
		TFD [%]	16.93	7.17	6.68	4.39	35.16
		FC [count]	7.21	4.35	3.91	2.47	17.94
		FC [%]	12.18	7.35	6.60	4.17	30.30
Pictorial	7	TFD [s]	0.79	0.87	1.48	2.03	5.18
		TFD [%]	6.85	7.54	12.81	17.57	44.77
		FC [count]	3.33	4.03	6.00	8.24	21.59
		FC [%]	6.42	7.77	11.57	15.89	41.64
	8	TFD [s]	1.95	0.85	3.12	0.64	6.54
		TFD [%]	18.01	7.82	28.84	5.88	60.56
		FC [count]	6.94	4.44	11.65	3.24	26.26
		FC [%]	15.03	9.61	25.23	7.01	56.88

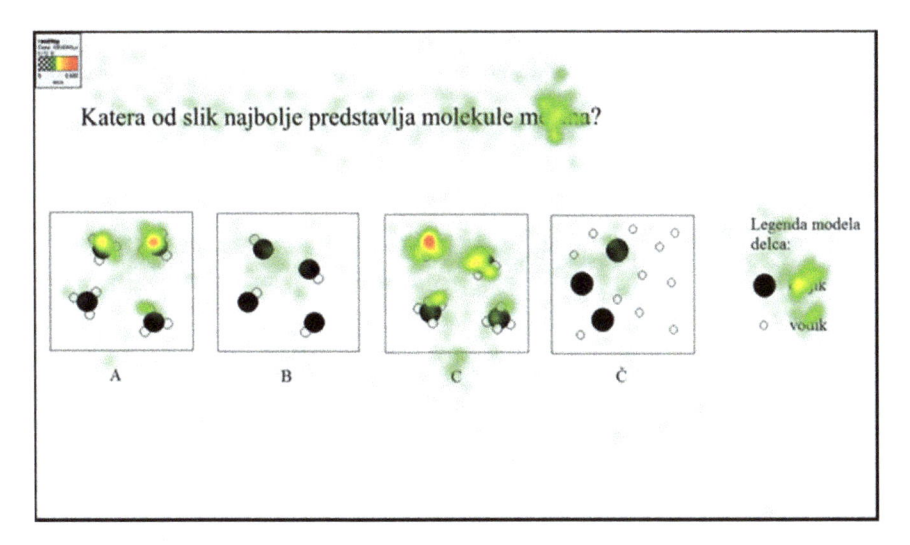

4 The Role of the Explanatory Key in Solving Tasks ...

Fig. 4.6 A heat map for Task 4, which includes pictorial explanatory key, shows the relative density of fixations using a colour gradient [Task 4 Instruction (above): Which of pictures best presents the methane molecule? Task 4 Instruction (right): Legend of the particles in the model: bigger black circle—carbon, smaller white circle—hydrogen]

where the choice Č as the correct answer of the Task 3, achieved the attention of most students (TFD 17.57%; FC 15.89%) with a response accuracy of 94.61%.

Figure 4.6 shows that the AOI of choice A and choice C have a higher relative density of fixations, since the students' gaze was more often fixed on these representations when solving the task. The observation is also evident from Table 4.3, where choice C as the correct answer of the Task 4 received the most attention from the students (TFD 17.57%; FC 15.89 with a student response accuracy of 87.45%.

Table 4.3 shows that the highest mean values of relative TFDs and FCs for the explanatory key were achieved in the task with a textual key (Task 1: TFD = 70.04%; FC = 69.78%), and the lowest mean values of relative TFDs and FCs were achieved in the task with a pictorial representation (Task 4: TFD = 39.44%; FC = 43.12%). This is also consistent with the highest response accuracy of students in Task 1 (100.00%) and the lowest response accuracy of Task 4 (87.45%).

To further examine how textual and pictorial explanatory key influence students' task solving the AOI for *All presented models*, AOI for *Instructions and explanatory key* and AOI for *Models, instructions and explanatory key* (Table 4.4) were examined by the use of the Wilcoxon Ranks Test. Thereby, AOI for the pictorial explanatory key in Tasks 7 and Task 4 also included task instruction in order to equalise the AOI with the textual explanatory key where the task instructions and key were jointly presented (Task 1 and Task 2). The common AOI for both kinds of tasks is therefore named *Instructions and explanatory key*. The analogical reasoning was used with AIO *Models, instructions and explanatory key*.

Table 4.4 Mean values of absolute and relative TFD and FC for tasks 5 to 8 with the focus on AOI for All presented models, Instructions and explanatory key and Models, instructions and explanatory key

Type of explanatory key	Task	Eye-movement measures	Area of interest (AOI)		
			All presented models	Instructions and explanatory key	Models, instructions and explanatory key
Textual	5	TFD [s]	3.11	7.27	10.38
		TFD [%]	29.96	70.04	100.00
		FC [count]	16.44	37.97	54.41
		FC [%]	30.22	69.78	100.00
	6	TFD [s]	4.29	7.91	12.20
		TFD [%]	35.16	64.84	100.00
		FC [count]	17.94	41.26	59.20
		FC [%]	30.30	69.70	100.00
	SUM	TFD [s]	7.40	15.18	22.58
		TFD [%]	32.77	67.23	100.00
		FC [count]	34.38	79.23	113.61
		FC [%]	30.36	69.74	100.00
Pictorial	7	TFD [s]	5.18	6.39	11.57
		TFD [%]	44.77	55.23	100.00
		FC [count]	21.59	30.26	51.85
		FC [%]	41.64	58.36	100.00
	8	TFD [s]	6.54	4.26	10.80
		TFD [%]	60.56	39.44	100.00
		FC [count]	26.26	19.91	46.17
		FC [%]	56.88	43.12	100.00
	SUM	TFD [s]	11.72	10.65	22.37
		TFD [%]	52.39	47.61	100.00
		FC [count]	47.85	50.17	98.02
		FC [%]	48.82	51.18	100.00

With regard to AOIs of **All presented models** (Table 4.4) the Wilcoxon Ranks Test indicated significant differences in the sum of the relative mean values of TFD and FC (TFD: $Z = -3.958$, $p < 0.001$; FC: $Z = -3.405$, $p < 0.001$). The significant differences in the sum of the mean values of TFD as well as the sum of the mean values of FC indicate, that students spent more time glancing on the AOIs of models that were accompanied by pictorial explanatory key in comparison to the AOIs of models accompanied with the textual explanatory key. Thereby, students also fixated on particular spots of the AOIs of models accompanied with the pictorial explanatory key more often than on AOIs of models accompanied with the textual explanatory key while solving the tasks.

The Wilcoxon Ranks Test showed also significant differences in the sum of the relative mean values of TFD and FC on AOIs of **Instructions and explanatory key** (Table 4.4) of the pictorial and textual explanatory key (TFD: $Z = -3.838$, $p < 0.001$; FC: $Z = -4.626$, $p < 0.001$). The significant differences in the sum of the mean values of TFD as well as the sum of the mean values of FC could be interpreted with the postulation, that students not only spent more time on the textual explanatory key in comparison with the pictorial explanatory key, but also fixated on particular spots (words) of the textual explanatory key more often while solving the task in comparison to a pictorial key.

On the other hand, Wilcoxon Ranks Test indicated, that the sum of the relative mean values of TFD on AOIs of **Models, instructions and explanatory key** (Table 4.4) of the pictorial versus textual explanatory key was not significantly different (TFD: $Z = -0.445$, $p = 0.657$). However interestingly, the number of FC on these AOIs was significantly different (FC: $Z = -2.360$, $p = 0.018$), which points to a difference in number of fixations on particular spots of AOIs. This could be interpreted with assumption, that the students in overall used the comparable amount of time for solving tasks either with a pictorial or textual explanatory key, but during the process of task solving students fixated their attention on more spots, when solving tasks with textual explanatory key. It would be interesting to further examine these results, especially with taking into consideration that the students' response accuracy in tasks with textual explanatory key yielded slightly higher (Task 1: 100% and Task 2: 97.01%) then tasks with pictorial explanatory key (Task 3: 94.61% and Task 4: 87.45%).

As the final part of the consideration about possible added value of a certain type of explanatory key to the SMRs, the attention was payed to students' perception of their value. Therefore, in the **interviews**, students were asked to compare two tasks (one having the textual other the pictorial explanatory key) and explain if there was any difference in difficulty between them. If students would observe differences, they would be asked to explain possible reasons from their perspective.

The majority of the students (91.43%; $N = 32$ from 35 students) claimed that from their perspective there was no difference in difficulty between task with various explanatory key (in terms of textual versus pictorial) accompanying the SMRs:

Typical students' comment:

"There are no differences, if you know the molecular formula, they're all easy".

"They are similar; the procedure of solving is the same".

Some students ($f = 8.57\%$; $N = 3$ from 35 students) explicitly pointed out that their difficulties in solving such tasks are due to their lack of chemistry knowledge.

Typical student's comment:

"The one with the methane was the hardest. I didn't know, if it was CH_3 or CH_4".

Conclusion

In recent decades, submicroscopic representations (SMRs) have been integrated into textbook sets, as research in chemistry education has shown that their meaningful integration can facilitate learning chemistry with understanding. The paper focuses on two perspectives of the explanatory key accompanying SMRs in chemistry learning, and attempts to combine the findings from both perspectives to provide recommendations that could be useful in the school practise of chemistry teaching. The paper firstly examines the state-of-art with respect to the presence of explanatory key accompanying SMRs in Slovenian chemistry textbook sets for the 8th and 9th grade of primary school, and secondly it aims at the role of the explanatory key in the processing of SMRs in solving chemistry tasks in which they are involved using the Eytracker method, followed by a short interview with the students.

The results of the first part indicate, that the number of SMRs in the Slovenian chemistry textbook for the 8th and 9th grade of primary school varies from curriculum topic to curriculum topic. SMRs with an explanatory key represent 14.80% of all SMRs integrated in analysed chemistry textbook sets, which means that the majority (85.20%) of SMRs in all chemistry curriculum topics are not accompanied with an explanatory key. It can be assumed that in order to support the representational competence of learners, the introduction of SMRs with their explanatory keys into the textbook sets from the beginning to the end of the textbook sets could probably have been more systematic.

In the second part, which examined the possible added value of the explanatory key that accompanies the SMRs, it was found that students use the explanatory key efficiently in solving chemistry tasks both in the pictorial and textual form of the explanatory key. Overall, the students spent the same amount of time solving tasks that were solved with the pictorial or textual explanatory key. However, the students spent more time glancing on the AOIs of models that were accompanied by pictorial explanatory key compared to the area of interest AOIs of models accompanied with the textual explanatory key. In solving the tasks, the students also fixated on particular spots of the AOIs of models that were accompanied by the pictorial explanatory key than on AOIs of models that were accompanied by the textual explanatory key. In the cases studied, the students not only spent more time on the textual explanatory key compared to the pictorial explanatory key, but also fixed themselves more often on particular spots (words) of the textual explanatory key compared to a pictorial key when solving the task. Despite the observed differences in the use of certain forms of the explanatory key, it can be concluded from the results that both forms of the explanatory key supported the students' ability to perceive SMRs correctly when solving tasks and enabled them to solve the tasks with approximately the same amount of time.

From the results it can be concluded that the explanatory key accompanying the SMRs plays an important role in information processing and that it should be an integral part of the SMRs in chemistry textbook sets. It would be beneficial for students if the SMRs were more systematically included in textbook sets from beginning to end to facilitate the development of their representational competence and to support students' learning.

In order to support the efficient use of SMRs by students, additional parts (e.g. annexes) of chemistry textbooks could also be proposed by textbook authors and/or editors, in which the possibilities to support students in the use of different types of SMRs, both in traditional and computer-based form (e.g. ball-and-stick, space-filling, wire-frame, valence-shell electron pair repulsion model) are elaborated, the colour conventions used to label the respective elements are presented and didactic explanations aimed at possible misunderstandings in the use of SMRs are addressed (e.g. the type of models, the rigidity of particles and bonds in models, the role of colours in labelling particles, the speed of processes at particle level).

Acknowledgements The reseach work was partialy supported by the Faculty of Education University of Ljubljana, project—framework "Interni razpis za financiranje raziskovalnih in umetniških projektov 2015/16 [Internal call for funding of research and art projects 2015/16]", project title "Pojasnjevanje uspešnosti reševanja kemijskih nalog na submikro ravni ter preučevanje kompetentnosti bodočih učiteljev kemije za njihovo poučevanje [The efficiency of students in solving chemical tasks at the submicroscopic level and investigating the ability of future teachers to use them in classroom practice]".

References

Akaygun, S., & Jones, L. L. (2014). Words or pictures: A comparison of written and pictorial explanations of physical and chemical equilibria. *International Journal of Science Education, 36*(5), 783–807.

Al-Balushi, S. M., & Al-Hajri, S. H. (2014). Associating animations with concrete models to enhance students' comprehension of different visual representations in organic chemistry. *Chemistry Education Research and Practice, 15*(1), 47–58.

Bačnik, A., Bukovec, N., Vrtačnik, M., Poberžnik, A., Križaj, M., Stefanovik, V., … Preskar, S. (2011). *Učni načrt. Program osnovna šola. Kemija* [Curicculum. Primary school. Chemistry.]. Ministrstvo za šolstvo in šport, Zavod RS za šolstvo. http://www.mizs.gov.si/fileadmin/mizs.gov.si/pageuploads/podrocje/os/prenovljeni_UN/UN_kemija.pdf.

Barke, H. D., Hazari, A., & Yitbarek, S. (2009). *Misconceptions in chemistry: Addressing perceptions in chemical education.* Springer Science & Business Media.

Barke, H. D., & Wirbs, H. (2002). Structural units and chemical formulae. *Chemistry Education Research and Practice, 3*(2), 185–200.

Corey, R. B., & Pauling, L. (1953). Molecular models of amino acids, peptides, and proteins. *Review of Scientific Instruments, 24*(8), 621–627.

Devetak, I., Vogrinc, J., & Glažar, S. A. (2010). States of matter explanations in Slovenian textbooks for students aged 6 to 14. *International Journal of Environmental and Science Education, 5*(2), 217–235.

Devetak, I., & Vogrinc, J. (2013). The criteria for evaluating the quality of the science textbooks. In M. Swe Khine (Ed.), *Critical analysis of science textbooks* (pp. 3–15). Springer.

Ferk Savec, V., Hrast, Š., Devetak, I., & Torkar, G. (2016). Beyond the use of an explanatory key accompanying submicroscopic representations. *Acta Chimica Slovenica, 63*(4), 864–873.

Ferk Savec, V., Vrtačnik, M., & Gilbert, J. K. (2005). Evaluating the educational value of molecular structure representations. In J. K. Gilbert (Ed.), *Visualization in Science Education* (pp. 269–297). Springer.

Ferk Savec, V., Sajovic, I., & Wissiak Grm, K. S. (2009). Action research to promote the formation of linkages by chemistry students between the macro, submicro, and symbolic representational levels. In J. K. Gilbert (Ed.), *Multiple representations in chemical education* (Models and Modeling in Science Education, vol. 4, pp. 309–331). Springer.

Francoeur, E. (1997). The forgotten tool: The design and use of molecular models. *Social Studies of Science, 27*(1), 7–40.

Furió-Más, C., Luisa Calatayud, M., Guisasola, J., & Furió-Gómez, C. (2005). How are the concepts and theories of acid–base reactions presented? Chemistry in textbooks and as presented by teachers. *International Journal of Science Education, 27*(11), 1337–1358.

Gilbert, J. K., Reiner, M., & Nakhleh, M. (2008). *Visualization: Theory and practice in science education.* Springer.

Gkitzia, V., Salta, K., & Tzougraki, C. (2011). Development and application of suitable criteria for the evaluation of chemical representations in school textbooks. *Chemistry Education Research and Practice, 12*(1), 5–14.

Gregory, A. (2000). *Plato's philosophy of science.* Bloomsbury.

Hardwicke, A. J. (1995). Using molecular models to teach chemistry. Part I : modelling molecules. *School Science Review, 77*(278), 59–64.

Harrison, A. G. (2001). How do teachers and textbook writers model scientific ideas for students? *Research in Science Education, 31*(3), 401–435.

Havanki, K. L., & Vanden Plas, J. R. (2014). Eye tracking methodology for chemistry education research. In D. M. Bunce & R. S. Cole (Eds.), *Tools of chemistry education research* (pp. 191–218). American Chemical Society.

Helmenstine, T. (2019). *Molecule atom colors—CPK colors.* https://sciencenotes.org/molecule-atom-colors-cpk-colors/.

Hinze, S. R., Rapp, D. N., Williamson, V. M., Shultz, M. J., Deslongchamps, G., & Williamson, K. C. (2013). Beyond ball-and-stick: Students' processing of novel STEM visualizations. *Learning and Instruction, 26,* 12–21.

Hrast, Š., & Ferk Savec, V. (2017a). Informational value of submicroscopic representations in Slovenian chemistry textbook sets. *Journal of Baltic Science Education, 16*(5), 694–705.

Hrast, Š., & Ferk Savec, V. (2017b). The integration of submicroscopic representations used in chemistry textbook sets into curriculum topics. *Acta Chimica Slovenica, 64*(4), 959–967.

Johnstone, A. H. (1991). Why is science difficult to learn? Things are seldom what they seem. *Journal of Computer Assisted learning, 7*(2), 75–83.

Jones, L. L. (2013). How multimedia-based learning and molecular visualization change the landscape of chemical education research. *Journal of Chemical Education, 90*(12), 1571–1576.

Jmol Colors. (n.d.). *Colors.* http://jmol.sourceforge.net/jscolors/.

Justi, R., & Gilbert, J. K. (2002). Models and modelling in chemical education. In J. K. Gilbert, O. De Jong, R. Justi, D. F. Treagust, & J. H. Van Driel (Eds.), *Chemical education: Towards research-based practice* (pp. 47–68). Springer.

Kahveci, A. (2010). Quantitative analysis of science and chemistry textbooks for indicators of reform: A complementary perspective. *International Journal of Science Education, 32*(11), 1495–1519.

Koltun, W. L. (1965). *Patent 3170246. U. S.* https://patents.google.com/patent/US3170246A/en.

Laçin-Şimşek, C. (2011). Women scientist in science and technology textbooks in Turkey. *Journal of Baltic Science Education, 10*(4), 277–284.

Majidi, S., & Mäntylä, T. (2011). Knowledge organization in physics text books: A case study of magnetostatics. *Journal of Baltic Science Education, 10*(4), 285–299.

Mason, M., Pluchino, P., Tornatora, M. C., & Ariasi, N. (2013). An eye-tracking study of learning from science text with concrete and abstract illustrations. *The Journal of Experimental Education, 81*(3), 356–384.

Merriam-Webster Dictionary. (n.d.). https://www.merriam-webster.com/dictionary.

Mumba, F., Chabalengula, V. M., Wise, K., & Hunter, W. J. (2007). Analysis of New Zambian high school physics syllabus and practical examinations for levels of inquiry and inquiry skills. *Eurasia Journal of Mathematics, Science & Technology Education, 3*(3), 213–220.

Nakhleh, M. B. (1992). Why some students don't learn chemistry: Chemical misconceptions. *Journal of Chemical Education, 69*(3), 191–196.

Nobel Assembly at Karolinska Institutet. (2019). *Press release: The Nobel Prize in Physiology or Medicine 2019*. https://www.nobelprize.org/prizes/medicine/2019/press-release.

Pavlin, J., Glažar, S. A., Slapničar, M., & Devetak, I. (2019). The impact of students' educational background, interest in learning, formal reasoning and visualisation abilities on gas context-based exercises achievements with submicro-animations. *Chemistry Education Research and Practice, 20*(3), 633–649.

Petersen, Q. R. (1970). Some reflections on the use and abuse of molecular models. *Journal of Chemical Education, 47*(1), 24–29.

Rayner, K. (2009). Eye movements and attention in reading, scene perception, and visual search. *The Quarterly Journal of Experimental Psychology, 62*(8), 1457–1506.

Slapničar, M., Tompa, V., Glažar, S., & Devetak, I. (2014). Fourteen-year-old students' misconceptions regarding the sub-micro and symbolic levels of specific chemical concepts. *Journal of Baltic Science Education, 17*(4), 620–632.

Slykhuis, D. A., Wiebe, E. N., & Annetta, L. A. (2005). Eyetracking students' attention to PowerPoint photographs in a science education setting. *Journal of Science Education and Technology, 14*(5–6), 509–520.

Torkar, G., Veldin, M., Glažar, S. A., & Podlesek, A. (2018). Why do plants wilt? Investigating students' understanding of water balance in plants with external representations at the macroscopic and submicroscopic levels. *Eurasia Journal of Mathematics, Science & Technology Education, 14*(6), 2265–2276.

Van Driel, J. H., & Verloop, N. (1999). Teachers' knowledge of models and modelling in science. *International Journal of Science Education, 21*(11), 1141–1153.

Yen, M. H., & Yang, F. Y. (2016), Methodology and application of eye-tracking techniques in science education. In M. H. Chiu (Ed.), *Science education research and practices in Taiwan* (pp. 249–277). Springer.

Vesna Ferk Savec, Ph.D. is a Full Professor of Chemical Education at the University of Ljubljana, Faculty of Education, Slovenia. Her research in the field of chemistry education focuses on the use of visualisation of the triple nature of chemical concepts, context-based learning, inquiry-based learning, micro-scale experimental work, and various aspects of green chemistry related to chemistry education. She is also interested in innovative approaches of the use of ICT to support the study process, especially in the education of future teachers. Since her employment at the University in Ljubljana in 1998, she has been involved in several national and international projects in the field of science education as a coordinator and/or collaborator. She has been a (co)author of more than 300 research publications. Since 2010, she has served as a member of Professional Development Group in Chemistry Education at National Education Institute Slovenia. She is also the Head of Center KemikUm—development and innovation school laboratory at the Faculty of Education, University of Ljubljana.

Špela Hrast Ph.D. student, is a Teaching Assistant for Chemical Education at the University of Ljubljana, Faculty of Education, Slovenia. Her research focuses on examining the relevance of the university–industry–school collaboration from the perspective of main stakeholders, using eye-tracking technology in explaining science learning, and supporting the development of the representational competence in learning and teaching Chemistry and Science.

Chapter 5
Investigating the Role of Conceptual Understanding on How Students Watch an Experimental Video Using Eye-Tracking

Sevil Akaygun and Emine Adadan

Introduction

Understanding chemistry has been challenging for students because it involves conceptualization of chemical phenomena at macroscopic, symbolic, and submicroscopic levels (Johnstone, 1991; Taber, 2013). Macroscopic level involves observable and tangible phenomena such as observing rusting of an iron nail; symbolic level includes the symbols, graphs, and mathematical formulas, such as writing a balanced chemical reaction; finally, submicroscopic level involves the structure and the processes of particles (atoms, ions, molecules, etc.) such as interaction among zinc atoms and copper ions in a solution. When learning chemistry, students should be able to make connection between these levels so that they could conceptually understand the processes. On the other hand, understanding submicroscopic level can be even more challenging for students because they need to visualize the motion and interaction of the particles that they can never see (Nakhleh, 1992; Smith et al., 2006). Thus, chemistry instructors make use of physical or computer-based models representing particles (National Research Council (NRC), 2012; Williamson, 2008). Considering the dynamic nature of chemical processes where there is ongoing interaction among the species, particulate level computer animations have been found to be helpful to students in improving their conceptual understandings (Ardac & Akaygun, 2004; Kelly & Jones, 2007; Williamson & Abraham, 1995). Because the animations depicting the motion and interaction of particles help students visualize the processes occurring at the submicroscopic level, and thus developmental models for chemical phenomena (Akaygun & Jones, 2013; Kelly, 2014).

Chemical reactions is one of the topics for which animations have been used because it involves understanding of how particles are rearranged during a reaction

---

S. Akaygun (✉) · E. Adadan
Bogazici University, Istanbul, Turkey
e-mail: sevil.akaygun@boun.edu.tr

process. If the students do not visualize these processes they may develop alternative conceptions (Ahtee & Varjola, 1998; Ben-Zvi et al., 1987; Yarroch, 1985). For instance, Yarroch (1985) stated that students who successfully balanced chemical equations showed no understanding of the relation between macroscopic and particulate levels. Similarly, Ben-Zvi et al. (1987) reported that students perceived chemical reactions as static phenomena; thus they had difficulties in understanding the dynamic processes, that is to say, how the atoms rearrange as the bonds break and form. Redox reactions have been found to be challenging for students to understand, thus various alternative conceptions including electrochemical and electrolytic cells, and the processes of oxidation and reduction have been identified in the previous studies (Acar & Tarhan, 2007; Brandriet & Bretz, 2014; Garnett & Treagust, 1992).

Eye-tracking technology have been used in chemistry education to investigate students' understanding of chemical phenomenon while they are working on a task because it provides information about the cognitive processes during the activity (Rayner, 1998). It has been asserted that there is a relation between where a person is starring at and what he or she is thinking (Just & Carpenter, 1984). The area where the person focuses on is where he or she pays attention (Hoffman & Subramaniam, 1995). To ensure the validity of their attention in terms of where and how much time they focus on the specific areas, the eye movements have been accompanied with verbal explanations through think-aloud protocols (Stieff et al., 2011).

By the inclusion of eye-tracking technology in chemistry education research, especially during the last decade, students' understanding of chemical structure such as the complexity of organic molecules (Havanki, 2012), nuclear magnetic resonance spectroscopic signals and molecular structure (Tang et al., 2012), molecular modeling (Williamson et al., 2013), stoichiometry (Baluyut & Holme, 2019; Tang et al., 2014); and the processes such as oxidation-reduction and double displacement reactions (VandenPlas, 2008), gas laws problems (Tang & Pienta, 2012), chemical change (Hansen, 2014) have been investigated. Even though the type of representations studied in the eye-tracking research varied from static images (Baluyut & Holme, 2019; Stieff et al., 2011) to dynamic computer visualizations (Hansen, 2014; VandenPlas, 2008), the main question to be answered has always been on how the eye movements can be related with understanding chemistry. Considering individual differences, eye-tracking has been used to explore the patterns observed due to the variations in the level of competency or expertise in chemistry. For instance, some researchers sought to identify the differences between the eye movement patterns of experts and novices when looking at different chemical representations (Havanki, 2012; VandenPlas, 2008). In terms of the academic achievement, few studies reported that less successful students spent more time, or fixated their eyes longer, on solving problems than more successful ones because less successful students tend to seek information from the given question (Tang & Pienta, 2012; Tang et al., 2014). Spatial ability which refers to the ability to generate, visualize, retain, and manipulate abstract visual images (Lohman, 1979) has been found to be related with understanding chemistry (Barke & Engida, 2001; Dori & Barak, 2001; Pribyl & Bodner, 1987). To this end, spatial ability was found to be effective on visual problem-solving when eye tracking patterns between high and low performers were compared on engineering-related computer games (Gomes et al., 2013). Eye-tracking studies have also helped researchers identify the mostly referred representations. For instance, Stieff et al. (2011) found that students answered the questions that could be answered only with the ball and stick models more accurately rather than the ones that involved the integration of ball and stick model and equations. Williamson et al. (2013) reported that students referred mostly on the more familiar, ball-stick images, rather than the electrostatic potential maps when solving a problem. Students' preference of using familiar or simpler representation might have served as a scaffold in learning from visualizations.

In this study, it was aimed to investigate the role of students' level of conceptual understanding on how they navigate while watching an experimental video of a redox reaction.

Method

This qualitative study adopted phenomenography (Ebenezer & Erickson, 1996) to understand the preservice teachers' level of conceptual understanding of a redox reaction occurring between zinc metal and copper(II) sulfate solution and how their understandings relate with the pattern of their eye movements while watching the experimental video of this reaction. Phenomenography explores how different individuals experience and comprehend a specific phenomenon (Marton, 1986). Thus, phenomenological perspective allowed researchers to make sense of how preservice teachers conceptualize redox reactions and how their understandings relate with the pattern of watching an experimental video of a redox reaction.

Participants

A total of 20 (12 chemistry and 8 science) preservice teachers (all female) voluntarily participated in the study. They were 3rd- or 4th-year students enrolled in a teaching chemistry/science program of a mid-size university located in a metropolitan city in Turkey. The preservice teachers completed general chemistry and analytical chemistry courses with the laboratory components where the redox reactions are covered. The study was approved by the Institutional Review Board, and all the ethical guidelines were followed throughout the study.

Procedure and Data Collection

In this study, in a 2-hour session, preservice teachers first watched a 4-minute experimental video of a redox reaction occurring between zinc metal and copper(II) sulfate solution. The experimental video was recorded and edited by the researchers at the university's teaching laboratory. The study was conducted in a computer laboratory where the participants worked individually on their own pace and were allowed to watch the video as many times as they wanted. The screenshots of the video and the link to access it were given in Fig. 5.1.

After watching the video, they were asked to write a reflection on determining the experimental evidences of the redox reaction taking place by guiding questions. They were expected to notice all the evidences for a redox reaction to happen, namely, the accumulation of copper metal on the zinc wire, paling color of the copper(II) sulfate solution and almost no change in the conductivity of the solution. After writing this reflection they were asked to represent this reaction at the submicroscopic level by preparing storyboards and generating ChemSense animations to link experimental evidences to the process occurring among the particles. On the storyboards, which are graphical representations that show the sequential flow of a story like a comic book (Lottier, 1986), they were also asked to write explanations to describe their drawing. Next, they were asked to prepare submicroscopic animations by using the software ChemSense, which was developed to prepare stop-motion animations (Schank & Kozma, 2002). One week after this session, they were interviewed to have further information about how they conceptualized the redox reaction. During the interview they were asked to explain the experimental evidences they determined in the video, how they represented the redox reaction at submicroscopic level via storyboarding and animations, and how these submicroscopic representations were connected with the experimental evidence. After the interview, they were invited to watch the experimental video one more time through an eye-tracker (ASL D-6) and think aloud at the same time. The data collection procedure is given in Fig. 5.2.

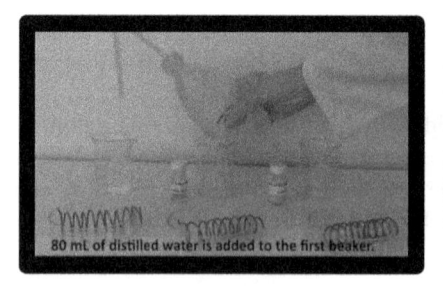

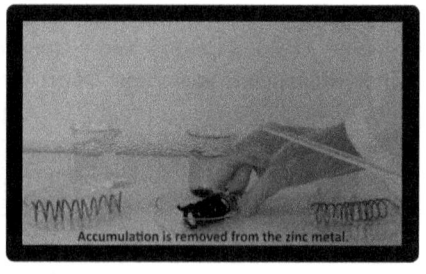

Fig. 5.1 Screenshots of the redox reaction video (Akaygun & Adadan, 2019). https://www.dropbox.com/s/o62epab4k0dkr06/Redoks_Video_06112017_ingilizce.avi?dl=0

5 Investigating the Role of Conceptual Understanding …

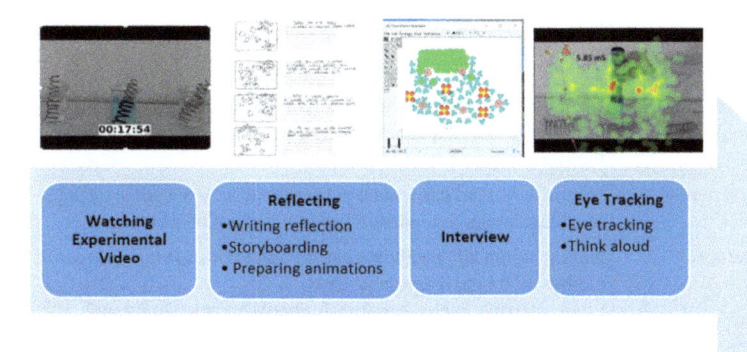

Fig. 5.2 Data collection procedure

Data Analysis

The data analysis was carried out in two parts: by analyzing *level of understandings* and *eye movements*. In the first part, participating preservice teachers' conceptual understandings were identified through their reflections, representations (storyboards and animation), and interviews. The participants' written, oral, and drawn responses were analyzed by utilizing the qualitative content analysis method (Hsieh & Shannon, 2005). During this process, the transcribed reflections and interviews were divided into meaningful units, and each unit was coded, and then the meaningfully similar codes were classified into the particular categories. Similarly, preservice teachers' representations were coded to convey their understandings and compared with the results obtained from the analysis of responses in the reflections and interviews. Thus, based on the preservice teachers' responses in these data resources their level of conceptualizations was determined through a scale ranging from 1 to 4; *alternative understanding, weak understanding, moderate understanding,* and *scientific understanding,* as reported earlier (Akaygun & Adadan, 2019; Akaygun et al., 2019). Table 5.1 describes the criteria for each level of understanding, and Figs. 5.3 and 5.4 illustrate sample representations for different levels of understanding.

In the second part, the eye-tracking data were analyzed qualitatively and confirmed by the transcribed think aloud protocols. At this stage, after the gaze plots embedded on the video were formed, the video was segmented with respect to the information displayed. The incidences displayed in the video segments follow:

1. Introductory text
2. Preparation for the experiment
3. Showing the chemicals
4. Preparing the solutions
5. Initial conductivity measurements
6. Immersion of zinc wire into the beakers
7. Course of reaction (speeded up)

Table 5.1 Criteria for the levels of understanding of redox reactions

Level of Understanding (LU)	Criteria
Scientific understanding	Including all particulate entities (electrons, ions, atoms, and molecules) and properly showing all the changes in entities along the reaction process
Moderate understanding	Not including all particulate entities, and showing some changes in entities along the reaction process, but not all of them
Weak understanding	Not including all particulate entities, and not showing changes in entities along the reaction process • not referring to electron transfer between zinc atoms and copper ions as part of the reaction process but not showing any alternative understanding
Alternative understanding	In addition to weak understanding criteria, the representations included one or more of the following alternative conceptions: • showing copper(II) sulfate in water as molecules not as ions • showing zinc sulfate as molecular precipitate

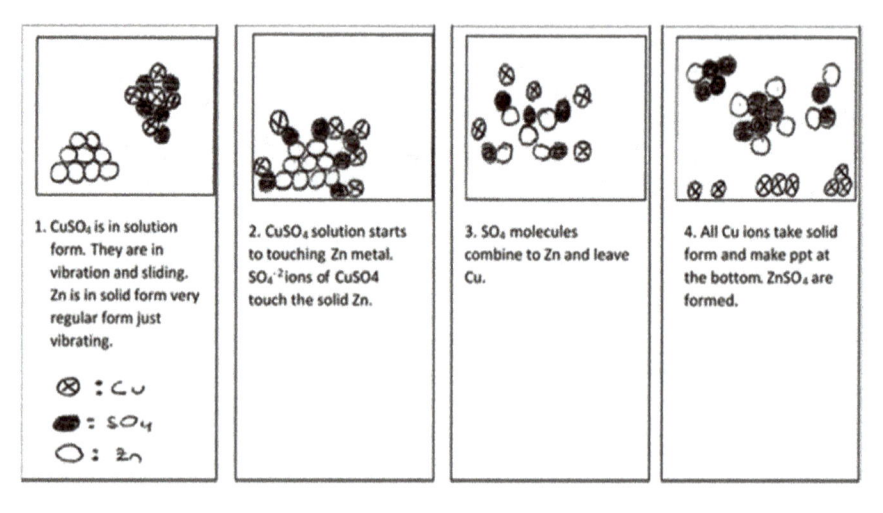

Fig. 5.3 Storyboard of a preservice teacher having an alternative understanding of redox reactions (Akaygun et al., 2019)

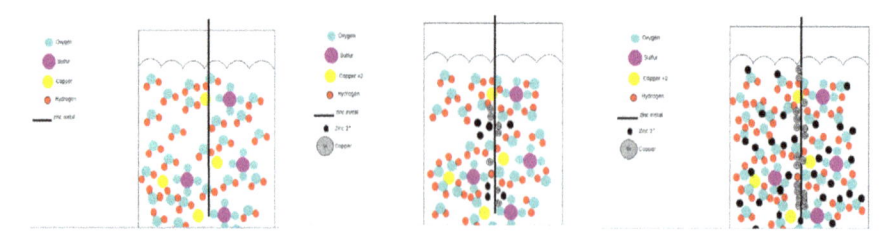

Fig. 5.4 Animation frames of a preservice teacher having scientific understanding of redox reactions

5 Investigating the Role of Conceptual Understanding ...

8. Removing of zinc wire from the beakers
9. Final conductivity measurements
10. Removal of accumulated copper metal from the zinc wire.

The experimental evidences were shown in the segments of 3, 5, and 7, therefore, it is important to pay attention to the conductivity measurements and the beaker containing $CuSO_4$ solution with the zinc wire.

Results

In Fig. 5.3, the storyboard of a preservice teacher with alternative understanding is given. This participant showed both zinc and copper(II) sulfate as a solid, did not indicate the charges of copper and sulfate ions, represented zinc sulfate in the molecular form and precipitated copper atoms.

In Fig. 5.4, the scenes from the animation prepared by a preservice teacher with scientific understanding is given. This participant showed both zinc, copper, and sulfate ions in the solution as well as water. After the zinc wire is immersed into the beaker, the preservice teacher represented the transfer of Zn^{2+} ions into the solution and deposition of Cu atoms on the wire. She did not represent the electron transfer occurring between the species but described it during the interview, so her level of understanding was identified as *scientific*.

Based on the qualitative analysis of the eye tracking data accompanied with the think aloud protocols, the characteristic features with respect to each level of understandings were outlined. Table 5.2 shows the eye movement patterns of preservice teachers with different levels of understanding of redox reactions.

The results of the analysis showed that the preservice teacher having a *scientific understanding* first just briefly gazed at all the materials and then focused only on the experimental evidences which is shown in the second beaker where redox reaction takes places, and the conductivity measurements. Then she looked through all the three beakers for comparison. This preservice teacher did not look at the background details such as laboratory instructor. Some of the scenes from the gaze plot of this preservice teacher are given in Fig. 5.5.

The preservice teacher with a moderate understanding of redox reactions briefly gazed through materials and mostly focused on the experimental evidences, to the beaker where the reaction takes place. She neither looked at the beakers at both sides nor the other details. Some of the scenes from the gaze plot of this preservice teacher are given in Fig. 5.6.

In general, preservice teachers with a weak understanding of redox reactions less focused on the experimental evidences and occasionally looked at the beakers. For instance, at the 34th minute, toward the end of the video, one of the preservice teachers did not look at solution or the wire inside the solution, but the metal outside the beaker as seen in Fig. 5.7. Some of the scenes from the gaze plot of this preservice teacher are given in Fig. 5.7.

Table 5.2 The eye movement patterns of preservice teachers with different levels of understanding of redox reactions

Level of understanding	N	Features of the eye movement patterns

Scientific	1	• Focusing **only** on the experimental evidences
		  – metal deposition
		  – paling solution
		  – conductivity measures
		• Very briefly gazing through the other beakers
		• **Not looking** at any other details

Moderate	1	• Focusing **mostly** on the experimental evidences
		  – metal deposition
		  – paling solution
		  – conductivity measures **briefly**
		• **Not looking** at any other details

Weak	5	• Focusing **less** on the experimental evidences
		  – metal deposition
		  – conductivity measures **very briefly**
		• **Occasionally looking** through the other beakers

Alternative	13	• Focusing **less** on the experimental evidences
		  – metal deposition
		  – conductivity measures **very briefly**
		• **Frequently looking** through the other beakers

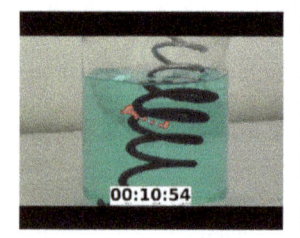

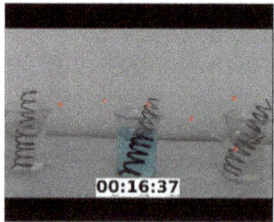

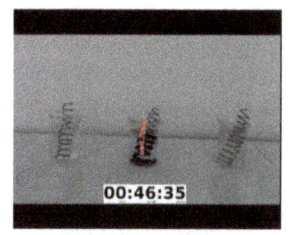

Fig. 5.5 Gaze plot of a preservice teacher having a scientific understanding of redox reactions

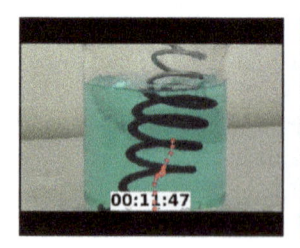

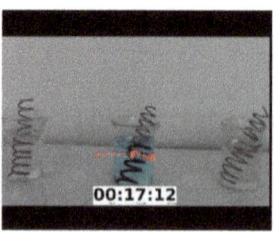

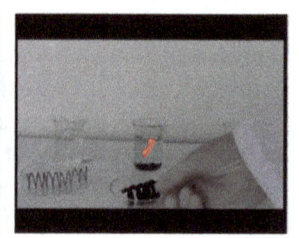

Fig. 5.6 Gaze plot of a preservice teacher having a moderate understanding of redox reactions

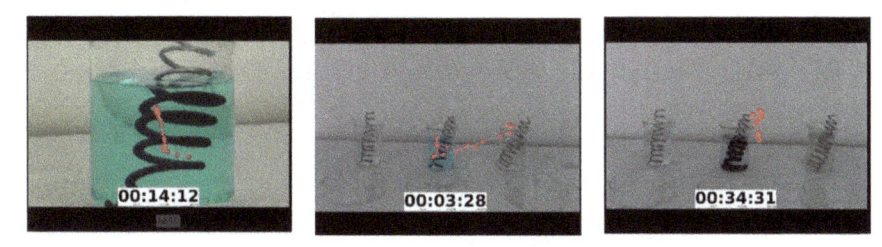

5 Investigating the Role of Conceptual Understanding ...   101

Fig. 5.7 Gaze plot of a preservice teacher having a weak understanding of redox reactions

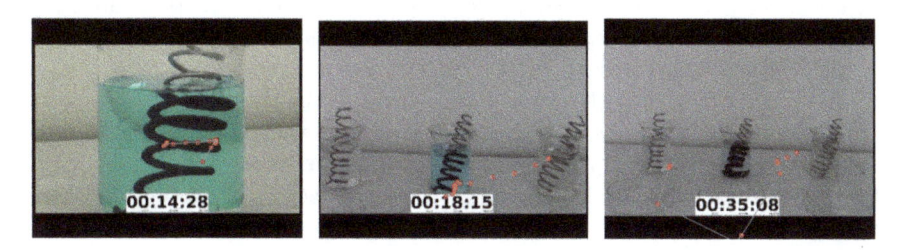

Fig. 5.8 Gaze plot of a preservice teacher having an alternative understanding of redox reactions

Generally, preservice teachers with a weak understanding of redox reactions less focused on the experimental evidences and frequently looked at the beakers nearby. For instance, at the 35th minute of the reaction, toward the end of the video, one of the preservice teachers mostly looked at all three beakers, not much at solution or the wire inside as seen in Fig. 5.8. Some of the scenes from the gaze plot of this preservice teacher are given in Fig. 5.8.

The analysis of eye-tracking data was confirmed with the analysis of think aloud protocols revealed the features preservice teachers demonstrated while watching an experimental video as shown in Table 5.2.

It was observed that the differences in their eye movement patterns can be grouped into three main categories: *looking where they need to look, staying focused*, and *ignoring the details*. First of all, it was observed that preservice teachers who had scientific or moderate understanding mostly focused on the experimental evidences, not the other beakers or parts. On the other hand, preservice teachers who had weak or alternative understanding, did not focus on the experimental evidences which were mostly observed in the beaker in the middle, instead they looked through the other beakers. So, a difference was observed in terms of looking where the experimental evidences were mostly observed. Secondly, preservice teachers with scientific or moderate understanding, stay focused and keep looking at the experimental evidences whereas preservice teachers with weak or alternative understanding, changed their focus occasionally or even some did frequently. Thirdly, preservice teachers with scientific or moderate understanding ignored the details given in the video, however, preservice teachers with weak or alternative understanding, also looked at these details such as the piece of metal outside the solution. Thus, it seemed that preservice teachers' levels of understanding of redox reactions had an effect on the patterns of their eye movements.

Discussion and Conclusion

Understanding chemical reactions, including redox reactions, has been challenging for students (Acar & Tarhan, 2007; Brandriet & Bretz, 2014) due to the difficulty of making connection between macroscopic and submicroscopic levels (Ahtee & Varjola, 1998; Ben-Zvi et al., 1987).

Eye-tracking technology has been used in chemistry education to better evaluate students' conceptualizations of chemical phenomenon while they are working on a particular task because it informs educators about students' cognitive processes during the activity (Rayner, 1998). In this study, eye-tracking technology was used to explore the role of students' level of understandings of redox reactions on how they navigate while watching an experimental video of a redox reaction. Specifically, the patterns they followed and the aspects they mostly focused were determined and compared across different levels of understandings.

The results of the study revealed that, based on the levels of understanding of redox reactions, preservice teachers exhibited different patterns of eye movements and focuses while watching an experimental video. These differences between the levels of understandings suggested three main categories: *looking where they need to look*, *staying focused*, and *ignoring the details*. First, preservice teachers who had scientific or moderate understanding mostly focused on the experimental evidences, not the other beakers or parts. On the other hand, preservice teachers who had weak or alternative understanding, did not focus much on the experimental evidences which were mostly observed in the beaker in the middle, instead they looked through the other beakers. So, it can be suggested that preservice teachers with higher levels of understanding could better make connections between macroscopic and submicroscopic levels as they paid more attention to the experimental evidences. Secondly, these preservice teachers also stayed focused throughout the video and kept looking at the experimental evidences whereas preservice teachers with weak or alternative understanding, changed their focus occasionally or even some did frequently. This might have happened due to being mentally engaged on what is happening in the reaction vessel. Finally, preservice teachers with scientific or moderate understanding ignored the details given in the video, whereas the ones with weak or alternative understanding, also looked at these details such as the piece of metal outside the solution. This might have been observed due to the lack of understanding what is happening and trying to look for evidence at other parts.

In conclusion, in this study, it can be suggested that the preservice teachers' levels of understanding of redox reactions played a role on the patterns of their eye movements as they looked at where they needed to look, stayed focused, and ignored the details, as they were more mentally engaged. In other words, as the level of understanding approach to an expert-like understanding, the eye movements conveyed this understanding as also captured in this study by eye-tracking. It was also reported in the previous studies that there existed differences between the eye movements of experts and novices (Havanki, 2012; VandenPlas, 2008).

It was also observed that when the second beaker where the redox reaction takes place zoomed in all the preservice teachers looked at the experimental evidence, the copper metal depositing on zinc wire. So, the action of zooming helped them better recognize the experimental evidences. When it was zoomed out, preservice teachers with a low level of understanding also looked at the other parts and lost their focus. It can be said that they needed more scaffolding such as zooming to catch important aspects.

In this study, the majors of the preservice teachers, as studying chemistry or science, were not considered because the level of understanding was the main cause of conveying differences in the eye movement patterns, that was also revealed by the findings. Even though they were majoring at different subject areas, they had taken similar chemistry classes and developed similar understandings independent of their major. Therefore, in this study they were all grouped and called as preservice teachers.

As an implication for teaching, experimental videos can be used in teaching chemistry in- or out-of-classroom setting as they help students make connection between macroscopic and submicroscopic phenomena. Yet, when the experimental videos are used in chemistry classes, teachers may help students direct their attentions to experimental evidences or important aspects, such that in this study, it was the beaker where the redox reaction took place. As an implication for research, visual tools can be used to scaffold students' attention to the main aspects and stay focused. For instance, zooming, labels, or other cues can be included in the video.

As a limitation, this study utilized qualitative analysis that might have missed some of the minor changes in the eye movements. Repeating the study with quantitative methods may provide further evidence about students' eye movement as they view experimental video recording.

References

Acar, B., & Tarhan, L. (2007). Effect of cooperative learning strategies on students' understanding of concepts in electrochemistry. *International Journal of Science and Mathematics Education, 5*(2), 349–373.
Ahtee, M., & Varjola, I. (1998). Students' understanding of chemical reaction. *International Journal of Science Education, 20*(3), 305–316.
Akaygun, S., & Adadan, E. (2019). Revisiting the understanding of redox reactions through critiquing animations in variance. In M. Schultz, S. Schmid, & G. Lawrie (Eds.), *Research and practice in chemistry education* (pp. 7–29). Singapore: Springer.
Akaygun, S., Adadan, E., & Kelly, R. (2019). Capturing preservice chemistry teachers' visual representations of redox reactions through storyboards. *Israel Journal of Chemistry, 59,* 493–503.
Akaygun, S., & Jones, L. L. (2013). Research-based design and development of a simulation of liquid-vapor equilibrium. *Chemistry Education Research and Practice, 14,* 324–344.

Ardac, D., & Akaygun, S. (2004). Effectiveness of multimedia-based instruction that emphasizes molecular representations on students' understanding of chemical change. *Journal of Research in Science Teaching, 41*(4), 317–337.

Barke, H.-D., & Engida, T. (2001). Structural chemistry and spatial ability in different cultures. *Chemistry Education: Research and Practice in Europe, 2*, 227–239.

Baluyut, J. Y., & Holme, T. A. (2019). Eye tracking student strategies for solving stoichiometry problems involving particulate nature of matter diagrams. *Chemistry Teacher International, 1*(1). https://doi.org/10.1515/cti-2018-0003.

Ben-Zvi, R., Eylon, B., & Silberstein, J. (1987). Students' visualization of a chemical reaction. *Education in Chemistry, 24*, 117–120.

Brandriet, A. R., & Bretz, S. L. (2014). The development of the redox concept inventory as a measure of students' symbolic and particulate redox understandings and confidence. *Journal of Chemical Education, 91*(8), 1132–1144.

Dori, Y. J., & Barak, M. (2001). Virtual and physical molecular modeling: Fostering model perception and spatial understanding. *Educational Technology & Society, 4*(1), 61–74.

Ebenezer, J. V., & Erickson, G. L. (1996). Chemistry students' conceptions of solubility: A phenomenography. *Science Education, 80*(2), 181–201.

Garnett, P. J., & Treagust, D. F. (1992). Conceptual difficulties experienced by senior high school students of electrochemistry: Electrochemical (galvanic) and electrolytic cells. *Journal of Research in Science Teaching, 29*(10), 1079–1099.

Gomes, J. S., Yassine, M., Worsley, M. A. B., & Blikstein, P. (2013). Analysing engineering expertise of high school students using eye tracking and multimodal learning analytics. In S. K. D'Mello, R. A. Calvo, & A. Olney (Eds.), *Proceedings of the 6th International Conference on Educational Data Mining, EDM 2013*. International Educational Data Mining Society.

Hansen, S. (2014). *Multimodal study of visual problem solving in chemistry with multiple representations*. Doctoral Dissertation, Columbia University.

Havanki, K. (2012). *A process model for the comprehension of organic chemistry notation*. Doctoral Dissertation, The Catholic University of America.

Hsieh, H. F., & Shannon, S. E. (2005). Three approaches to qualitative content analysis. *Qualitative Health Research, 15*(9), 1277–1288.

Hoffman, J. E., & Subramaniam, B. (1995). The role of visual attention in saccadic eye movements. *Perception and Psychophysics, 57*(6), 787–795.

Johnstone, A. H. (1991). Why is science difficult to learn? Things are seldom what they seem. *Journal of Computer Assisted learning, 7*, 75–83.

Just, M. A., & Carpenter, P. A. (1984). *Reading skills and skilled reading in the comprehension of text*. Erlbaum, Hillsdale, NJ: Learning and Comprehension of Text.

Kelly, R. M., & Jones, L. L. (2007). Exploring how different features of animations of sodium chloride dissolution affect students' explanations. *Journal of Science Education and Technology, 16*(5), 413–429.

Kelly, R. M. (2014). Using variation theory with metacognitive monitoring to develop insights into how students learn from molecular visualizations. *Journal of Chemical Education, 91*(8), 1152–1161.

Lohman, D. F. (1979). *Spatial ability: A review and reanalysis of the correlational literature*. School of Education, Stanford, CA: Stanford University.

Lottier, L. F. (1986). Storyboarding your way to successful training. *Public Personnel Management, 15*(4), 421–427.

Marton, F. (1986). Phenomenography—A research approach to investigating different understandings of reality. *Journal of Thought, 21*(3), 28–49.

Nakhleh, M. B. (1992). Why some students don't learn chemistry. *Journal of Chemical Education, 69*(3), 191–196.

National Research Council. (2012). *A framework for K-12 science education: Practices, crosscutting concepts, and core ideas*. Washington, DC: National Academies Press.

Pribyl, J. R., & Bodner, G. M. (1987). Spatial ability and its role in organic chemistry: A study of four organic courses. *Journal of Research in Science Teaching, 24,* 229–240.

Rayner, K. (1998). Eye movements in reading and information processing: 20 years of research. *Psychological Bulletin, 124*(3), 372–422.

Schank, P., & Kozma, R. (2002). Learning chemistry through the use of a representation-based knowledge building environment. *Journal of Computers in Mathematics and Science Teaching, 21*(3), 253–279.

Smith, C. L., Wiser, M., Anderson, C. W., & Krajcik, J. (2006). Implications of research on children's learning for standards and assessment: A proposed learning progression for matter and the atomic molecular theory. *Measurement, 4*(1–2), 1–98.

Stieff, M., Hegarty, M., & Deslongchamps, G. (2011). Identifying representational competence with multi-representational displays. *Cognition and Instruction, 29*(1), 123–145.

Taber, K. S. (2013). Revisiting the chemistry triplet: Drawing upon the nature of chemical knowledge and the psychology of learning to inform chemistry education. *Chemistry Education Research and Practice, 14,* 156–168.

Tang, H., & Pienta, N. (2012). Eye-tracking study of complexity in gas law problems. *Journal of Chemical Education, 89,* 988–994.

Tang, H., Topczewski, J., Topczewski, A., & Pienta, N. (2012). Permutation test for groups of scan-paths using normalized Levenshtein distances and application in NMR Questions. In *Proceedings of the Symposium on Eye Tracking Research and Applications* (pp. 169–172).

Tang, H., Kirk, J., & Pienta, N. (2014). Investigating the effect of complexity factors in stoichiometry problems using logistic regression and eye tracking. *Journal of Chemical Education, 91*(7), 969–975.

VandenPlas, J. (2008). *Animations in chemistry learning: Effect of expertise and other user characteristics*. Doctoral Dissertation, The Catholic University of America.

Williamson, V. M. (2008). The particulate nature of matter: An example of how theory-based research can impact the field. In D. Bunce & R. S. Cole (Eds.), *Nuts and Bolts of chemical education research* (pp. 67–78). Washington, DC: American Chemical Society.

Williamson, V. M., & Abraham, M. R. (1995). The effects of computer animation on the particulate mental models of college chemistry students. *Journal of Research in Science Teaching, 32*(5), 521–534.

Williamson, V. M., Hegarty, M., Deslongchamps, G., Williamson, K. C., & Shultz, M. J. (2013). Identifying student use of ball-and-stick images versus electrostatic potential map images via eye tracking. *Journal of Chemical Education, 90*(2), 159–164.

Yarroch, W. L. (1985). Student understanding of chemical equation balancing. *Journal of Research in Science Teaching, 22*(5), 449–459.

Sevil Akaygun, Ph.D. is an Associate Professor of Chemical Education at Bogazici University, Faculty of Education, Turkey. Her research interests include visualisations in chemistry education, integration of ICT in chemistry education, nanotechnology education, STEM education, and socio-scientific issues in science education. Dr. Akaygun has coordinated various national and international research projects, including IRRESISTIBLE (funded by European Commission), Turkey–USA bilateral collaboration project (funded by TUBITAK), given seminars and workshops to STEM teachers, and served on the national chemistry curriculum development programmes organised by Turkish Ministry of Education. She (co)authored chapters in international books (published by Springer, American Chemical Society, IGI Global) and published numerous papers in respected international and national journals. Dr. Akaygun coedited the themed issue on Visualizations and Representations in Chemistry Education published in Chemistry Education Research and Practice (CERP), in 2019. She has been vice editor of Journal of Turkish Chemical Society Section C: Chemical Education (JOTCSC), chemistry field editor of Teaching Science Journal (published by Turkish Science Education and Research Association), guest co-editor of the special issue, "STEM Education" to be published in Boğaziçi University

Journal of Education (BUJE), and guest co-editor of the special issue, "Science Education for Citizenship through Socio-Scientific Issues" to be published in Frontiers in Education.

Emine Adadan, Ph.D. is a Professor of Chemistry Education at Bogazici University, Faculty of Education, Turkey. Her research focuses on students' learning progressions, using multiple (and visual) representations in chemistry instruction, role of metacognitive awareness in chemistry learning, pedagogical content knowledge, using e-portfolios for promoting preservice teachers' reflective thinking skills, and nanoscience education. Dr. Adadan was involved in several national and international research projects, including IRRESISTIBLE (funded by European Commission), Turkey–USA bilateral collaboration project (funded by TUBITAK). She published numerous papers in respected refereed international journals (e.g. Journal of Research in Science Teaching, International Journal of Science Education, etc.). Dr. Adadan has served as an associate editor of Boğaziçi University Journal of Education (BUJE) and the advisory board member of Science and Technology Journal, a popular science journal, published by the Scientific and Technological Research Council of Turkey. Dr. Adadan is also the recipient of the Excellence in Research Award from Bogazici University Foundation in 2017.

Chapter 6
Using an Eye-Tracker to Study Students' Attention Allocation When Solving a Context-Based Problem on the Sublimation of Water

Miha Slapničar, Valerija Tompa, Iztok Devetak, Saša Aleksij Glažar, and Jerneja Pavlin

Introduction

Science knowledge is of utmost importance for understanding the processes and phenomena around us. Appropriate basic education in science enables critically evaluating the deluge of information from various and diverse media. The learning content of science subjects is perceived as abstract and difficult to understand and learn. One of the challenges of modern science education is learning to understand science concepts, phenomena and processes and to apply them when solving authentic tasks (Wu et al., 2001). The way students process information in science subjects can be supported in solving problems by observing a person with an eye tracker, which is the subject of our research, the theoretical background of which includes the triple nature of chemical concepts, the understanding of submicroscopic representations, and the properties of the eye-tracker method.

Learning the Triple Nature of Chemical Concepts

The complexity of teaching and learning chemistry is related to the representation of chemical concepts on three different levels (macroscopic, submicroscopic and symbolic). The connection between these levels was initially represented by a triangle (model) in which each level complements the other without giving superiority to any of the levels (Johnstone, 1982, 1991). This model has been further developed by

---

M. Slapničar (✉) · V. Tompa · I. Devetak · S. A. Glažar · J. Pavlin
Faculty of Education, University of Ljubljana, Ljubljana, Slovenia
e-mail: miha.slapnicar@pef.uni-lj.si

© Springer Nature Switzerland AG 2021
I. Devetak and S. A. Glažar (eds.), *Applying Bio-Measurements Methodologies in Science Education Research*,
https://doi.org/10.1007/978-3-030-71535-9_6

several authors: Mahaffy (2004) added the human element; Devetak, Vogrinc et al. (2009) implemented a visualisation method and a mental model.

The macroscopic level of the chemical concepts is illustrated by the observed chemical phenomena, while at the submicroscopic level the interpretation of the observed interaction of the particles is explained. For the representation of chemical concepts at the submicroscopic level, submicroscopic representations (SMRs) can be used in static and dynamic representations (Devetak & Glažar, 2010). SMRs could be translated into different symbols to represent the symbolic level, such as symbols of elements, formulas, equations, mathematical equations, as well as graphical and schematic representations (Devetak, 2012; Johnstone, 2001; Levy & Wilinsky, 2009; Taber, 2013).

Johnstone (2001) found that chemical concepts are difficult to learn and are, therefore, often misunderstood by students, which may be related to an inability to connect all three levels of chemical concept representation because their working memory is overloaded during the learning process (Johnstone & El-Banna, 1986). The age and stage of education of the students influence their ability to represent chemical concepts (e.g., states of water). Older students, for example, used the macroscopic level of representation to explain 3D dynamic SMRs (Slapničar et al., 2017). The ten-year longitudinal study by Löfgren and Hellden (2009) shows that students were unable to transfer the acquired knowledge of the particulate nature of matter to situations of everyday life.

In this context, one of the approaches to teaching chemistry (context-based approach) attempts to improve students' understanding of chemistry. The context-based approach usually starts with contexts (e.g., topics, questions) that are close to their everyday lives (authentic contexts). The application of this approach has an impact on raising pupils' interest, activating their prior knowledge and providing situations for the application of newly developed knowledge by linking it to basic concepts (Parchmann et al., 2017).

Understanding the Particle Representations of Chemical Concepts

Particle representations (e.g., 2D, 3D dynamic SMRs, computer animations of particles, etc.) play an essential role in chemistry lessons, as they enable students to visualise phenomena (e.g., atoms, ions, molecules) that cannot be directly observed (Phillips et al., 2010; Wu & Shah, 2004).

SMRs are the essential elements of various chemistry-learning materials that can be successfully implemented in chemistry education and can contribute significantly to developing an adequate understanding of chemical concepts at all three levels (Ferk Savec et al., 2016; Russell et al., 1997). However, researchers (Kelly & Jones, 2008; Tien et al., 2007) reported that students exposed to SMRs during the learning process understood the nature of particle interactions better than students who learned concepts only by reading textbooks. In contrast, much research has shown that students had difficulty in understanding SMRs, including the states of matter, the states of water (Bunce & Gabel, 2002; Devetak, Drofenik Lorber et al., 2009; Kind, 2004; Slapničar et al., 2017) and the transitions between them (Devetak, Vogrinc et al., 2009; Kind, 2004).

Eye-Tracking Measurements

Understanding the learning process is related to measurements of the individual's eye movements, which can be used to interpret processes that occur during context-based problem solving due to the connection between the direction of gaze and the focus of attention (Hyönä et al., 2003; Just & Carpenter, 1980; Rayner, 2009).

Eye movements consist of two components called 'fixations' and 'saccades'. During fixations, visual attention is focused on the specific location (periods of eye stability). Saccades are quick eye movements between fixations and indicate how the eyes move from one location to another. Fixations generally last about 200 to 300 ms, while saccades last 30 to 80 ms (Holmqvist et al., 2011).

The most commonly used eye movement measurement is called fixation duration, which is a measure of the total time a person spends within an area of interest (AOI). Within a single AOI, the fixation duration may be represented as a total of all fixations. A longer total fixation duration (TFD) in an AOI could indicate the degree of detail (deeper and more complex information processing), the difficulty in processing this area (Goldberg & Kotval, 1999; Holmqvist et al., 2011; Hyönä et al., 2002) or a lower efficiency in finding the information on the computer screen (Green et al., 2007). Research has shown that experts had shorter fixations on irrelevant information than novices did, while there were longer fixations on relevant information compared to novices (Gegenfurtner et al., 2011).

A second eye-tracking measure, a VC is related to the number of times a person returns to an AOI, which could provide insight into the attraction, usefulness and confusion of a particular AOI by the person. Repeated visits to an AOI could indicate which characteristics are important or interesting (West et al., 2006).

In problem-solving in chemistry, for example, it might be useful to track the number of visits to a particular resource (e.g., periodic table) to measure how often the person has to use it for the given problem (Cullipher & VandenPlas, 2018).

As a third eye-tracking measure, pupil dilation is a commonly researched area because of the relationship between pupil dilation and cognitive load (Bassok, 1990). Research has shown that pupil size is not an accurate indicator of cognitive load during reading problems, but that it is sensitive to task difficulty in certain subproblems (Schultheis & Jameson, 2004). During mental effort, the pupil diameter increases, which correlates with the difficulty of the task (Just et al., 2003). One of the most commonly used variables is mean pupil dilation, which is calculated by averaging all dilations over a period of interest (Beatty & Lucero-Wagoner, 2000).

Eye-tracking research studies have shown that unsuccessful problem solvers had difficulty distinguishing between relevant and irrelevant factors and focusing on relevant factors to solve the problem. Success in selecting the right information is critical to successful problem solving and is similar to observing 3D SMRs (Koning et al., 2009). Successful problem solvers focused more on relevant factors, while unsuccessful problem solvers had difficulty deciphering the problem and identifying the relevant factors (Tsai et al., 2012).

The integration of the triple nature of chemical concepts, diverse educational materials and teaching approaches contribute to problem-solving ability (Slapničar et al., 2017, 2018). Success in solving chemical problems depends on the simultaneous and properly associated three levels of chemical concepts (Taber, 2013), the students' previous knowledge, and their experience in a particular field (Avramiotis & Tsaparlis, 2013).

Aims and Research Questions

This research aims to investigate differences in attention allocation (TFD and VC) to specific AOIs and the cognitive load (average pupil size) of successful and unsuccessful students at three different stages of education (lower secondary, upper secondary and university education). The context-based problem includes questions related to the sublimation of water, which are presented with macroscopic representations (photos) and 3D SMR.

Concerning the research aim, two research questions can be addressed:

1. Are there significant differences in attention allocation (TFDs and VCs) to specific AOIs in context-based problems between successful and unsuccessful students in different age groups and within the groups of successful or unsuccessful students?
2. Are there significant differences in cognitive load (average pupil size) between successful and unsuccessful students in different age groups?

Method

For the research, a quantitative research approach with descriptive and non-experimental methods was used.

Participants

Seventy-nine students from three different age groups (primary, upper secondary and university educational stages) participated in the research. The first group consisted of 30 seventh-grade students ($Mdn = 12.0$ years, $IQR = 0.0$ year); the second group consisted of 29 students in the first year of upper secondary school (*Mdn* = 16.0 years, *IQR* = 1.0 year) and the third group consisted of 20 pre-service two subject teachers (chemistry and biology) (*Mdn* = 23.0 years, *IQR* = 2.0 years).

The participants were selected from a mixed urban population in the Ljubljana region. Each student was assigned a code.

The students participated in the study on a voluntary basis. Consent was obtained for primary and upper secondary students (from school boards, teachers and parents) in accordance with the judgement of the Ethics Committee for Pedagogy Research of the Faculty of Education.

Instrument and Procedures

During the research, data were collected with a context-based problem related to water sublimation and an eye-tracking apparatus (participants' measures of TFD, VC and average pupil size).

Context-Based Problem of Sublimation of Water

The analysis of students' eye-tracking measurements while solving a context-based problem related to the sublimation of water is presented in this paper.

This specific context-based problem is one of eleven science-based problems investigated in research funded by the Slovenian Research Agency (ARRS) entitled 'Explaining Problem Solving of the Triplet Relationship in Science Concepts Representations' (JS-6814). These context-based problems were developed according to the Slovenian students' achievements on TIMMS (Trends in International Mathematics and Science Study), PISA (The Programme for International Student Assessment) and the Slovenian national external assessment for chemistry, physics and biology, and modified for this research (Pavlin et al., 2019).

The 3D dynamic SMR of the sublimation of water was designed by science educators and completed by a computer expert; it was developed solely for this research and is not available for general use. There was no time limit for viewing the selected SMR. When the participants needed more time to solve the task, the dynamic 3D SMR started again. Participants could not control the display of the 3D dynamic SMR. The text of the context-based problem was in the Slovene language. Only for the purpose of this research paper, the text of the context-based problem has been translated into English (see Fig. 6.1).

The context-based problem consisted of two questions and one justification three questions related to the recognition and explanation of the transition between states of water characteristic for frost formation (presented on the photos and the 3D dynamic SMR). The first question was related to defining the substance included in the frost formation. The second question was linked to the aggregate states of matter involved

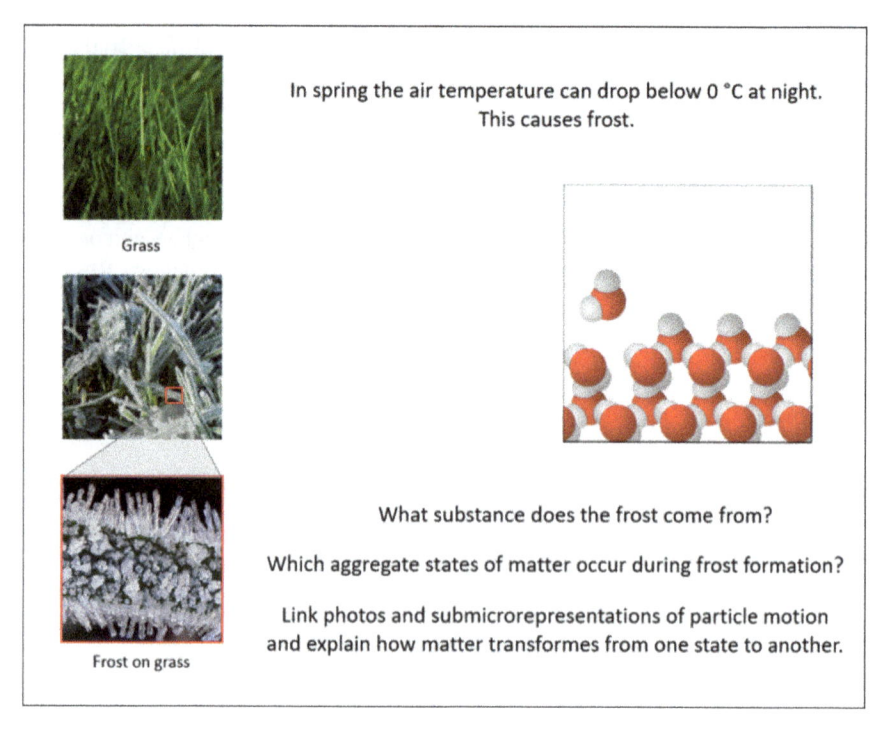

Fig. 6.1 Screen image of the context-based problem related to the sublimation of water in English

in the frost formation (solid and gaseous state of water). The justification was related to the explanation of the transformation of matter from one state to another, based on both representations of chemical concepts (three photos and the 3D dynamic SMR).

The 3D dynamic SMR was displayed on the computer screen simultaneously to present the submicroscopic level of transition between the gaseous and solid state of water during frost formation. The participants could use both levels of the chemical concept representation to form the answers (macroscopic and submicroscopic).

The context-based problem consisted of six AOIs (see Fig. 6.2), four of which were selected for research purposes (Photos 1, 2, 3 and 3D dynamic SMR).

The correct answers to the context-based problem are:

1. What substance does the frost come from?
   From water.
2. Which aggregate states of matter occur during frost formation?
   Solid and gaseous states of matter (water).
3. Link the photos and submicro-representations of particle motion and explain how matter transforms from one state to another
   Matter (water) transforms from the gaseous state of water (when the air temperature is below 0°C) to the solid state of water. On the 3D dynamic SMR, only sublimation is shown without the process of resublimation (transformation of water from the solid to the gaseous state).

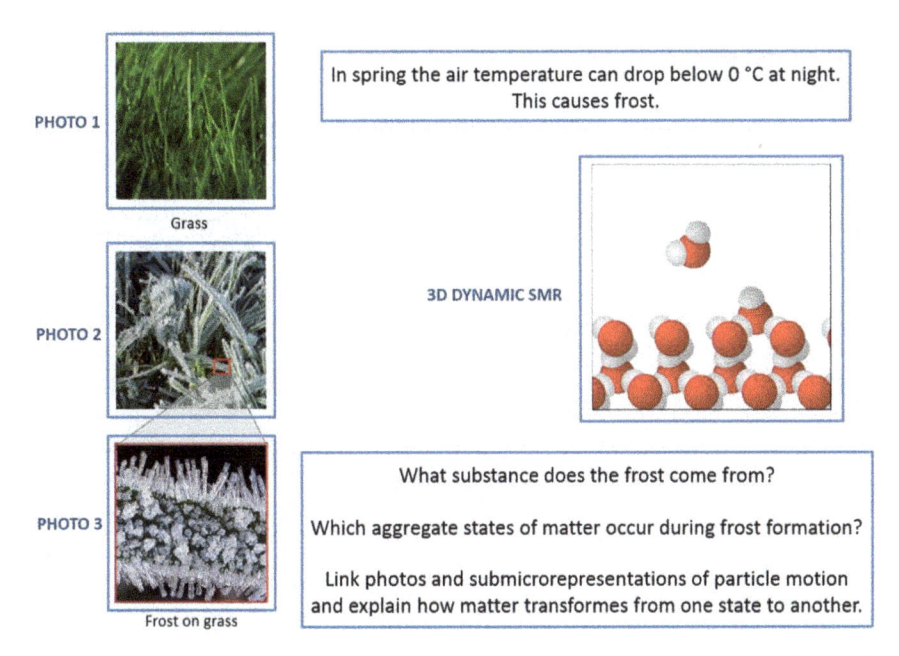

Fig. 6.2 Screen image of AOIs marked in blue rectangles in the context-based problem of the sublimation of water

Eye-Tracking Apparatus

The screen-based EyeLink 1000 (35 mm lens, horizontal orientation) eye tracker apparatus with associated software (Experiment Builder) for the preparation of the experiment and a connection with EyeLink was used in this research. Data Viewer was suitable for obtaining the data and basic analysis, recording and analyses of students' eye movement during problem-solving. The data were collected from the right eye (monocular data collection following corneal reflection and pupil responses) at 500 Hz (Torkar, 2018).

The context-based problem displayed on the computer screen was separated into clearly and carefully divided AOIs according to the placement elements. The eye-tracker measured eye movements, such as TFD, average pupil size and VC on the specific AOI (on the computer screen) while solving a context-based problem.

The identification of fixations was related to the motion of gaze during each sample collected. When the velocity and acceleration thresholds (in our case: 30 degrees per second and 8000 degrees per second squared) are exceeded, a saccade begins. In other cases, the sample is marked as fixation.

Data Collection

The eye-tracking research was conducted from November 2016 to March 2017 in the laboratory of the Department of Psychology at the Faculty of Arts, University of Ljubljana. Each participant was tested with eye tracker individually while measuring eye movements. The participants had no time limit when solving eleven context-based problems.

Before the context-based problems solving, all participants (in all age groups) were individually informed about the purpose of the study, the method used and their role in the research. According to the most optimal recordings gathered, students sat approximately 60 cm from the computer screen (distance to the eyes) with their heads stabilised in the head-supporting stand. After calibration and validation through the algorithm of nine points, the individual solved all the tasks by verbally providing the answers. The experimenter wrote down the individual's answers to each problem. The description of the screen images of the task is presented in the subsection context-based problem of water sublimation.

Data Analysis

Participants' answers obtained during the solving of the context-based problem were graded based on the correct model answer. The reliability of the grading (97%) was ensured by the grading of two researchers (authors of this article).

The eye movement measures were obtained with an EyeLink 1000 eye tracker and analysed with the associated software. Experiment Builder was used to prepare the experiment and connect to the EyeLink, as well Data Viewer as for data selection and basic data analysis.

All data were statistically processed in SPSS (Statistical Package for the Social Sciences). Basic descriptive statistics (medians Mdn, interquartile range IQR) of numerical variables were determined, whose distribution was non-normal (TFD, VC and average pupil size) were determined. Three groups of students were classified according to their success in solving the context-based problem related to frost formation.

The non-parametric Mann-Whitney U test was used to explain the relationship between groups of successful and unsuccessful students in each age group in terms of average pupil size, VCs and TFDs. This test compares the medians between two independent groups on a continuous measure, the results of which are then converted into ranks.

The Kruskal-Wallis test was selected to determine differences between groups of successful and unsuccessful students in different age groups in terms of average pupil size, VCs and TFDs. Post hoc Dunn's tests were used to define the differences between different pairs of age groups. Statistical hypotheses for all tests were tested at an alpha error rate of 5%.

The Mann-Whitney U test and Kruskal-Wallis tests were used due to the small sample size and the non-normal data distribution (Pallant, 2011).

The measure of effect size eta squared η^2 was used to describe the strength of a phenomenon in relation to the definition of whether the effects have a relevant magnitude. Benchmarks for defining small (.01), medium (.06) and large (.14) magnitude effects were provided by Cohen (1988).

Results and Discussion

The interpretation of results and discussion are presented according to the research questions.

Differences in Visual Attention on the AOIs

The first research question relates to differences in attention allocation (variables TFD and VC) between groups of successful and unsuccessful students in different age groups (e.g., primary, upper secondary and university stages of education). Successful students achieved two or three out of three points in the context-based problem, with each part of them scoring one point.

Table 6.1 shows the absolute and relative frequencies for the groups of successful and unsuccessful students in each age group and shows that success improves with the stages of education.

Table 6.2 presents the successful students' TFDs by *Mdns* and *IQRs* at macroscopic (Photos 1, 2 and 3) and submicroscopic levels (3D dynamic SMR) AOIs.

All age groups of students show the highest values of TFDs for the 3D dynamic SMR, which decrease with the age of the students.

The students of all three stages of education, spent most of their time at the macroscopic level while observing Photo 2, which shows the frost formation (water in the solid state on the grass).

Table 6.1 Absolute and relative frequencies for the groups of successful and unsuccessful students in each age group

Age group	Absolute (relative) frequencies	
	Successful students	Unsuccessful students
Primary school students	7 (23.3%)	23 (76.7%)
Upper secondary school students	8 (27.6%)	22 (75.9%)
University students	6 (30.0%)	14 (70.0%)

Table 6.2 TFD spent on different AOIs in context-based problem related to sublimation of water in all age groups of successful students

AOI	Primary school students (Group 1) Mdn [s]	IQR [s]	Upper secondary school students (Group 2) Mdn [s]	IQR [s]	University students (Group 3) Mdn [s]	IQR [s]
Photo 1	1.4	2.8	2.6	2.6	1.7	4.7
Photo 2	8.2	7.7	8.9	6.6	6.0	3.7
Photo 3	3.7	7.0	7.8	5.5	2.8	3.1
3D SMR	35.9	26.0	36.5	34.6	20.0	15.4

No statistical differences in TFDs were found between the groups of successful students for certain AOI (Kruskal-Wallis Test: $p > .05$ in all cases).

Table 6.3 shows the TFDs of unsuccessful students by *Mdns* and *IQRs* at the macroscopic (photos 1, 2 and 3) and submicroscopic (3D dynamic SMR) levels of AOIs.

The unsuccessful students of all three stages of education spent most of their time on the 3D dynamic SMR. Photo 2 shows the AOI with the highest attention allocation (TFD values), which decreases with the age of the students.

The differences between Groups 1 (primary school students) and 3 (university students) are statistically significant for Photos 2 and 3 (size effects are small or large—η^2 in Table 6.3).

Statistical differences between the groups of successful and unsuccessful primary school students were found in the TFDs on the AOI—Photo 2 ($U = 46.000$; $p = .049$; $\eta^2 = .129$). The TFD value of mean ranks of the unsuccessful students ($MR = 17.41$) was statistically higher than that of the successful students ($MR = 10.25$). No statistical differences in the TFD values on the AOIs were found between the

Table 6.3 TFD spent on different AOIs in context-based problem related to sublimation of water in all age groups of unsuccessful students

AOI	Primary school students (Group 1) Mdn [s]	IQR [s]	Upper secondary school students (Group 2) Mdn [s]	IQR [s]	University students (Group 3) Mdn [s]	IQR [s]	Kruskal-Wallis test χ^2	p	η^2
Photo 1	3.1	5.1	1.7	2.9	3.4	5.2	3.739	.154	–
Photo 2	11.0	5.4	8.0	6.6	5.7	4.7	12.134	.002[a]	.184
Photo 3	7.9	7.5	4.1	4.9	2.8	2.8	13.914	.001[b]	.217
3D SMR	27.8	27.7	18.8	22.4	25.1	28.6	4.997	.082	–

Results of post hoc Dunn's tests
[a]Group 1–2: $p = .036$; group 2–3: $p = .864$; group 1–3: $p = .003$; 1 > 2, 2 = 3, 1 > 3
[b]Group 1–2: $p = .013$; group 2–3: $p = 1.000$; group 1–3: $p = .002$; 1 > 2, 2 = 3, 1 > 3

Table 6.4 VCs on different AOIs in context-based problem related to sublimation of water in all age groups of successful students

AOI	Primary school students (Group 1)		Upper secondary school students (Group 2)		University students (Group 3)	
	Mdn	*IQR*	*Mdn*	*IQR*	*Mdn*	*IQR*
Photo 1	4.5	5.5	5.0	3.0	4.0	4.3
Photo 2	9.5	13.0	23.0	22.0	14.0	7.0
Photo 3	6.5	9.8	13.0	20.0	9.5	6.0
3D SMR	22.5	13.0	23.0	35.0	15.0	12.8

groups of successful and unsuccessful students of upper secondary education, upper secondary students and university students.

Table 6.4 shows the *Mdns* with *IQRs* of VCs for AOIs at the macroscopic and submicroscopic levels for all three age groups of successful students.

Table 6.4 shows that all three groups of successful students have the highest *Mdn* of VCs on Photo 2 at the macroscopic level. It is interesting to note that among the successful upper secondary students who had similar *Mdns* of TFD as successful primary students, higher Mdns of VC are found on this AOI.

No statistically significant differences were found between groups of successful students of all ages on any of the AOIs (Kruskal-Wallis test: $p > .05$ in all cases).

Table 6.5 presents the VC' *Mdns* with *IQRs* for the AOIs at macroscopic and submicroscopic levels for different age groups of unsuccessful students.

For the unsuccessful upper secondary and university students, the highest values of VCs were on the 3D dynamic SMR. Photo 2 represented the AOI with the highest *Mdns* of VCs for the group of primary school students even if the value of TFD was higher for the 3D dynamic SMR.

Table 6.5 VCs on different AOIs in context-based problem related to sublimation of water in all age groups of unsuccessful students

AOI	Primary school students (Group 1)		Upper secondary school students (Group 2)		University students (Group 3)		Kruskal-Wallis test		
	Mdn	*IQR*	*Mdn*	*IQR*	*Mdn*	*IQR*	χ^2	*p*	η^2
Photo 1	7.0	8.5	3.5	5.0	8.0	9.8	6.502	.039[a]	.082
Photo 2	20.5	10.0	14.5	10.8	13.0	11.8	8.195	.017[b]	.113
Photo 3	13.0	10.5	9.0	9.5	9.5	6.3	7.373	.025[c]	.098
*3D SMR	18.0	19.3	17.0	10.5	17.5	14.5	1.251	.535	–

Results of post hoc Dunn's tests
[a]Group 1–2: $p = .047$; group 2–3: $p = .218$.; group 1–3: $p = 1.000$; $1 > 2, 2 = 3, 1 = 3$
[b]Group 1–2: $p = .039$; group 2–3: $p = 1.000$; group 1–3: $p = .052$; $1 > 2, 2 = 3, 1 = 3$
[c]Group 1–2: $p = .086$; group 2–3: $p = .1.000$; group 1–3: $p = .048$; $1 = 2, 2 = 3, 1 > 3$

Kruskal-Wallis tests and post hoc tests showed statistically significant differences between unsuccessful primary school students and upper secondary students (AOIs—Photos 1 and 2), and primary school students and university students (AOI—Photo 3) in VCs. The size effects of Kruskal-Wallis tests were small or medium.

Between groups of successful and unsuccessful primary school students, statistically significant differences were detected in VCs on Photo 2 ($U = 36.500$; $p = .016$; $\eta^2 = .194$) and Photo 3 ($U = 44.500$; $p = .041$, $\eta^2 = .139$). Unsuccessful students' values of mean ranks were higher (Photo 2—$MR = 17.84$; Photo 3—$MR = 17.48$) than those of successful students (Photo 2—$MR = 9.06$; Photo 3—$MR = 10.06$). Successful and unsuccessful upper secondary school students differentiated between themselves in VCs on the 3D dynamic SMR ($U = 38.000$; $p = .046$; $\eta^2 = .095$). Successful students ($MR = 13.23$) had higher mean rank values than unsuccessful students did ($MR = 20.57$).

Between groups of successful and unsuccessful university students, no statistically significant differences were detected.

The results of the research showed statistically significant differences between the VCs: (1) of unsuccessful primary and upper secondary school students and (2) of primary and university students only on the AOIs representing the macroscopic level (Photos 1, 2 and 3).

Among the successful and unsuccessful students, the highest values of TFDs were on the AOI representing the 3D dynamic SMR, which is related to the results of the research studies that indicate difficulties in understanding the SMRs representing the states of matter (water) and the transitions between them (Bunce & Gabel, 2002; Devetak, Vogrinc et al., 2009; Kind, 2004; Slapničar et al., 2017) and difficulties in dealing with this area (Goldberg & Kotval, 1999; Holmqvist et al., 2011; Hyönä et al., 2002) or less efficiency in finding the information on the computer screen (Green et al., 2007). With regard to small groups of successful students, we can also connect the research results to the fact that students as a whole were unable to transfer the knowledge gained about the particulate nature of matter to situations of everyday life (frost formation) (Löfgren & Hellden, 2009).

Unsuccessful primary school students had the highest values of VCs at the macroscopic level (Photo 2), while upper secondary school students and university students were at the submicroscopic level. Successful primary and university students had the highest values of VCs on 3D dynamic SMR, while upper secondary students had the same amount of VCs on Photo 2 and 3D dynamic SMR. Based on these results, it is obvious that in most groups, successful and unsuccessful students needed more time to process the submicroscopic level (SMR) information associated with the higher values of VCs (West et al., 2006).

Successful primary school students had lower values of VCs on the AOIs at the macroscopic level (Photos 2 and 3) than unsuccessful students did. Successful upper secondary school students had lower values of VCs on the AOIs at the macroscopic level (Photos 2 and 3) on 3D dynamic SMR than unsuccessful students did. These results indicate that primary school students are more focused on the macroscopic level when they are unsuccessful in solving problems, while successful upper secondary school students use more of the submicroscopic level representation than unsuccessful students did.

Moreover, primary school students are more familiar with the macroscopic level of representation, which is a consequence of the fact that, due to the Slovenian science curriculum, primary school teachers in the sixth year introduce matter at the submicroscopic level with the help of static 2D one-particle SMRs and that more complex 3D dynamic SMR s are applied in chemistry and physics lessons in the classroom and used in further science lessons (Bačnik et al., 2009, 2011; Planinšič et al., 2009; Verovnik et al., 2011).

Average Pupil Size in Relation to Cognitive Load

The second research question concerned differences between successful and unsuccessful students in average pupil size and the ability of this variable to indicate cognitive load. Areas with higher *Mdns* indicate the AOIs of the problem that are more difficult to process.

Table 6.6 shows the values of average pupil size on different AOIs for all age groups of successful students.

For all three groups of successful students, the *Mdns* are similar to the average pupil size on all AOIs. For primary school students, the *Mdns* on Photo 1 are the highest, while for upper secondary and university students, the *Mdns* on Photos 2, 3 and the 3D dynamic SMR are highest.

Table 6.7 shows the average pupil size values for all age groups of unsuccessful students on the AOIs at the macroscopic and submicroscopic levels.

From Table 6.7, it is evident that the AOIs on the macroscopic level (Photo 3) have the highest value of average pupil size for the unsuccessful upper secondary school students, primary school students for Photo 2 and 3D SMR and university students for the 3D dynamic SMR.

The Kruskal-Wallis test and post hoc test showed statistically significant differences between unsuccessful primary school and university students in average pupil

Table 6.6 Average pupil size on different AOIs in context-based problem related to sublimation of water in all age groups of successful students

AOI	Primary school students (Group 1)		Upper secondary school students (Group 2)		University students (Group 3)	
	Mdn [mm]	*IQR* [mm]	*Mdn* [mm]	*IQR* [mm]	*Mdn* [mm]	*IQR* [mm]
Photo 1	1.8	.9	1.8	.7	1.5	.4
Photo 2	1.7	.5	1.9	.7	1.6	.5
Photo 3	1.6	1.0	1.9	.6	1.6	.5
3D SMR	1.7	1.0	1.9	.6	1.6	.7

Table 6.7 Average pupil size on different AOIs in context-based problem related to sublimation of water in all age groups of unsuccessful students

AOI	Primary school students (Group 1) Mdn [mm]	IQR [mm]	Upper secondary school students (Group 2) Mdn [mm]	IQR [mm]	University students (Group 3) Mdn [mm]	IQR [mm]	Kruskal-Wallis test χ^2	p	η^2
Photo 1	1.9	.8	1.8	1.1	1.3	.7	6.930	.031[a]	.097
Photo 2	2.0	.8	1.8	.9	1.3	.5	8.371	.015[b]	.116
Photo 3	1.9	.7	2.0	.9	1.3	.7	8.366	.015[c]	.116
3D SMR	2.0	.8	1.8	1.0	1.4	.5	8.125	.017[d]	.111

Results of post hoc Dunn's tests
[a]Group 1–2: $p = .150$; group 2–3: $p = 1.000$; group 1–3: $p = .047$; 1 = 2, 2 = 3, 1 > 3
[b]Group 1–2: $p = .065$; group 2–3: $p = 1.000$; group 1–3: $p = .028$; 1 = 2, 2 = 3, 1 > 3
[c]Group 1–2: $p = .101$; group 2–3: $p = 1.000$; group 1–3: $p = .021$; 1 = 2, 2 = 3, 1 > 3
[d]Group 1–2: $p = .084$; group 2–3: $p = 1.000$; group 1–3: $p = .028$; 1 = 2, 2 = 3, 1 > 3

size on all of the AOIs; therefore, the primary school students had higher values of mean ranks. The size effects of Kruskal-Wallis tests were medium.

Mann-Whitney U tests did not show any statistically significant differences between successful and unsuccessful students in average pupil size ($p > .05$ in all cases).

From the results presented, it can be seen that unsuccessful primary school students, who could be regarded as novices had higher average pupil size values on all AOIs than university students did, who could be considered experts, which points to a higher effort in information processing by novices. It is interesting to note that, in the groups of successful and unsuccessful students, the average pupil size values did not differ; thus, the success in problem-solving is not related to the average pupil size values.

Only successful and unsuccessful university students had the highest average pupil size values on the 3D dynamic SMR, even though the TFD values were the highest for all groups of successful and unsuccessful students on this AOI, indicating difficulty in processing this area. Nevertheless, this suggests that pupil dilation is not an accurate indicator of cognitive load during problem-solving, which is related to the results of the research (Schultheis & Jameson, 2004). Nor can this variable be used to define task difficulty elements.

Conclusions and Implications

This research aimed to investigate differences in visual attention (VCs and TFDs) and cognitive load (average pupil size) in different AOIs (macroscopic and submicroscopic representations) between successful and unsuccessful students in three age groups. Because of the relationship between the above variables that might influence the occurrence of differences between the age groups of successful and unsuccessful students, two sets of conclusions are presented.

Differences in Attention Allocation on the AOIs

The first set of results represents the differences in attention allocation to the selected AOIs between successful and unsuccessful students in different age groups and within the groups of successful or unsuccessful students. All age groups of successful and unsuccessful students spent most of their time observing the 3D dynamic SMR, indicating difficulties in information processing and consequently even comprehension problems.

Unsuccessful primary school and university students differ in the time they spend on the macroscopic representation of frost on grass and magnified frost on grass. The differences are also evident when looking at the time spent on the macroscopic representation of frost on grass between successful and unsuccessful primary school students.

Successful primary school and university students most often visit the AOI with the 3D dynamic SMR, in which the context-based problem is solved, while upper secondary school students similarly often visit the AOI with the macroscopic representation of frost on grass and the 3D dynamic SMR. The unsuccessful students in primary school had the highest values of visits on the macroscopic representation of frost on grass, while the other two age groups had the highest values of visits on 3D dynamic SMR. The visits are not related to the time spent at specific AOIs, because for successful and unsuccessful students the time spent at the AOI was highest for the 3D dynamic SMR in all age groups. From the results of the number of visits for unsuccessful students, it can be concluded that primary students process information at the macroscopic rather than submicroscopic level, while other unsuccessful students process information at the submicroscopic level (higher value of visits).

Unsuccessful primary and secondary school students (AOIs—the macroscopic representation of grass and frost on grass) and primary and university students (AOI—magnified frost on grass) differed in the number of visits. Between groups of successful and unsuccessful primary students, there are differences in visits in the macroscopic representation of frost on grass and magnified frost on grass, while between successful and unsuccessful upper secondary students there are differences in the 3D dynamic SMR.

The unsuccessful primary students had the highest number of visits at the macroscopic level (macroscopic representation of frost on grass), while upper secondary school and university students had the highest number of visits at the submicroscopic level.

Differences in Average Pupil Size and Cognitive Load of Defining the Problem

The second set of results show differences in average pupil size between or within groups of successful and unsuccessful students.

Unsuccessful primary school and university students differ in average pupil size on all AOIs (macroscopic and submicroscopic representations). The average pupil size was higher for the primary school students, indicating a higher effort in information processing.

Successful and unsuccessful university students had the highest values of average pupil size only on the 3D dynamic SMR, while other age groups of successful and unsuccessful students had the highest values on different AOIs, which suggests that, due to the similar values of average pupil size for different AOIs, pupil dilation, in this case, is not relevant for the definition of cognitive load during problem-solving.

It was identified that there was no difference in average pupil size between groups of successful and unsuccessful students, which means that the success of problem-solving in this case is not related to average pupil size.

Limitations of the Research

The research limitation is a small sample of participants in each age group.

Screen images should be developed in such a way that clearly defined eye-tracking measurements are guaranteed. It is especially important to minimise the number of variables on a specific 3D dynamic SMR and to focus on the number of representations at the macroscopic level. The triple nature of chemical concepts should be used for context-based problems, if applicable. In this research, the context-based problem was designed to test the understanding of the macroscopic and submicroscopic levels, while the symbolic level was not necessary.

Implications for the Educational Process

Based on the research results, it is helpful for the teacher to identify the background of chemistry knowledge of the students when teaching the selected learning content. The results obtained help the teacher to form heterogeneous groups of students and to adapt the teaching to specific content for students with different pre-knowledge.

When teaching phase transitions, it is important to stress that when teaching content with 3D dynamic presentations, it is essential to choose the appropriate process photo at the macroscopic level.

Further Research

Further research is needed to investigate how students solve context-based problems, including other transitions between the states of matter and the triple nature of chemical concepts, in order to explore students' full understanding of chemical concepts. The research should include a larger sample of groups to define the differences between them in the TFD and VC values in various context-based problems related to working memory capacity to determine the relationship between these two variables. Average pupil size should be studied during problem-solving to identify the difficulty of a particular problem or its elements.

It would be interesting to obtain insight into the impact of motivation, formal-reasoning thinking and visualisation abilities on problem-solving.

Acknowledgements This research was supported by the project Explaining Effective and Efficient Problem Solving of the Triplet Relationship in Science Concepts Representations (J5-6814), financed by the Slovenian Research Agency (ARRS).

References

Avramiotis, S., & Tsaparlis, G. (2013). Using computer simulations in chemistry problem solving. *Chemistry Education Research and Practice, 14*(3), 297–311.

Bačnik, A., Bukovec, N., Poberžnik, A., Požek Novak, T., Keuc Z., Popič, H., & Vrtačnik M. (2009). *Učni načrt, Program srednja šola, Kemija: gimnazija: klasična, strokovna gimnazija* [Curriculum, program of secondary school, chemistry: gymnasium: classical, professional gymnasium]. Ljubljana: National Education Institute Slovenia.

Bačnik, A., Bukovec, N., Vrtačnik, M., Poberžnik, A., Križaj, M., Stefanovik, V., Sotlar, K., Dražumerič, S., & Preskar, S. (2011). *Učni načrt, Program osnovna šola, Kemija* [Curriculum, program of primary school, chemistry]. Ljubljana: National Education Institute Slovenia.

Bassok, M. (1990). Transfer of domain-specific problem solving procedures. *Journal of Experimental Psychology. Learning, Memory, and Cognition, 16*(3), 522–533.

Beatty, J., & Lucero-Wagoner, B. (2000). The pupillary system. *Handbook of psychophysiology* (2nd ed., pp. 142–162). New York: Cambridge University Press.

Bunce, D. M., & Gabel, D. (2002). Differential effects in the achievement of males and females of teaching the particulate nature of chemistry. *Journal of Research in Science Teaching, 39*(10), 911–972.

Cohen, J. (1988). *Statistical power analysis for the behavioral sciences* (2nd ed.). Hillsdale: Lawrence Erlbaum Associates.

Cullipher, S., & VandenPlas, J. R. (2018). Using fixations to measure attention in eye tracking for the chemistry education researcher. In J. R. VadenPlas, S. J. R. Hansen, & S. Cullipher (Eds.), *Eye tracking for the chemistry education researcher* (pp. 53–72). Washington, DC: American Chemical Society.

Devetak, I. (2012). *Zagotavljanje kakovostnega znanja naravoslovja s pomočjo submikroreprezentacij, Analiza ključnih dejavnikov zagotavljanja kakovosti znanja v vzgojno – izobraževalnem sistemu* [The analysis of the key factors in ensuring the quality of knowledge in educational system]. Ljubljana: Faculty of Education, University of Ljubljana.

Devetak, I., Drofenik Lorber, E., Juriševič, M., & Glažar, S. A. (2009). Comparing Slovenian year 8 and year 9 elementary school pupils' knowledge of electrolyte chemistry and their intrinsic motivation. *Chemistry Education Research and Practice, 10*(4), 281–290.

Devetak, I., & Glažar, S. A. (2010). The influence of 16-year-old students' gender, mental abilities, and motivation on their reading and drawing submicrorepresentations achievements. *International Journal of Science Education, 32*(12), 1561–1593.

Devetak, I., Vogrinc, J., & Glažar, S. A. (2009). Assessing 16-year-old students' understanding of aqueous solution at submicroscopic level. *Research in Science Education, 39*(2), 157–179.

Ferk Savec, V., Hrast, Š., Devetak, I., & Torkar, G. (2016). Beyond the use of an explanatory key accompanying submicroscopic representations. *Acta Chimica Slovenica, 63*(4), 864–873.

Gegenfurtner, A., Lehtinen, E., & Saljo, R. (2011). Expertise differences in the comprehension of visualisations: A meta-analysis of the eye-tracking research in professional domains. *Educational Psychology Review, 23*(2), 523–552.

Goldberg, J. H., & Kotval, X. P. (1999). Computer interface evaluation using eye movements: Methods and constructs. *International Journal of Industrial Ergonomics, 24*(2), 631–645.

Green, H. J., Lemaire, P., & Dufau, S. (2007). Eye movement correlates of younger and older adults' strategies addition. *Acta Psychologica, 125*(12), 257–278.

Holmqvist, K., Nyström, M., Andersson, R., Dewhurst, R., Jarodzka, H., & Van de Weijer, J. (Eds.). (2011). *Eye tracking: A comprehensive guide to methods and measures*. Oxford: Oxford University Press.

Hyönä, J., Lorch, R. F., & Kaakinen, J. K. (2002). Individual differences in reading to summarise expository text: Evidence from eye fixation patterns. *Journal of Education Psychology, 94*(1), 44–55.

Hyönä, J., Lorch, R. F., & Rinck, M. (2003). Eye movement measures to study global text processing. In *The mind's eye, cognitive and applied aspects of eye movement research* (pp. 313–334). Elsevier: North Holland.

Johnstone, A. H. (1982). Macro- and micro-chemistry. *School Science Review, 64*(227), 377–379.

Johnstone, A. H. (1991). Why is science difficult to learn? Things are seldom what they seem. *Journal of Computer Assisted Learning, 7*(2), 75–83.

Johnstone, A. H. (2001). Teaching of chemistry-logical or psychological? *Chemical Education: Research and Practice in Europe, 1*(1), 9–15.

Johnstone, A. H., & El-Banna, H. (1986). Capacities, demands and processes—A predictive model for science education. *Education in Chemistry, 23*(3), 80–84.

Just, M. A., & Carpenter, P. A. (1980). A theory of reading: From eye fixations to comprehension. *Psychological Review, 87*(4), 329–354.

Just, M. A., Carpenter, P. A., & Miyake, A. (2003). Neuroindices of cognitive workload: Neuroimaging, pupillometric and event-related potential studies of brain work. *Theoretical Issues in Ergonomics Science, 4*(1), 56–88.

Kelly, R. M., & Jones, L. L. (2008). Investigating students' ability to transfer ideas learned from molecular animations of the dissolution process. *Journal of Chemical Education, 85*(2), 303–309.

Kind, V. (2004). *Beyond appearances: Students' misconceptions about basic chemical ideas* (2nd ed.). Durham: Durham University, School of Education.

Koning, B. B., Tabbers, H. K., Rikers, R. M. J. P., & Paas, F. (2009). Towards a framework for attention cueing in instructional animations: Guidelines for research and design. *Educational Psychology Review, 21*(3), 113–140.

Levy, S. T., & Wilinsky, U. (2009). Crossing levels and representations: The connected chemistry (CC1) curriculum. *Journal of Science Education and Technology, 18*(3), 224–242.

Löfgren, L., & Hellden, G. (2009). A longitudinal study showing how students use a molecule concept when explaining everyday situations. *International Journal of Science Education, 31*(4), 1631–1655.

Mahaffy, P. (2004). The future shape of chemistry education. *Chemistry Education Research and Practice, 5*(3), 229–245.

Pallant, J. (2011). *SPSS survival manual: A step by step guide to data analysis using SPSS* (4th ed.). Crows Nest: Allen & Unwin.

Parchmann, I., Blonder, R., & Broman, K. (2017). Context-based chemistry learning: The relevance of chemistry for citizenship and responsible research and innovation (RRI). In L. Leite, L. Dourado, A. S. Afonso, & S. Morgado (Eds.), *Contextualising teaching to improve learning the case of science and geography* (pp. 25–39). New York: Nova Science Publishers.

Pavlin, J., Glažar, S. A., Slapničar, M., & Devetak, I. (2019). The impact of students' educational background, interest in learning, formal reasoning and visualisation abilities on gas context-based exercises achievements with submicro-animations. *Chemistry Education Research and Practice, 20*(3), 633–649.

Phillips, L. M., Norris, S. P., & Macnab, J. S. (2010). *Visualisation in mathematics, reading and science education*. Dordrecht: Springer.

Planinšič, G., Belina, R., Kukman, I., & Cvahte, M. (2009). *Učni načrt, Program srednja šola, Fizika: gimnazija: klasična, strokovna gimnazija* [Curriculum, Program of secondary school, physics: Gymnasium: Classical, professional gymnasium]. Ljubljana: National Education Institute Slovenia.

Rayner, K. (2009). Eye movements and attention in reading, scene perception, and visual search. *The Quarterly Journal of Experimental Psyhology, 62*(8), 1457–1506.

Russell, J., Kozma, R., Jones, T., Wykoff, J., Marx, N., & Davis, J. (1997). Use of simultaneous-synchronised macroscopic, microscopic, and symbolic representations to enhance the teaching and learning of chemical concepts. *Journal of Chemical Education, 74*(3), 330–334.

Schultheis, H., & Jameson, A. (2004). Assessing cognitive load in adaptive hypermedia systems: Physiological and behavioral methods. In P. M. E. De Bra & W. Nejdl (Eds.), *Adaptive hypermedia and adaptive web-based systems* (pp. 225–234). Berlin: Springer.

Slapničar, M., Devetak, I., Glažar, S. A., & Pavlin, J. (2017). Identification of the understanding of the states of water and air among Slovenian students aged 12, 14 and 16 years through solving authentic exercises. *Journal of Baltic Science Education, 16*(3), 308–323.

Slapničar, M., Tompa, V., Glažar, S. A., & Devetak, I. (2018). Fourteen-year-old students' misconceptions regarding the sub-micro and symbolic levels of specific chemical concepts. *Journal of Baltic Science Education, 17*(4), 620–632.

Taber, K. S. (2013). Revisiting the chemistry triplet: Drawing upon the nature of chemical knowledge and the psychology of learning to inform chemistry education. *Chemistry Education Research and Practice, 14*(2), 156–168.

Tien, L. T., Teichert, M. A., & Rickey, D. (2007). Effectiveness of a MORE laboratory module in prompting students to revise their molecular-level ideas about solutions. *Journal of Chemical Education, 84*(1), 175–180.

Torkar, G., Veldin, M., Glažar, S. A., & Podlesek, A. (2018). Why do plants wilt? Investigating students' understanding of water balance in plants with external representations at the macroscopic and submicroscopic levels. *Eurasia Journal of Mathematic Science Technology & Education, 14*(6), 2265–2276.

Tsai, M., Hou, H., Lai, M., Liu, W., & Yang, F. (2012). Visual attention for solving multiple-choice science problem: An eye-tracking analysis. *Computer Education, 58*(4), 375–385.

Verovnik, I., Bajc, J., Beznec, B., Božič, S., Brdar, U. V., Cvahte, M., Gerlič, I., & Munih S. (2011). *Učni načrt, Program osnovna šola, Fizika* [Curriculum. Program of primary school. Physics]. Ljubljana: national education institute Slovenia.

West, J. M., Haake, A. R., Rozanski, E. P., & Karn, K. S. (2006). *EyePatterns: Software for identifying patterns and similarities across fixation sequences.* Paper presented at the Proceedings of the 2006 symposium on eye tracking research & applications (pp. 149–154). New York: ACM Press.

Wu, H. K., Krajcik, J. S., & Soloway, E. (2001). Promoting understanding of chemical representations: Students' use of a visualisation tool in the classroom. *Journal of Research in Science Teaching, 38*(7), 821–832.

Wu, H. K., & Shah, P. (2004). Exploring visuospatial thinking in learning. *Science Education, 88*(3), 465–492.

Miha Slapničar Ph.D. student, is a teaching assistant of chemical education at University of Ljubljana, Faculty of Education, Department of Biology, Chemistry and Home Economics. His research focuses on students' redox reaction comprehension in connection with triple nature of chemical concepts using eye-tracking technology, experimental work in the field of general, inorganic and organic chemistry, spectroscopy aspects of the chemistry of natural compounds, and the chemical knowledge assessment.

Valerija Tompa is a teacher of chemistry and home economics at the primary school in Ljubljana and project worker at the University of Ljubljana, Faculty of Education, Slovenia. Her research focuses on how students understand certain chemical concepts and how the eye-tracking measurements identify information processing and visual attention of students when solving science problem tasks.

Iztok Devetak, Ph.D. is a Professor of Chemical Education at University of Ljubljana, Faculty of Education, Slovenia. His research focuses on how students, from elementary school to university, learn chemistry at macro-, submicro- and symbolic level, how chemistry in context and active learning approaches stimulate learning, using eye-tracking technology in explaining science learning, aspects of environmental chemistry education, developing teachers' health-managing competences etc. He has been involved in research projects in the field of science education and he was the national coordinator of PROFILES project (7th Framework Program) for 4.5 years. He co-edited a Springer monograph about active learning approaches in chemistry. He (co)authored chapters in international books (published by Springer, American Chemical Society, Routledge…) and published papers in respected journals (altogether about 400 different publications). He was a Fulbright scholar in 2009. He is a member of ESERA (European Science Education Research Association) and Vice-chair for Eastern Europe of EuChemMS DivCEd (European Association for Chemical and Molecular Sciences Division of Chemical Education). He is a Chair of Chemical Education Division in Slovenian Chemical Society and president of the national Subject Testing Committee for chemistry in lower secondary school. Dr. Devetak is Editor-in-Chief of CEPS Journal and editorial board member of respected journals, such as Chemistry Education Research and Practice, International Journal of Environmental and Science Education, and Eurasian Journal of Physics and Chemistry Education.

Saša Aleksij Glažar, Ph.D. is a Professor Emeritus at University of Ljubljana, Slovenia. His research is focussed mainly on the development of methodology for the organisation of chemical data into networks of knowledge and building chemical relational information systems to be applied in the transfer of knowledge in education and industrial development. Saša A. Glažar is involved in defining, classifying and categorisation of science concepts in building relational systems for various levels of education, structuring information into knowledge maps, i.e. hierarchically built relational systems in science education. These systems have been tested through evaluation of knowledge of students at various levels, thus identifying misconceptions. The same approach was also applied in designing study programmes for all levels of education. New research area of S. A. Glažar is aimed at developing teaching methods based on macroscopic observation and linking it to the microscopic and symbolic interpretations of science phenomena, it is also aimed at implementation of new approaches using computer technology to develop motivation for learning and guidance in the science professions. As an expert, he participates in the development and evaluation of science curricula and in international studies (TIMSS, PISA).

Jerneja Pavlin, Ph.D. is an Assistant Professor of Physics Education at University of Ljubljana, Faculty of Education, Slovenia. Her research focuses on the investigation of different approaches to physics and science teaching (e.g. using 3D sub-microscopic representations, didactic games, outdoor teaching and learning, peer instruction etc.), aspects of scientific literacy and the introduction of contemporary science into physics teaching, from the development and optimization of experiments in the field of modern materials to the evaluation of the developed of learning modules, etc. She was involved in several research projects in the field of science education. She is a member of GIREP (International Research Group on Physics Teaching) and ESERA (European Science Education Research Association). In total she has published about 150 different publications. Dr. Pavlin is a member of the editorial board of CEPS Journal.

Chapter 7
Using an Eye-Tracking Approach to Explain Students' Achievements in Solving a Task About Combustion by Applying the Chemistry Triplet

Iztok Devetak

Introduction

The application of the macro, submicro and symbolic levels of chemical concept representations (chemistry triplet) at all levels of education is an essential part of teaching and learning chemistry. It is necessary to understand how students can translate the chemistry triplet in solving specific problems. Chemical reaction is one of the fundamental concepts in chemistry education and combustion is a specific example of it. Eye-tracking technology can provide opportunities to monitor cognitive processes based on positive correlations between eye movements and the individual's cognitive process during a specific task. The duration and frequency of gaze fixations are related to the ongoing mental processes associated with fixed information. Research also shows that students who choose an inaccurate animation of a chemical reaction are often attracted by a model that is easier to explain and fits their understanding of the reaction equations.

This chapter aimed to examine students' performance in solving a chemistry context-based task about chemical reactions, more specifically about the combustion of methane. The task displayed on the computer screen included a photograph of the methane combustion, an animated 3D submicron representation (SMR) and chemical equations.

I. Devetak (✉)
Faculty of Education, University of Ljubljana, Ljubljana, Slovenia
e-mail: iztok.devetak@pef.uni-lj.si

Understanding Chemical Reaction in the Context of the Chemistry Triplet

Chemical reaction is one of the basic concepts in chemistry education, and combustion is a specific example of it. However, the complexity of teaching and learning chemistry can be explained by the presentation of the concepts on three levels: the macroscopic, the (sub)microscopic and the symbolic level, which could be imagined as the corners of a triangle in which no form of presentation is superior to the others, but rather complements each other (Johnstone, 1982, 1991). The original triangular model, which represents the chemistry triplet, has been further established, and a teaching and learning model for chemistry has been developed (see Fig. 7.1).

Chemical phenomena (macroscopic level) can be explained on the submicroscopic or particulate level; this is where chemistry shows its complexity, which must be dealt with in the learning material if teachers attempt different visualisation methods and adequate language in their lessons (without confusing the macro, submicro and symbolic levels of concepts), and the whole setting of the classroom situation should be situated in the social context (adequate teacher–student and student–student interactions) in the classroom. During this process, students should develop a certain

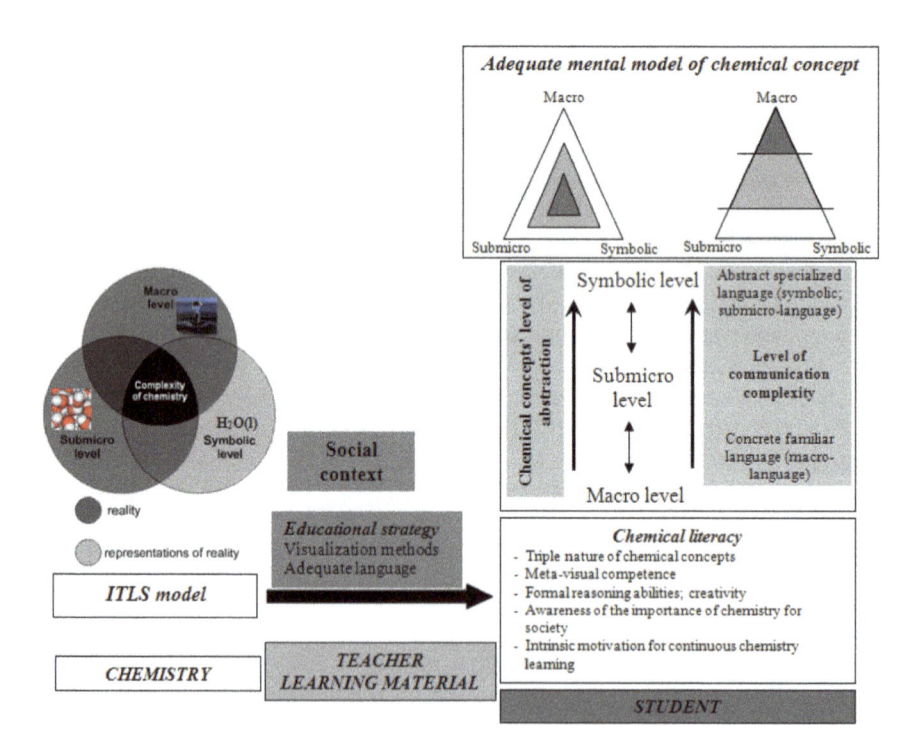

Fig. 7.1 Teaching and learning chemistry model (Adapted from Devetak & Glažar, 2014a; Chittleborough, 2014)

level of chemical literacy, understanding the specific level of complexity of chemical concepts. In this way, a suitable mental model can be created, representing an expanding triangle or a rising iceberg, depending on the students' level of education (Chittleborough, 2014). Submicro representations (SMRs) can be used to represent chemical concepts at the particle level, which can be represented as static or dynamic representations (Devetak & Glažar, 2010). For this study, dynamic 3D SMRs of chemical change (methane combustion) were generated. These SMRs could be translated into established symbols (symbolic level of concept representation), such as symbols of elements, formulas and equations, mathematical equations and various graphical and schematic representations (Levy & Wilinsky, 2009). Johnstone (2001) noted that chemistry concepts are difficult to learn and are often misunderstood by students. The reason for this could be the inability of students to combine the three levels of chemical concepts adequately. Chemistry is inherently conceptual, abstract and difficult to understand without adequate pre-knowledge, which can be understood as a level of chemical literacy adequately embedded in students' long-term memory as specific and adequate mental models.

As Reid (2014) points out, almost by definition, if a concept is to be understood, many facts must be stored simultaneously by the learner in the working memory, whose capacity is very limited. These efforts can lead to learners overloading their working memory capacity (Johnstone & El-Banna, 1986). For this reason, this study also monitored the working memory capacity. Medical research had shown that human memory has several components, one of which is working memory. It is a mental and physical space in the brain in which incoming information is temporarily stored, in which information can be extracted from long-term memory and in which information can be manipulated. The capacity of working memory, the part of the brain that processes information, is quite small. It is known to grow with age, but the final capacity is genetically determined. An average adult (16 years or older) can store seven pieces of information simultaneously, and almost all adults have a capacity between five and nine. It has been found that this part of the brain not only stores information temporarily but is also the location where thinking, understanding information and solving problems takes place, from an educational point of view. It is a "hold thinking" space. However, because it is limited in its capacity and, if there is too much to hold, little space is left for thinking and understanding; therefore, it is a controlling phase for all learning with understanding. Information from the working memory could be transferred to long-term memory, leaving the working space free for further processing (Reid, 2014).

Another important aspect of learning chemistry is the attitude of the learner towards the subject to be learned. Reid (2008) reported that attitudes have a powerful and continuous influence on the learning process. Attitudes can influence the filter for information perception and control which information enters the learner's working memory. The research (Jung & Reid, 2009) also confirms that those students who have shown high working memory capacity are more interested in science, try to understand it, and do not intend to learn it by memorising the concepts, as is the case for students with intermediate and low working memory performance.

However, the attitude of the students is an important aspect that guides the learning of chemistry triplets on their abstract level, but the interest and motivation of the students to learn chemistry is also important. As Taber (2014) points out, even if the students have adequately developed formal reasoning abilities, learning abstract chemical concepts will be fairly minimal if a student can see little sense in a lesson and has no interest in being attentive. Stipek (1998) argues that a higher level of intrinsic motivation for learning a specific content has a positive effect on the students' success in understanding new concepts of the specific content. Various research studies (Devetak & Glažar, 2010, 2014b; Devetak et al., 2009; Juriševič et al., 2008; Patrick et al., 2007; Cavas, 2011) have concluded that the motivation of primary and secondary school and university students to learn chemistry or science, in general, is moderately correlated with their performance in chemistry. Concerning the chemistry triplet, it can be concluded that students tend to be more motivated to learn concepts at the macro level and less motivated to learn at the submicroscopic or symbolic levels (Devetak & Glažar, 2014b; Devetak et al., 2009; Juriševič et al., 2008).

Another aspect of students' cognition of the chemistry triplet can be recognised in their formal reasoning abilities. Students need to acquire formal reasoning abilities to understand 3D animated SMRs correctly and to solve context-based exercises (Pavlin et al., 2019). The fact that some students' formal reasoning abilities never reach the formal operational stage must be taken into account (Labinowicz, 1989). Devetak and Glažar (2010) showed that, at the submicroscopic level, there are statistically significant correlations between formal reasoning abilities and students' chemical knowledge. On average, 28% of the students' achievement variance for items requiring reading 2D-SMRs can be explained by the TOLT (Test of Logic Thinking) score. Similar results were reported by Valanides (1996), who found that formal operation scores correlate significantly with performance in chemistry; Lewis and Lewis (2007) reported that TOLT might predict chemistry exam success, based on a large sample study. In this study, students' formal reasoning abilities were also controlled.

The chemistry triplet can be understood if adequate and specific visualisation methods are used in chemistry teaching. The use of particulate representations plays an essential role in chemistry teaching, as they can enable students to visualise phenomena that cannot be directly observed due to the size of the particles (e.g. atoms, ions, molecules and subatomic particles) (Phillips et al., 2010). For this chapter, visualisation abilities, as proposed by Gilbert (2004), are described as students' abilities to recognise and manipulate visual objects. Research shows (Gilbert, 2008) that students have difficulty constructing relevant information from dynamic visual representations when the represented particles move too fast. Students' visualisation abilities correlate with scientific achievement, but this relationship is influenced by exercise requirements and learning strategies (Hinze et al., 2013). However, Raiyn and Rayan (2015) report that there is a positive correlation between students' visuo-spatial abilities and their problem-solving performance, and good 3D visualisation tools improve students' understanding of molecular structures. Ferk Savec et al. (2005) investigated the usefulness of concrete three-dimensional models, virtual computer-molecular models and their combination as tools for students to solve spatial chemical tasks involving three-dimensional perception, rotation and reflection. Students' perception of the three-dimensional structure was better when a stereo-chemical formula was used in comparison to the formula supported by a computer image. The results suggest that both types of molecular models used as auxiliary tools can facilitate the solution of chemical tasks that require three-dimensional thinking.

Animations improved students' conceptual understanding by helping them to create dynamic mental models of particulate phenomena (Williamson & Abraham, 1995). Research also shows that students who select inaccurate animations of chemical reactions are often attracted to a model that is easier to explain and fits their understanding of the chemical equations. Research has also been conducted to identify misconceptions of chemical reactions at three levels of chemical concept representations (e.g. Barker & Millar, 1999; Chandrasegaran et al., 2007; Bergliot Øyehaug & Holt, 2013; Kelly & Hansen, 2017; Cheng, 2018). Robertson and Shaffer (2014) noted that the literature on students' understanding of combustion has reported that students often rely on a descriptive (what they see during combustion) rather than an explanatory (why something happens) characterisation of combustion. Research indicates (e.g. Meheut et al., 1985; Löfgren & Hellden, 2009; Robertson & Shaffer, 2014) that students usually say that oxygen is necessary for combustion and that water and/or carbon dioxide is produced. However, the most significant problem that can be identified is that they are not specifically associated with the presence of carbon and/or hydrogen in the substance being burned (specific reactants). Robertson and Shaffer (2014) summarised research on students' explanations of certain aspects of combustion. They identified three students' misconceptions regarding the production of water vapour during combustion: (1) the water condenses from the air or environment; (2) the water comes out of the flame and (3) the water is displaced from the wood. However, Barker and Millar (1999) speculated that there are different problems in understanding the change (including combustion) when it is explained as a closed or open system. They concluded that students are inconsistent in including the mass of gas in chemical changes, which leads to confusing thinking. They explained the reasons for these misconceptions by saying that many reactions in open systems involve atmospheric oxygen (such as combustion). Although students experience atmosphere constantly, they do not measure its mass, so the authors speculate that many students omit the mass of oxygen that enters the chemical reaction of gasoline combustion in the reaction of the car engine. Furthermore, Bergliot Øyehaug and Holt (2013) argued that the students' idea that oxygen binds to the reactant during combustion reactions is so strong that they use it as a focal point to integrate other ideas. Research (Christian & Yezierski, 2012) also shows that only 17% of 11- and 12-year-old students correctly estimate the gases released when burning wood.

Eye Movement Measurements in Chemical Education Research

Eye-tracking (eyeT) technology can offer opportunities for monitoring cognitive processes due to the links between eye movements and mental processes of cognition. In addition, to understand these mental processes, the eye movements of individuals can be measured and, after careful consideration, used to interpret processes during task completion, since the direction of the human gaze is closely related to the focus of attention when individuals process the observed visual information (Just & Carpenter, 1980; Hyönä et al., 2003). Research (e.g. Slykhuis et al., 2005; Mason et al., 2013; Havanki & VandenPlas, 2014; Ho et al., 2014; Yen & Yang, 2016; Torkar et al., 2018) has shown that the use of eye-tracking technology in various fields, for example, to study how students process text, data diagrams, relevant photos, explanatory keys, SMRs and similar, can provide relevant information on how students learn and solve different tasks and problems. Havanki and VandenPlas (2014) investigated how the previous knowledge of the students and additional indications in the material guide the allocation of attention, since material with scientific content usually consists of several representations (e.g. text, illustrations). Eye-tracking technologies provide real-time information for understanding students' cognitive activities when processing information encoded in different formats. The combination of quantitative methods and eye-tracking technologies can explain how students interact with different presentation formats in multimedia learning environments (Chuang & Liu, 2012).

Eye-tracking technology can provide measurements of different eye movements. The most common measurements used in chemical education research are total fixation duration (TFDs) and frequency of fixation or fixation counts (FCs) in the specific area of interest (AOI). Other eye movements are the visit count (VC) to the specific area of interest, the average pupil size (APS) obtained by pupillometry, and the spontaneous blink rate (SBR) associated with the ongoing mental processes associated with the processing of information (see Table 7.1).

The eye tracker was used to measure specific eyeT measurements and also other eye movements for each participant in each area of interest, which were determined in a specific material used to collect research data. Fixations occur when the eye is stabilised over an area of interest. Fixations are separated by saccades or "jumps" between fixations. Research suggests that learners fixate on features that are conspicuous, interesting or important through experience (Goldberg & Kotval, 1999; Henderson, 1992). However, fixation counts (FCs) typically indicate the focus of attention, with areas of high fixation count being the most prominent. It can also be understood that the higher the fixation count, the less efficient the viewer's search for information on the computer screen (Chuang & Liu, 2012). When the eyes fixate on an area with high salience, the duration of fixation (absolute TFD) is determined by the time it takes to process the perceptual and cognitive information in the area. For example, the duration of a subject's fixation within an area of interest is a measure of processing difficulty (Goldberg & Kotval, 1999) or the observer has encountered a

Table 7.1 Eye movement measurements and their correlations to cognitive process

Measure at the specific Area Of Interest (AOI)	Description	Correlation to cognitive process
Absolute total fixation durations (TFD)	The total time someone fixes his/her gaze to a specific AOI	Longer fixation durations are indications that the visual and cognitive information requires more complex processing or it is a measure of processing difficulty
Fixation counts (FC)	The number of fixation of someone's gaze to a specific AOI	Typically indicates the focus of attention on the specific AOI and higher number can indicate less efficiency in searching the relevant information or how important the information in that region is
Visit count (VC)	The number of times someone returns to a specific AOI	Higher number of visits can indicate how attractive/useful a particular AOI is
Average pupil size (APS)	The average size of a pupil while someone is looking at a specific AOI	Greater pupil size indicates higher cognitive load during a specific task
Spontaneous eye blink rate (SEBR)	Frequency of eye blinks per minute while someone is looking at a specific AOI	Higher blink rate is related to lower distractibility on tasks that place higher demands on working memory

more difficult element of the task (Chuang & Liu, 2012). Therefore, longer fixation times are usually an indication that the visual information requires more complex processing (Cook et al., 2008).

A visit begins when the eyes first fixate on a particular AOI and ends when it moves away. Visit count (VC) is a measure of how often the person returns to a particular AOI, which can indicate how attractive a particular AOI is, how useful or confusing this AOI can be for the person (Cullipher & VandenPlas, 2018).

Average pupil size (APS) has long been used to identify the cognitive load or mental effort when a person performs various tasks. Pupil dilation increases when the task being performed is cognitively demanding (Beatty, 1982; Dionisio et al., 2001). The larger pupil size when solving specific tasks of wind formation over land or sea suggests that pupil size has been a useful indicator for measuring the cognitive load of learners (Chuang & Liu, 2012). The fact that pupil diameter scales with task requirements makes it a valuable tool for objectively measuring the intensity of cognitive processing in participants of all ages (Eckstein et al., 2017). Research also shows that adults with higher scores on intelligence tests had lower pupil dilation for a range of cognitive tasks (mental multiplication, digit span, sentence comprehension) than those with lower scores (Beatty, 1982), suggesting that more skilled participants exerted less effort to complete the task. This study demonstrates the relationship between pupil response and individual differences in cognitive processing (Ecksteina et al., 2017).

The last eye movement used in this chapter to examine student activity in working memory is the spontaneous eye blink rate (SEBR), which indicates the number of blinks per minute. Spontaneous blinking is one of the most common human movements, with 14,000 spontaneous eye blink per day and an average rate of 14 eye blink per minute when looking straight ahead (Kaminer et al., 2011). Blinking fulfils various functions ranging from maintaining eye health to non-verbal communication. There are three main types of blinking: voluntary, reflexive and spontaneous. Spontaneous blinking occurs without choice and is characterised by a highly synchronised and temporary closing and reopening of the eyelids, a movement that helps to distribute the tear film evenly across the eye (Cruz et al., 2011, cited in Eckstein et al., 2017). It has been suggested that the spontaneous eye blink rate or the frequency with which the eyelids open and close can serve as a non-invasive, indirect measure of dopamine activity in the central nervous system (Eckstein et al., 2017), since dopamine regulates eyelid blinking (Jongkees & Colzato, 2016). Dopamine is an important neurotransmitter involved in learning, working memory functions and goal-oriented behaviour (Kaminer et al., 2011). These functions keep a person focused until a solution is found. Functions related to the frontal lobe, including working memory, are responsible for maintaining a high degree of concentration on a task (Duncan et al., 2000). The basal ganglia, which are connected to the cerebral cortex (Bostan et al., 2013), have a critical function for memory, attention and consciousness. They regulate the release of dopamine in the striatum and thus influence spontaneous eye blinking (Evinger et al., 1993). The basal ganglia also control the input of working memory and have the ability to manipulate information in short-term memory and use it to control actions (Baddeley, 1998). They also filter what enters working memory and modulate its focus by modifying dopamine levels (Schroll & Hamker, 2013), acting as perception filters. In fact, dopamine is released in the prefrontal cortex during higher executive functions such as learning, remembering and recalling memories (Puig et al., 2014), which indicates the importance of dopaminergic mechanisms for cognitive performance (Paprocki & Lenskiy, 2017).

However, eyeT, pupillometry and spontaneous blink data can be misinterpreted. For example, a long fixation on a particular area of interest may be caused by a variety of factors: (1) the participants find that the information in that region is important or relevant to the problem; (2) the material in the area is interesting; (3) the material is difficult or (4) the participants simply stared at the location or object without any associated mental activity (Knoblich et al., 2001).

Research Problem and Research Questions

The main purpose of this chapter was to present the students' achievements in solving the context-based exercise on chemical reaction, more specifically methane combustion and the use of the chemistry triplet in the process. This study aims to show the differences between the students who chose the correct chemical equation to represent methane combustion (G1) and those who had not (G2) in specific eye movement measurements on the specific AOIs (e.g. absolute total fixation duration (TFD), fixation counts (FC), visit count (VC), average pupil size (APS) and spontaneous eye blink rate (SEBR)) and their pre-knowledge, formal reasoning abilities, level of motivation to learn science, working memory capacity and visualisation abilities.

According to the research problem, four research questions can be addressed:

1. Do students who correctly applied the symbolic level of the natural gas combustion process in solving the context-based exercise show statistically significantly different levels of pre-knowledge, motivation to learn the science and some mental abilities (such as formal reasoning abilities, working memory capacity and visualisation abilities) than those who incorrectly applied the symbolic level?
2. Do students who have correctly applied the symbolic level of the natural gas combustion process in solving the context-based exercise describe methane combustion at the submicroscopic level more successfully than those who have misapplied the symbolic level?
3. Do students who correctly applied the symbolic level of the natural gas combustion process in solving the context-based exercise differ statistically significantly in the eye movement measurements on-screen Images 1 and 2 from those who misapplied the symbolic level?
4. Do students who correctly applied the symbolic level of the natural gas combustion process while solving the context-based exercise differ statistically significantly in the eye movement measurements in the macro, submicro and symbolic representations from those students who misapplied the symbolic level?

Method

In this study, a cross-sectional and non-experimental pedagogical research design with a quantitative approach was used.

Participants

The non-random sample of this study comprises 49 participants. All participants came from the Ljubljana Region and participated in this study voluntarily. Secondary school students ($n = 29$) were on average 15.6 years old ($SD = 8.4$ months) and two subject pre-service teachers in bachelor's degree programmes studying biology or physics and chemistry ($n = 20$) at the Faculty of Education of the University of Ljubljana (at the age of $M = 23.18$ years, $SD = 12.0$ months). The consent for secondary school students was obtained from school authorities, teachers and parents, according to the opinion of the Ethics Committee for Pedagogy Research of the Faculty of Education of the University of Ljubljana. All participants had normal or corrected to normal vision, and all were competent readers. To ensure anonymity, each student was assigned a code.

Instruments

Various instruments were used to collect data to answer the research questions, such as pre-knowledge achievement test, eye-tracking apparatus with context-based natural gas combustion exercise on three levels of presentation of chemical concepts (Figs. 7.2 and 7.3 presenting screen images), a Test of Logical Thinking (TOLT) for the students' formal reasoning abilities, a Science Motivation Questionnaire (SMQ) for the students' motivation to learn science, a test of working memory capacity and a Pattern Comparison Test (PCT) for the students' visualisation abilities.

Pre-knowledge Achievement Test

Nine items (with 22 sub-items) pre-knowledge achievement test was developed by the researchers. The construct validity of the instrument was confirmed by three independent experts in science and chemical education. The items include concepts related to chemical and physical changes, as context-based exercises include these concepts.

The Context-Based Natural Gas Combustion Exercise

The selected context-based combustion exercise was presented on a computer screen in the form of a text describing the environment, visualisations of the combustion process of natural gas at all three levels of the presentation of chemical concepts and six tasks on two PowerPoint slides (Figs. 7.2 and 7.3). Both slides showed the same photo of a blue natural gas flame burning, as well as a dynamic 3D animation created specifically for this study. For on-screen Image 1 (Fig. 7.2), the students had to answer three questions and provide a justification for the second question. Three specific areas of interest (AOIs) were defined, representing the combustion of methane, as shown in Fig. 7.3, with green rectangles displayed: macro level, submicroscopic dynamic 3D-SMR, and the symbolic level. The symbolic level was used either for the correct selection of the chemical equation or for the wrong selection of the equations (more specific aspects of selecting different wrong chemical equations were not discussed).

7 Using an Eye-Tracking Approach to Explain …   139

Majority of European homes are heated by burning natural gas. The burning process is presented also at the particulate level with the animation below.

Compare photo and animation of natural gas burning and answer the questions.

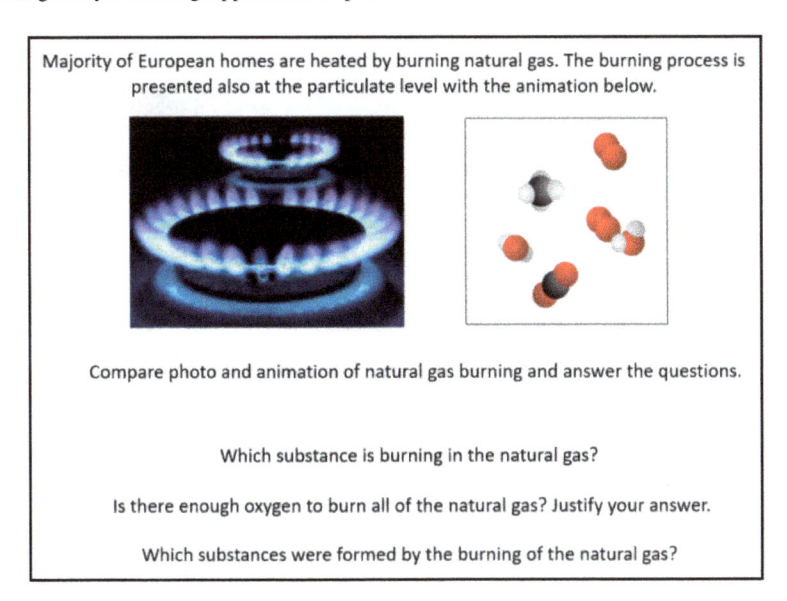

Which substance is burning in the natural gas?

Is there enough oxygen to burn all of the natural gas? Justify your answer.

Which substances were formed by the burning of the natural gas?

Fig. 7.2 Screen Image 1 of the burning contextual task, which includes macro and submicro level of representations of chemical concepts

Majority of European homes are heated by burning natural gas. The burning process is presented also at the particulate level with the animation below.

Macro   Submicro

Choose the correct equation representing the chemical reaction of natural gas burning. Justify your answer.

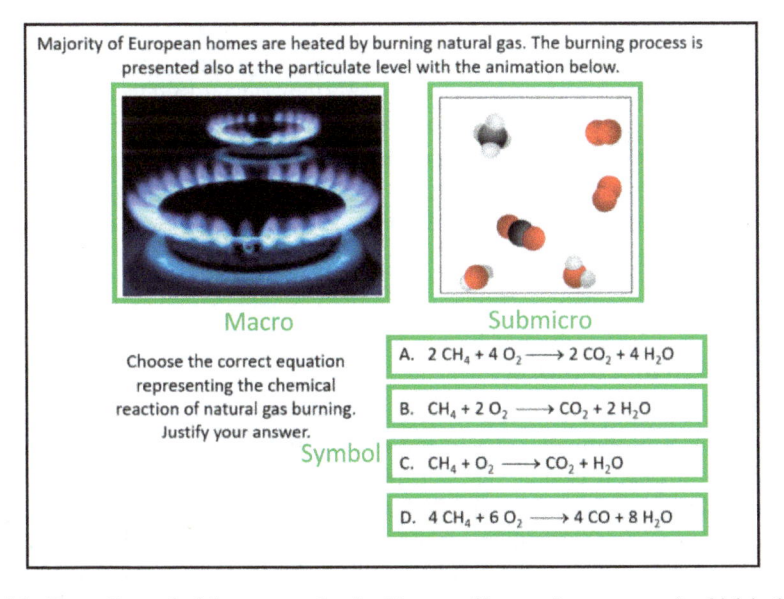

Symbol

A. $2\,CH_4 + 4\,O_2 \longrightarrow 2\,CO_2 + 4\,H_2O$

B. $CH_4 + 2\,O_2 \longrightarrow CO_2 + 2\,H_2O$

C. $CH_4 + O_2 \longrightarrow CO_2 + H_2O$

D. $4\,CH_4 + 6\,O_2 \longrightarrow 4\,CO + 8\,H_2O$

Fig. 7.3 Screen Image 2 of the contextual task with areas of interest (green squares), which includes macro-, submicro and symbolic levels of representations of chemical concepts (equation B is correct)

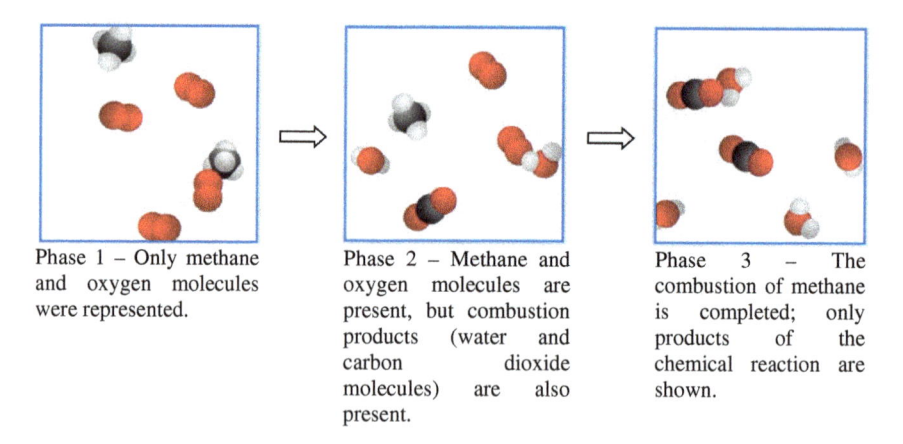

| Phase 1 – Only methane and oxygen molecules were represented. | Phase 2 – Methane and oxygen molecules are present, but combustion products (water and carbon dioxide molecules) are also present. | Phase 3 – The combustion of methane is completed; only products of the chemical reaction are shown. |

Fig. 7.4 Phases of the 3D animation of natural gas combustion on both slides

The particulate level of natural gas combustion is shown in the animation in three stages (see Fig. 7.4). The students had to analyse the course of the animation so that a correct symbolic representation (chemical equation) and a correct justification of the choice were provided.

For this chapter, eye movement measurements were used for screen Image 1 and 2. The screen-based EyeTracker (eyeT) device EyeLink 1000 (35 mm lens, horizontal orientation) and associated software (Experiment Builder to prepare the experiment and connect to EyeLink; Data Viewer to collect data and basic analysis) to measure and analyse the eye movements of the participants while solving context-based exercises was used. Data were collected at 500 Hz in the right eye (monocular data acquisition following corneal reflection and pupil responses) (Torkar et al., 2018).

Test of Logical Thinking

The test of logical thinking (TOLT) test is a multiple-choice test which assesses five skills of logical reasoning relevant for science teaching. The test contains ten problems that require some consideration and the use of problem-solving strategies in different areas (i.e. controlling variables, as well as proportional, correlational, probabilistic and combinatorial reasoning). Participants were given a point for a correct answer and its explanation (in Exercises 1–8) and for the correct combinations and their correct number (in Exercises 9–10). These points were added to an overall score (maximum 10 points), which was used as the main result of the test. The students had 38 min to solve the test (Tobin & Capie, 1984).

Science Motivation Questionnaire

To measure motivation for science in our study, we used an adapted Slovenian version of the self-assessment of science motivation questionnaire (SMQ) (Glynn et al., 2009). The term "science" included chemistry, biology and physics. Participants answered each of the 30 items of the SMQ on a five-point Likert scale ranging from 1-never to 5-always. The questionnaire consists of six five-item scales: (1) intrinsically motivated science learning, (2) extrinsically motivated science learning, (3) relevance of learning science to personal goals, (4) responsibility (self-determination) for learning science, (5) confidence (self-efficacy) in learning science and (6) anxiety about science assessment. The students were given 35 min to complete the questionnaire. We calculated the average answers in all six scales and overall (anxiety was coded in reverse) in order to compare and differentiate all aspects of motivation for science.

Visualisation Ability Test

The pattern-based approach was used to assess students' visual processing (visualisation) skills using the Visual Pattern Comparison Test (VPCT) from the Psychology Experiment Building Language (PEBL) test battery, a series of psychological tests for researchers and clinicians. In the PCT, there were 60 pairs of two grid patterns, 30 of which were equal and 30 different. The participants had to compare the stimuli in pairs and answer as quickly as possible whether the patterns were the same or different. The reaction time and the correctness of the answers were measured. The maximum score was 100 and participants had 15 min to complete the test (Perez et al., 1987).

Test of Working Memory Capacity

A simple digit span task (DST) from the PEBL test battery was used to measure the participants' working memory capacity. This test stores and processes lists of digits that are visually displayed on a screen and spoken through headphones, providing both visual and auditory stimuli. Participants were shown a series of digits, presented one by one on the screen and instructed to enter (recall) items on the next screen. The participants had to recall the numbers in the order in which they were presented. This variable indicates the participants' working memory capacity. However, participants were also instructed to recall the numbers in reverse order, and this mental reversal of the numbers requires both the storage and processing of information in working memory. If the participants were successful in recalling a three-digit list (first step), the list was extended by one digit in the next step (up to a maximum of 10 digits). The number of steps depends on the number of successful attempts until the participant unsuccessfully recall the list of digits without success (Averett, 2017; Croschere et al., 2012).

Research Design

This research was conducted from November 2016 to March 2017. The data were collected in the Department of Psychology Laboratory, at the Faculty of Arts, University of Ljubljana. The participants were informed about the purpose of the study and the methods used before the data collection. The participants had no time limit to solve 11 context-based tasks, but they spend approximately 30 min to solve them. Their eye movements were measured with the eye tracker apparatus. One (natural gas combustion) context-based task was selected for this chapter. The participants sat about 60 cm away from the computer screen on which the context-based tasks were displayed. They placed their heads in a special head-supporting stand to ensure stability and to record the most optimal eye movement data. Prior to recording eye movement data, each participant's gaze had to be calibrated and validated using a nine-point algorithm. Participants solved the context-based exercises using the method (in the same order for all participants). The researcher collected the data by writing down the participants' answers.

TOLT, SMQ, DST and VPCT were applied to all participants in the group before participating in the eye-tracking study. All data were collected in the Slovenian language.

All collected data were statistically processed in SPSS (Statistical Package for the Social Sciences). Basic descriptive statistics (median *Md* and an interquartile range *IQR*) of the numerical variables were determined. EyeLink Data Viewer was used to draw specific heat maps. The text of the heat maps shown in Figs. 7.5 and 7.6 is in Slovenian, since the data were collected in Slovenian, as it is usually appropriate that the data are analysed in the language in which they were collected (Taber, 2018).

The participants were divided into two groups according to their correct choice of the chemical equation: correct chemical equation (G1) and incorrect chemical equation (G2).

The Mann-Whitney U test was used to explain the differences between students who chose the chemical equation of natural gas combustion correctly or incorrectly and their formal reasoning abilities, visualisation abilities, motivation, working memory and eye movement measurements (e.g. absolute total fixation durations (TFD), fixation counts (FC), frequency of returning back (FRB), average pupil size (APS) and spontaneous eye blink frequency rate (sEBR)). The frequency distribution of students' responses to different questions on-screen Fig. 7.1 (Fig. 7.2) on two levels (i.e. the correct response and application of the triple nature of presenting chemical concepts) was also analysed.

7 Using an Eye-Tracking Approach to Explain … 143

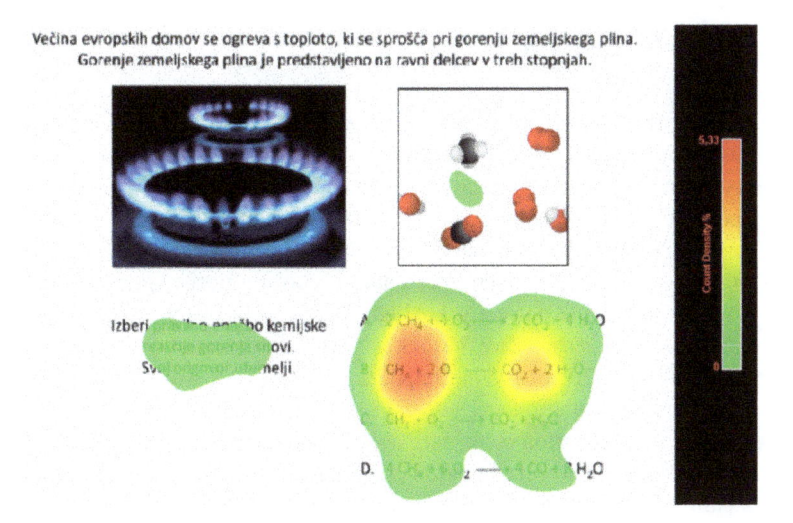

Fig. 7.5 Average heat map for students who correctly solve the context-based natural gas-burning exercise (G1)

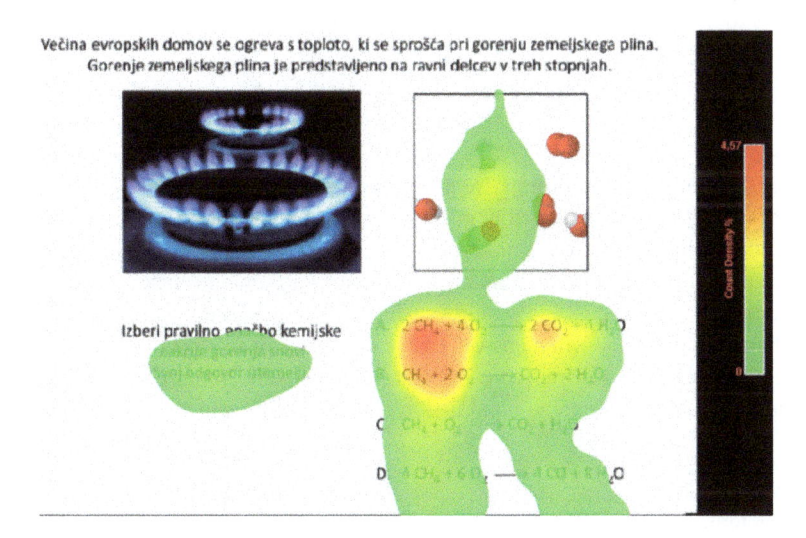

Fig. 7.6 Average heat map for students who incorrectly solve the context-based natural gas-burning exercise (G2)

Statistical hypotheses were tested at an alpha error rate of 5%. To describe whether the effects have a relevant magnitude, the effect size measure eta squared (η^2) was used to describe the strength of a phenomenon. Benchmarks for defining small (.01), medium (.06) and large (.14) effect sizes were provided by Cohen (1988).

Results and Discussion

The purpose of this chapter is to show the differences in the students' achievements and information processing in solving the context-based natural gas combustion exercise, depending on their success in choosing the correct chemical equation: the symbolic level of chemical concepts related to natural gas combustion on the screen image 2; the correct chemical equation (G1) and the wrong chemical equation (G2) (see Fig. 7.3). The results are presented according to the research questions; 65.3% of all participants correctly applied the symbolic level of the natural gas combustion process when solving the context-based natural gas combustion exercise (G1 students), but 34.7% of the participants (G2 students) were not successful in choosing the correct equation.

The first research question deals with control variables so that students' achievements in identifying the correct equation of methane combustion is influenced by the macro picture or dynamic submicro-representation.

The differences in controlling variables, such as pre-knowledge (G1 ($Md = 79.0$; $IQR = 73.0–88.0$); G2 ($Md = 75.0$; $IQR = 62.0–83.5$); Mann-Whitney $U = 191.0$; $p = .088$), motivation (G1 ($Md = 3.7$; $IQR = 3.3–4.0$); G2 ($Md = 3.4$; $IQR = 2.8–3.7$); Mann-Whitney $U = 184.5$; $p = .066$), formal reasoning abilities (G1 ($Md = 8.0$; $IQR = 6.0–7.6$); G2 ($Md = 7.0$; $IQR = 6.0–8.0$); Mann-Whitney $U = 246.0$; $p = .578$) and visualisation abilities (G1 ($Md = 57.0$; $IQR = 55.0–58.0$); G2 ($Md = 57.0$; $IQR = 56.5–59.0$); Mann-Whitney $U = 205.5$; $p = .155$) and working memory capacity (G1 ($Md = 6$; $IQR = 5–6$); G2 ($Md = 5$; $IQR = 4–6$); Mann-Whitney $U = 215.5$; $p = .221$), between the two groups are not significant.

The second research question is about the students' description of methane combustion at the submicroscopic level. Table 7.2 shows the students' (for groups G1 and G2) achievements in answering the questions about screen Image 1 regarding the combustion of natural gas. It also shows what kind of explanations triplet they used regarding the chemistry when answering the questions or justifying their answer.

It can be concluded from Table 7.2 that those students who correctly chose the chemical equation of methane combustion also generally achieved better results in

Table 7.2 Students' achievements in answering the questions for screen Image 1 (see Fig. 7.2) according to the correctly (G1) or incorrectly (G2) selected symbolic representation (chemical equation) on-screen Image 2 (see Fig. 7.3)

Question about screen Image 1		G1 students' answer [%] ($n = 32$)		G2 students' answer [%] ($n = 17$)	
		Correct	Incorrect/No. answer	Correct	Incorrect/No. answer
1. Which substance is burning in the natural gas?		53.1	46.9	47.1	52.9
2. Is there enough oxygen to burn all of the natural gas?		93.8	6.3	94.1	5.9
a. Justify your answer for Question 2.		56.3	43.8	41.2	58.8
3. Which substances were formed by the burning of the natural gas?	CO_2	96.9	3.1	70.6	29.4
	H_2O	100	0	88.2	11.8

answering the questions about screen Image 1. Both groups of students responded similarly, stating that there is enough oxygen to react with methane, but the group that chose the chemical equation correctly better justified its answer. The major difference between the two student groups (G1 and G2) was the prediction that carbon dioxide is the product of this chemical reaction. More than 26% of the students in G1 also correctly predicted both products of the chemical reaction.

Most students (84.4%) who correctly chose the chemical equation used the macro level to explain what the substance that burns in natural gas is. Similar results (82.4%) were obtained by students who did not choose the chemical equations correctly; others did not provide any explanation. Only one student mentioned methane molecules and two used the methane formula to express their reasoning. In explaining whether there is enough oxygen to burn all the natural gas, 65.6% of the students used only macroscopic explanations and 41.2% of those who chose the wrong symbolic representation. In total, 25.1% of the students explained sufficient amount of oxygen using the submicroscopic level or the combination of the submicroscopic and macroscopic levels, and 41.2% of those who chose the wrong chemical equation. Others did not provide any information or explanation. More students explained that the combustion of methane produces carbon dioxide molecules (34.4%) and water molecules (28.1%).

In contrast, no student who was unsuccessful in choosing the correct chemical equation for methane combustion explained that carbon dioxide and water molecules are formed. They provided their explanations at the macro level, stating that carbon dioxide (76.5%) and water (82.4%) are formed as substances. Others used the formula of these substances (CO_2 and H_2O, 11.8 and 5.8%, respectively) or did not provide an explanation.

The third research question relates to the differences in specific eye movement measurements (e.g. absolute total fixation durations (TFD) (in sec.), fixation counts (FC), average pupil size (APS) (in arbitrary units), spontaneous eye blink rate (SEBR)) for screen Images 1 and 2 of the context-based natural gas combustion exercise as between those who chose the chemical equation incorrectly and those who did not.

From Table 7.3, it can be concluded that there were no significant differences between the two groups of students, which may indicate that those students who were more successful in selecting the correct chemical equation were generally (the entire screen Image 1 or 2) less successful in focusing their attention or processing visual information than those who were less successful. Similar results were also obtained by comparing the students in both groups in terms of the cognitive load that the task imposed on the students, which is also consistent with the non-significant difference in blinking rate between the students who were correct in choosing the symbolic level representing methane combustion and those who were not successful.

The final research question focuses on the differences in student eye movement measurements for macro, submicro and symbolic representations between those who correctly and incorrectly chose the chemical equation. A more detailed data analysis of eye movement measurements between the two groups of students is presented in Table 7.4.

Table 7.3 Differences in eye movement measurements between students according to the correctly (G1) or incorrectly (G2) selected symbolic representation (chemical equation) on the screen image 1—SE1 (see Fig. 7.2) and 2—SE2 (see Fig. 7.3)

Eye movement measurement		G1 students' answer [%] ($n = 32$)		G2 students' answer [%] ($n = 17$)		Mann-Whitney U test	
		Mdn	IQR	Mdn	IQR	U	p
TFD	SE1	75.2	65.9–104.4	76.4	49.7–108.6	256.0	.737
	SE2	57.6	4.6–82.4	80.5	56.7–93.7	199.0	.125
FC	SE1	281	250.8–375.8	290.0	214.0–397.5	266.0	.900
	SE2	234	179.3–278.0	251.0	222.5–319.5	212.5	.211
APS	SE1	1643.7	1340.1–2006.9	1416.4	1271.0–1761.1	216.0	.240
	SE2	1482.1	1275.7–1893.4	1330.0	1161.3–1625.6	225.0	.324
SEBR	SE1	28.0	12.0–41.0	20.0	14.5–55.0	265.0	.883
	SE2	19.5	7.0–34.8	17.0	13.5–48.5	263.0	.449

TFD—absolute total fixation durations (in sec.), FC—fixation counts, APS—average pupil size (in arbitrary units), SEBR—spontaneous eye blink rate

In the screen Image 2 (see Fig. 7.3), three significant elements were covered in the analysis of the eye movement measurements of the participating students. The absolute total fixation durations (TFD) indicate that the 3D animation of the chemical change (combustion of methane) requires statistically significantly more complex processing of the presented visual information at the SMR for low-performance students. In contrast, students who correctly selected the chemical equation (underlined symbol B in Table 7.4) also find the correct symbolic representation of methane combustion more complex when processing visual stimuli, which is understandable: they must analyse the chemical equation to ensure that it is correct. The duration of visual information processing is also consistent with the second eye movement measurement, the number of fixation (FC) on the specific area of interest. It is clear that more successful students paid more attention to the correct chemical equation than to the other areas of interest identified by the researchers on the second screen image. Both significant differences in the average total absolute fixation durations and the fixation counts for both AOIs (3D-SMR and correct equation) show medium effect size, which means that about 10% of the variability in the TFD and FC for both AOIs is accounted for successfully selected the correct chemical equation in the context-based natural gas combustion exercise. Those students who chose the wrong chemical equation spent the longest time (longest average TFD and highest average fixation count) on the first (incorrect) chemical equation.

The number of visits (VC) to the specific AOI can be an indication of how attractive or useful a particular AOI is for students in solving a particular task. Students who chose the wrong chemical equation to represent methane combustion visit the 3D animated submicron representation of the chemical reaction statistically significantly more often than students who have chosen the right equation do. A higher number of visits to 3D-SMR by low-performing students may indicate that this animation

Table 7.4 Differences in eye movements between students according to the correctly (G1) or incorrectly (G2) selected symbolic representation (chemical equation) on specific AOIs at the screen image 2 (see Fig. 7.3)

Eye movement measurement		G1 students' answer [%] ($n = 32$)		G2 students' answer [%] ($n = 17$)		Mann-Whitney U test		
		Mdn	IQR	Mdn	IQR	U	p	η^2
TFD	Macro	.413	.266–.983	.594	.398–1.458	200.0	.130	
	Submicro	**2.1**	**.518–5.1**	**14.0**	**1.7–27.0**	**163.0**	**.022**	**.107**
	Symbol A	10.3	4.4–18.6	18.3	7.6–24.2	209.0	.186	
	Symbol B	**19.5**	**11.4–26.7**	**5.7**	**3.6–28.1**	**165.0**	**.025**	**.103**
	Symbol C	6.8	3.7–12.6	5.1	1.6–7.6	199.0	.125	
	Symbol D	2.4	1.2–4.5	2.8	1.3–7.2	252.0	.674	
FC	Macro	2.0	1.0–5.0	4.0	2.0–8.0	195.0	.101	
	Submicro	**9.0**	**3.0–12.0**	**50.0**	**4.5–68.5**	**171.5**	**.034**	**.091**
	Symbol A	36.0	21.5–65.3	67.0	31.0–85.0	200.0	.130	
	Symbol B	**63.5**	**47.3–97.3**	**20.0**	**16.5–87.0**	**162.0**	**.021**	**.109**
	Symbol C	24.5	16.3–48.8	18.0	7.5–29.5	183.0	.061	
	Symbol D	10.0	5.3–19.5	13.0	6.0–28.5	244.0	.556	
VC	Macro	2.0	1.0–3.0	3.0	1.0–4.5	197.5	.107	
	Submicro	**5.0**	**2.3–6.8**	**18.0**	**3.0–28.5**	**174.5**	**.040**	**.086**
	Symbol A	12.0	7.0–20.5	19.0	9.0–22.5	205.0	.159	
	Symbol B	**18.0**	**12.0–29.5**	**11.0**	**6.5–21.5**	**167.5**	**.028**	**.098**
	Symbol C	11.5	5.0–17.0	7.0	4.0–11.0	195.0	.105	
	Symbol D	5.0	2.0–8.3	6.0	3.0–7.5	236.5	.451	
APS	Macro	1490.9	1155.4–1817.0	1373.5	1219.2–1508.1	240.0	.612	
	Submicro	1603.0	1221.0–2072.2	1536.6	1272.0–1660.1	245.0	.690	
	Symbol A	1675.8	1252.7–1916.1	1517.9	1250.9–1747.4	233.0	.413	
	Symbol B	1612.6	1324.3–1930.5	1483.8	1235.6–1777.9	241.0	.515	
	Symbol C	1583.5	1189.9–1927.4	1523.7	1216.4–1748.7	259.0	.785	
	Symbol D	1637.9	1264.6–2050.5	1477.9	1256.1–1812.3	249.0	.629	

TFD—absolute total fixation durations, FC—fixation counts, VC—visit count, APS—average pupil size [in arbitrary units]
Underlined alternative indicates a correct symbolic representation
Significant differences are written in bold

was more attractive or useful to them than to high-performing students. However, high-performing students found the correct chemical equation more attractive than students who did not choose the correct equation. For these students, again, the first equation (see above) was most attractive. Both significant differences on average VC show medium effect size, which means that about 9% of the variability in VC 3D-SMR and the correct chemical equation is due to the fact that the correct chemical equation was successfully selected in the context-based natural gas combustion exercise. The final eye movement measurement indicates the cognitive load of the students during a given task, which means that a larger pupil size indicates a higher cognitive load. The results show that both groups of students experienced similar cognitive load during the context-based natural gas combustion exercise. However, when comparing the pupil size, it can be speculated that students who solved the exercise correctly experienced higher cognitive loads, but this is due to the greater effort required to solve the exercise correctly.

These results are also supported by the visual attention distribution (on average) between the two groups of students (see Figs. 7.5 and 7.6).

The average heat map for both groups of students shows that the macro level does not sufficiently draw their attention for them to choose the correct chemical equation; although the flame of methane combustion can indicate the chemical reaction with oxygen, this was obvious to the students after reading the text of the task (Translation: Choose the correct chemical equation representing the burning of the substance) and by looking at (G1) or processing (G2) the visual information represented by the animated 3D submicro-representation. Overall, the students showed little interest in the macro level, which is consistent with other research, as it can be assumed that students are less focused on the macro level when they are familiar with the task context (Chittleborough, 2014). For those students who chose the chemical equation correctly, other levels of conceptual representations (macro and submicro levels) were less relevant. Other studies described similar findings that experts spend less time on information that is irrelevant for the successful solution of the task (Gegenfurtner et al., 2011).

Conclusions

The main purpose of this chapter was to present the students' achievements in solving the context-based exercise on chemical reaction, more specifically, methane combustion and the use of the chemistry triplet in the process. This research aims to determine the differences between the students who chose the correct chemical equation and those who were unsuccessful in certain eye movement measurements in the specific area of interest, on two screen images where the context-based task was presented.

The differences between the two groups of students in pre-knowledge, motivation for learning science, formal reasoning abilities, visualisation abilities and working memory are not significant, which means that these variables do not statistically significantly affect the solving of the context-based combustion exercise, but the ability of the students to solve the exercise correctly may be due to their ability to process and determine relevant information presented in the two screen images and how familiar they were with the task.

Students' explanations, regardless of whether they are attending an upper secondary school or university, remain at the macro level.

It can be summarised that students who were more successful in selecting the correct chemical equation do not focus their attention overall or process visual information displayed on the whole screen Image 1 or 2 as less successful.

From this, it can be concluded that students who chose the correct chemical equation when solving the context-based natural gas combustion exercise spent less time processing 3D animations and photos representing methane combustion. However, more successful students spend more time mentally analysing the correct chemical equation without searching for much information at the macro or submicro levels. They also find it more relevant to the solution of the task, but for both groups of students, the context-based natural gas combustion exercise presents similar cognitive load while solving the task.

Some implications for chemistry teaching can be suggested. Teachers should stimulate students to provide explanations using correct language at the submicroscopic level and describe chemical phenomena not only at the macroscopic level. Poorly performing students should be encouraged to use all three levels of presentation of chemical concepts when attempting to solve the specific chemical exercise or problem successfully. In addition, the results can enable teachers to encourage students to develop successful problem-solving strategies, which means that teachers could focus on the analysis of those textual or visual elements of the exercise or problem that could lead the students to effectively use all three levels of chemical concept representations in finding the right solution.

There are some limitations to the research presented in this chapter. The sample size is relevant for this type of research, but it is difficult to obtain a large number of participants. It is also important to emphasise that students solve different tasks during eyeT measurements (some of which are also used in the chapters of this book), so it can be speculated that some students may not have made enough effort to solve the tasks. The eye-tracking technology can provide useful data, but we should be aware that this is not a standalone research method and that a triangulation of methods should be applied to a similar research design, which may also reduce the possibility of misinterpretation of eyeT, pupillometric and spontaneous blink data, as this can be a major problem when attempting to draw conclusions from eyeT research.

The conclusions of this study also suggest some further research. More attention should be paid to the analysis of eye movement measurements as a function of the abilities of certain students, which may influence learning and solving specific tasks. In addition, complex chemical problems involving the chemistry triplet should be used in similar studies to determine students' strategies for solving these tasks.

Acknowledgements This chapter is the result of a research project "Explaining Effective and Efficient Problem Solving of the Triplet Relationship in Science Concepts Representations" (J5-6814), which was supported by the Slovenian Research Agency (ARRS).

References

Averett, A. A. (2017). *The effects of high intensity interval training on working memory performance in sendentary young adults.* A Master Thesis, Northern Arizona University.

Baddeley, A. (1998). Recent developments in working memory. *Current Opinion in Neurobiology, 8*(2), 234–238.

Barker, V., & Millar, R. (1999). Students' reasoning about chemical reactions: What changes occur during a context-based post-16 chemistry course? *International Journal of Science Education, 21*(6), 645–665.

Beatty, J. (1982). Task-evoked pupillary responses, processing load, and the structure of processing resources. *Psychological Bulletin, 91*(2), 276–292.

Bergliot Øyehaug, A., & Holt, A. (2013). Students' understanding of the nature of matter and chemical reactions—A longitudinal study of conceptual restructuring. *Chemistry Education Research and Practice, 14*(4), 450–467.

Bostan, A. C., Dum, R. P., & Strick, P. L. (2013). Cerebellar networks with the cerebral cortex and basal ganglia. *Trends in Cognitive Sciences, 17*(5), 241–254.

Cavas, P. (2011). Factors affecting the motivation of Turkish primary students for science learning. *Science Education International, 22*(1), 31–42.

Chandrasegaran, A. L., Treagust, D. F., & Mocerino, M. (2007). The development of a two-tier multiple-choice diagnostic instrument for evaluating secondary school students' ability to describe and explain chemical reactions using multiple levels of representation. *Chemistry Education Research and Practice, 8*(3), 293–307.

Cheng, M. M. W. (2018). Students' visualisation of chemical reactions—Insights into the particle model and the atomic model. *Chemistry Education Research and Practice, 19*(1), 227–239.

Chittleborough, G. (2014). The development of theoretical frameworks for understanding the learning of chemistry. In I. Devetak & S. A. Glažar (Eds.), *Learning with understanding in the chemistry classroom* (pp. 25–40). Dordrech: Springer.

Christian, B. N., & Yezierski, E. J. (2012). Development and validation of an instrument to measure student knowledge gains for chemical and physical change for grades 6–8. *Chemistry Education Research and Practice, 13*(3), 384–393.

Chuang, H.-H., & Liu, H.-C. (2012). Effects of different multimedia presentations on viewers' information-processing activities measured by eye-tracking technology. *Journal of Science Education and Technology, 21*(2), 276–286.

Cohen, J. (1988). *Statistical power analysis for the behavioural sciences* (2nd ed.). Hillsdale, NJ: Lawrence Erlbaum Associates.

Cook, M., Carter, G., & Wiebe, E. N. (2008). The interpretation of cellular transport graphics by students with low and high prior knowledge. *International Journal of Science Education, 30*(2), 239–261.

Croschere, J., Dupey, L., Hilliard, M., Koehn, H., & Mayra, K. (2012). *The effects of time of day and practice on cognitive abilities: Forward and backward Corsi block test and digit span* (PEBL Technical Report Series [On-line], #2012–03). http://sites.google.com/site/pebltechnicalreports/home/2012/pebl-technical-report-2012-03.

Cullipher, S., & VandenPlas, J. R. (2018). Using fixations to measure attention. In J. R. VandenPlas, S. J. R. Hansen, & S. Cullipher (Eds.), *Eye tracking for the chemistry education researcher* (pp. 53–72). ACS Symposium Series 1292. Washington, DC: American Chemical Society.

Devetak, I., Drofenik Lorber, E., Juriševič, M., & Glažar, S. A. (2009). Comparing Slovenian year 8 and year 9 elementary school pupils' knowledge of electrolyte chemistry and their intrinsic motivation. *Chemistry Education Research and Practice, 10*(4), 281–290.

Devetak, I., & Glažar, S. A. (2010). The influence of 16-year-old students' gender, mental abilities, and motivation on their reading and drawing submicrorepresentations achievements. *International Journal of Science Eduation, 32*(12), 1561–1593.

Devetak, I., & Glažar, S. A. (2014a). Learning with understanding in the chemistry classroom constructing active learning in chemistry: Concepts, cognition and conceptions. In I. Devetak

& S. A. Glažar (Eds.), *Learning with understanding in the chemistry classroom* (pp. 5–23). Dordrecht: Springer.

Devetak, I., & Glažar, S. A. (2014b). Educational models and differences between groups of 16-year-old students in gender, motivation, and achievements in chemistry. In I. Devetak & S. A. Glažar (Eds.), *Learning with understanding in the chemistry classroom* (pp. 103–126). Dordrech: Springer.

Dionisio, D. P., Granholm, E., Hillix, W. A., & Perrine, W. F. (2001). Differentiation of deception using pupillary responses as an index of cognitive processing. *Psychophysiology, 38*(2), 205–211.

Duncan, J., Seitz, R. J., Kolodny, J., Bor, D., Herzog, H., Ahmed, A., Newell, F. N., & Emslie, H. (2000). A neural basis for general intelligence. *Science, 289*, 457–460.

Eckstein, M. K., Guerra-Carrillo, B., Miller Singley, A. T., & Bunge, S. A. (2017). Beyond eye gaze: What else can eyetracking reveal about cognition and cognitive development? *Developmental Cognitive Neuroscience, 25*, 69–91.

Evinger, C., Basso, M. A., Manning, K. A., Sibony, P. A., Pellegrini, J. J., & Horn, A. K. (1993). A role for the basal ganglia in nicotinic modulation of the blink reflex. *Experimental Brain Research, 92*(3), 507–515.

Ferk Savec, V., Vrtačnik, M., & Gilbert, J. K. (2005). Evaluating the educational value of molecular structure representations. In *Visualisation in science education* (pp. 269–297). Dordrecht, Netherlands: Springer.

Gegenfurtner, A., Lehtinen, E., & Saljo, R. (2011). Expertise differences in the comprehension of visualisations: A meta-analysis of the eye-tracking research in professional domains. *Educational Psychology Review, 23*(2), 523–552.

Gilbert, J. K. (2004). Models and modelling: Routes to more authentic science education. *International Journal of Science and Mathematics Education, 2*(2), 115–130.

Gilbert, J. K. (2008). Visualisation: An emergent field of practice and enquiry in science education. In J. K. Gilbert, M. Reiner & M. Nakhlem (Eds.), *Visualisation: Theory and practice in science education* (pp. 3–24). Dordrecht: Springer.

Glynn, S. M., Taasoobshirazi, G., & Brickman, P. (2009). Science motivation questionnaire: Construct validation with nonscience majors. *Journal of Research in Science Teaching, 46*(2), 127–146.

Goldberg, J. H., & Kotval, X. P. (1999). Computer interface evaluation using eye movements: Methods and constructs. *International Journal of Industrial Ergonomics, 24*(6), 631–645.

Havanki, K. L., & VandenPlas, J. R. (2014). Eye tracking methodology for chemistry education research. In D. M. Bunce & R. S. Cole (Eds.), *Tools of chemistry education research* (pp. 191–218). Washington, DC: American Chemical Society.

Henderson, J. M. (1992). Visual attention and eye movement control during reading and picture viewing. In K. Rayner (Ed.), *Eye movements and visual cognition: Scene perception and reading* (pp. 260–283). New York: Springer-Verlag.

Hinze, S. R., Williamson, V. M., Shultz, M. J., Williamson, K. C., Deslongchamps, G., & Rapp, D. N. (2013). When do spatial abilities support student comprehension of STEM visualisations? *Cognitive Processing, 14*(2), 129–142.

Ho, H. N. J., Tsai, M.-J., Wang, C.-Y., & Tsai, C.-C. (2014). Prior knowledge and online inquiry-based science reading: Evidence from eye tracking. *International Journal of Science and Mathematics Education, 12*(3), 525–554.

Hyönä, J., Lorch, R. F., & Rinck, M. (2003). Eye movement measures to study global text processing. In *The mind's eye, cognitive and applied aspects of eye movement research* (pp. 313–334). Elsevier: North Holland.

Johnstone, A. H. (1982). Macro- and micro-chemistry. *School Science Review, 64*(227), 377–379.

Johnstone, A. H. (1991). Why is science difficult to learn? Things are seldom what they seem. *Journal of Computer Assisted Learning, 7*(2), 75–83.

Johnstone, A. H. (2001). Teaching of chemistry-logical or psychological? *Chemical Education: Research and Practice in Europe, 1*(1), 9–15.

Johnstone, A. H., & El-Banna, H. (1986). Capacities, demands and processes—A predictive model for science education. *Education in Chemistry, 23*(3), 80–84.

Jongkees, B. J., & Colzato, L. S. (2016). Spontaneous eye blink rate as predictor of dopamine-related cognitive function-A review. *Neuroscience and Biobehavioral Review, 71,* 58–82.

Jung, E. S., & Reid, N. (2009). Working memory and attitudes. *Research in Science and Technology Education, 27*(2), 205–224.

Juriševič, M., Devetak, I., Razdevšek Pučko, C., & Glažar, S. A. (2008). Intrinsic motivation of pre-service primary school teachers for learning chemistry in relation to their academic achievement. *International Journal of Science Education, 30*(1), 87–107.

Just, M. A., & Carpenter, P. A. (1980). A theory of reading: From eye fixations to comprehension. *Psychological Review, 87*(4), 329–354.

Kaminer, J., Powers, A. S., Horn, K. G., Hui, C., & Evinger, C. (2011). Characterising the spontaneous blink generator: An animal model. *The Journal of Neuroscience, 31*(31), 11256–11267.

Kelly, R. M., & Hansen, S. J. R. (2017). Exploring the design and use of molecular animations that conflict for understanding chemical reactions. *Quimica Nova, 40*(4), 476–481.

Knoblich, G., Ohlsson, S., & Raney, G. E. (2001). An eye movement study of insight problem solving. *Memory & Cognition, 29*(7), 1000–1009.

Labinowicz, E. (1989). *Izvirni Piaget* [The Piaget Primer: Thinking, learning, teaching]. Ljubljana: Državna založba Slovenije.

Levy, S. T., & Wilinsky, U. (2009). Crossing levels and representations: The connected chemistry (CC1) curriculum. *Journal of Science Education and Technology, 18*(3), 224–242.

Lewis, S., & Lewis, J. (2007). Predicting at-risk students in general chemistry: Comparing formal thought to a general achievement measure. *Chemistry Education Research and Practice, 8*(1), 32–51.

Löfgren, L., & Hellden, G. (2009). A longitudinal study showing how students use a molecule concept when explaining everyday situations. *International Journal of Science Education, 31*(12), 1631–1655.

Mason, L., Pluchino, P., & Tornatora, M. C. (2013). Effects of picture labeling on illustrated science text processing and learning: Evidence from eye movements. *Reading Research Quarterly, 48*(2), 199–214.

Meheut, M., Saltiel, E., & Tiberghien, A. (1985). Pupils' (11–12 year olds) conceptions of combustion. *European Journal of Science Education, 7*(1), 83–93.

Paprocki, R., & Lenskiy, A. (2017). What does eye-blink rate variability dynamics tell us about cognitive performance? *Frontiers in Human Neuroscience, 11,* 620.

Patrick, A. O., Kpangban, E., & Chibueze, O. O. (2007). Motivation effects on test scores of senior secondary school science students. *Studies on Home and Community Science Education, 1*(1), 57–64.

Pavlin, J., Glažar, S. A., Slapničar, M., & Devetak, I. (2019). The impact of students' educational background, interest in learning, formal reasoning and visualisation abilities on gas context-based exercises achievements with submicro-animations. *Chemistry Education Research and practice, 20*(3), 633–649.

Perez, W. A., Masline, P. J., Ramsey, E. G., & Urban, K. E. (1987). *Unified tri-services cognitive performance assessment battery: Review and methodology.* http://www.dtic.mil/cgi-bin/GetTRDoc?Location=U2&doc=GetTRDoc.pdf&AD=ADA181697. Accessed 8 March 2020.

Phillips, L. M., Norris, S. P., & Macnab, J. S. (2010). *Visualisation in mathematics, reading and science education.* Dordrecht: Springer.

Puig, M. V., Rose, J., Schmidt, R., & Freund, N. (2014). Dopamine modulation of learning and memory in the prefrontal cortex: Insights from studies in primates, rodents, and birds. *Frontiers in Neural Circuits, 8,* 93.

Raiyn, J., & Rayan, A. (2015). How chemicals' drawing and modelling improve chemistry teaching in colleges of education. *World Journal of Chemical Education, 3*(1), 1–4.

Reid, N. (2008). A scientific approach to the teaching of chemistry (A Royal Society of Chemistry Nyholm Lecture, 2006-2007). *Chemistry Education Research and Practice, 9*(1), 51–59.

Reid, N. (2014). The learning of chemistry the key role of working memory. In I. Devetak & S. A. Glažar (Eds.), *Learning with understanding in the chemistry classroom* (pp. 77–101). Dordrecht: Springer.

Robertson, A. D., & Shaffer, P. S. (2014). "Combustion always produces carbon dioxide and water": A discussion of university chemistry students' use of rules in place of principles. *Chemistry Education Research and Practice, 15*(4), 763–776.

Schroll, H., & Hamker, F. H. (2013). Computational models of basal-ganglia pathway functions: Focus on functional neuroanatomy. *Frontiers in Systems Neuroscience, 7*, 122.

Slykhuis, D. A., Wiebe, E. N., & Annett, L. A. (2005). Eye-tracking students' attention to PowerPoint photographs in a science education setting. *Journal of Science Education and Technology, 14*(5/6), 509–520.

Stipek, D. (1998). *Motivation to learn: From theory to practice.* Boston: Allyn and Bacon.

Taber, K. S. (2014). Constructing active learning in chemistry: Concepts, cognition and conceptions. In I. Devetak & S. A. Glažar (Eds.), *Learning with understanding in the chemistry classroom* (pp. 5–23). Dordrecht: Springer.

Taber, K. S. (2018). Lost and found in translation: Guidelines for reporting research data in an 'other' language. *Chemistry Education Research and Practice, 19*(3), 646–652.

Tobin, K., & Capie, W. (1984). The test of logical thinking. *Journal of Science and Mathematics Education in Southeast Asia, 7*(1), 5–9.

Torkar, G., Veldin, M., Glažar, S. A., & Podlesek, A. (2018). Why do plants wilt? Investigating students' understanding of water balance in plants with external representations at the macroscopic and submicroscopic levels. *Eurasia Journal of Mathematics, Science and Technology Education, 14*(6), 2265–2276.

Valanides, N. (1996). Formal reasoning and science teaching. *School Science and Mathematics, 96*(2), 99–107.

Williamson, V. M., & Abraham, M. R. (1995). The effects of computer animation on the particulate mental models of college chemistry students. *Journal of Research in Science Teaching, 32*(5), 521–534.

Yen, M. H., & Yang, F. Y. (2016). Methodology and application of eye-tracking techniques in science education. In M. H. Chiu (Ed.), *Science education research and practices in Taiwan* (pp. 249–277). Springer: Singapore.

Iztok Devetak, Ph.D. is a Professor of Chemical Education at University of Ljubljana, Faculty of Education, Slovenia. His research focuses on how students, from elementary school to university, learn chemistry at macro-, submicro- and symbolic level, how chemistry in context and active learning approaches stimulate learning, using eye-tracking technology in explaining science learning, aspects of environmental chemistry education, developing teachers' health-managing competences etc. He has been involved in research projects in the field of science education and he was the national coordinator of PROFILES project (7th Framework Program) for 4.5 years. He co-edited a Springer monograph about active learning approaches in chemistry. He (co)authored chapters in international books (published by Springer, American Chemical Society, Routledge…) and published papers in respected journals (altogether about 400 different publications). He was a Fulbright scholar in 2009. He is a member of ESERA (European Science Education Research Association) and Vice-chair for Eastern Europe of EuChemMS DivCEd (European Association for Chemical and Molecular Sciences Division of Chemical Education). He is a Chair of Chemical Education Division in Slovenian Chemical Society and president of the national Subject Testing Committee for chemistry in lower secondary school. Dr. Devetak is Editor-in-Chief of CEPS Journal and editorial board member of respected journals, such as Chemistry Education Research and Practice, International Journal of Environmental and Science Education, and Eurasian Journal of Physics and Chemistry Education.

Chapter 8
Pre-service Teachers' Determination of Butterflies with Identification Key: Studying Their Eye Movements

Tanja Gregorčič and Gregor Torkar

Introduction

Science curricula that integrate more research and research like experiences are getting internationally increased support from preschool to university level (e.g., Bybee, 2011; NAAEE, 2019; NRC, 2012; OECD, 2018). The National Research Council (NRC, 2012) is emphasizing process skills, i.e., asking questions, defining problems, conducting investigations, interpreting and using evidence, constructing explanations and designing solutions. These ways children and adults develop their understanding of surrounding environments, it helps them build a strong foundation of skills and knowledge for further exploration of the world and develop deeper conceptual understandings that environmental literacy needs (NAAEE, 2019). Research into student's development of science concepts exposed a great need for studies on how observation and other scientific process skills develop. Similarly, as intuitive (spontaneous) concepts are gradually replaced by scientific concepts, intuitive observations are replaced with more selective, sophisticated, and theory-driven observations that lead to the development of scientific explanations (Duschl, 2000; Tomkins & Tunnicliffe, 2007).

Skills of Observation

Dallwitz et al. (2002) wrote that the observation and identification of organisms is a process in determining which taxon species belong to and represents a basic skill for understanding nature. The way that students observe differences in nature, in

---

T. Gregorčič · G. Torkar (✉)
Faculty of Education, University of Ljubljana, Ljubljana, Slovenia
e-mail: gregor.torkar@pef.uni-lj.si

observed behavior and perceived function is fundamental to developing scientific thinking (Tomkins & Tunnicliffe, 2007). Observations are motivated and guided by and acquire meaning about questions or problems about natural phenomena (Lederman, 2018). For an experienced observer, observation plays a key role throughout the entire investigation process, whereas everyday observer uses observation mainly for data collection (Eberbach & Crowley, 2009). Garcia Moreno-Esteva et al. (2020) provided a couple of examples of how personal identification processes of a biologist and non-biologist vary when they are carrying out a task concerning the observation of species-specific characteristics of two bird species. Observations and investigations of the environment enable students to identify and answer questions that trigger their curiosity about the world around them. Observation is an important initial skill in early years (Johnston, 2009) that helps students find and organize patterns in the observed natural world, which is crucial for scientific activity (Klemm & Neuhaus, 2017). Student's skills of observation develop with age (Johnston, 2011) and influence the development of other scientific skills (Johnston, 2009). Observations are more than just seeing things. Most students initially start observing using multiple senses at once (Johnston, 2011, 2013). Once they develop perceptions of objects, using their sight, sound, smell, feel, and/or taste, they rapidly start to construct a concept of identity for an object (Tomkins & Tunnicliffe, 2015). Simple explanations of observations gradually develop into complex interpretations (Johnston, 2009). They start recognizing similarities and differences between objects, observing patterns, identifying sequences and events in their surroundings, and interpreting observations (Johnston, 2011).

Identification Keys

In biology, an identification key is a tool that aids the identification of biological entities (e.g., plants, animals, animal tracks). Thus, one of the most fundamental objectives in biology teaching is to strive for developing students' skills and abilities to use biological identification keys (Randler & Zehender, 2006). For these purposes simplified identification keys are used. Bajd (2016) explains that with simplified identification keys primary and secondary school students are taught to observe organisms closely and classify them, to use biological terminology, to use the identification keys (the skill that can help them later use scientific identification keys) and about biodiversity.

Each identification step in simplified identification key requires from learner to choose between two options (dichotomous keys) or multiple ones (polytomous keys); consisting of text, graphics, or both. Simplified keys mostly contain organisms from the student's local environment and which allows them to get to know these organisms more closely through identification (Bajd, 2016). There is an increasing number of studies (e.g., Anđić et al., 2019; Laganis et al., 2017; Stagg et al., 2015; Randler, 2008; Bromham & Oprandi, 2006; Randler & Bogner, 2006) investigating the effectiveness of identification keys. Most of these studies were done with primary and secondary school students.

The results of Randler and Bogner (2006) indicate that using identification keys in biology teaching is an effective educational tool to explain scientific principles. Through the process of determination with identification key students improve their observational skills and terminology (Laganis et al., 2017), and can work independently, without the teacher's help (Bromham & Oprandi, 2006). Students are more motivated for learning about plants when working with identification keys, although botanical content is not attractive for students (Silva et al., 2011).

Nowadays, loads of identification keys are available, usually using words, illustrations, and/or photographs that guide a learner through the identification of the organism. For a review on the use of images in field guides and identification keys see Leggett and Kirchoff (2011). Authors exposed best practices in an image used in guides and keys, based on their review e.g., multiple images should be included to illustrate the taxon descriptions (characters indicated with arrows to direct the user's attention); an observed organism in the photograph should be pointed out from the backgrounds, where possible, the background should be of a standard color; the use of drawings is more reasonable than photographs when representing a typical example of an organism; when used, illustrations should be prepared by professional botanical illustrators, and clearly labeled. However, little is known what type of visual representations (illustrations or photographs) users prefer to use.

Eye-Tracking Technique

Eye-tracking is a technique for studying various visualizations because it makes it possible to monitor cognitive processes due to the links between eye movements and cognition. Eye movements indicate where attention is being directed and total fixation time (i.e., cumulative duration of fixations within a region) is considered as a sign of the amount of total cognitive processing engaged with the fixated information. Data gathered with eye tracing provides information about the cognitive process of the student, such as reading, scene perception, visual search, and other information on processing the problem (Rayner, 1998, 2009). The eye-tracking technique has been used in many studies, analyzing the learning process and problem-solving (e.g., Lai et al., 2013; Pavlin et al., 2019; Torkar et al., 2018). For detecting information about students' procession of various visualizations, the eye-tracking is also efficiently used (e.g., Ferk Savec et al., 2016).

Research Problem and Research Questions

Concerning this fact, collecting eye movements can provide important information in investigating students' identification of organisms with simplified identification keys. In the present study, we focus our attention on pre-service teachers, who are going to teach students to use simplified identification keys in primary and lower secondary school and their task is also to develop student's scientific skills. The goal of the present research was to observe pre-service teacher's eye movements during the determination of butterfly species with a simplified dichotomous key containing illustrations, photographs, and written descriptions that guide a learner through the identification step.

The research questions underlying this research were:

1. Do pre-service teachers use statistically significantly more illustrations or photographs in identification keys?
2. Are there statistically significant differences in the identification process between students who identified organisms correctly or incorrectly?
3. Are there statistically significant differences in the identification process between students with more or less prior experiences with identification keys and butterflies?
4. Are there statistically significant differences in results about the identification process, success in the identification process and experiences with identification keys and butterflies between biology major and non-major pre-service teachers?

Method

Participants

Slovenian pre-service teachers ($n = 58$) participated in the study: pre-service biology teachers ($n = 26$) and pre-service primary school teachers ($n = 32$). Seven males and fifty-one females. They are going to teach science or biology subjects in primary or lower secondary school. Education staff at the school level have to hold relevant educational qualification (ISCED 7 for primary school and lower secondary school teachers) and they have to pass the state professional examination for education staff (Eurydice, 2019). Even though the curriculum varies among universities, pre-service teachers typically take content knowledge courses, pedagogical content knowledge courses, and general education courses. Pre-service teachers learn how to create and use simplified identification keys in general biology or science courses. Pre-service biology teachers also learn to use scientific identification keys in zoology and botany courses. Participating pre-service teachers were in the second, third, or fourth year of studies at the University of Ljubljana's Faculty of Education. They all learned how to make and use simplified dichotomous identification keys before the study.

Through the curriculum for the pre-service primary school teachers and pre-service biology teachers at the Faculty of Education, University of Ljubljana, there is a great emphasis on learning how to use and create simplified identification keys. Pre-service primary school teachers, in the first year of study, at the subject of Natural Sciences—biology, learn about the importance of observational skills for classification of organisms. Throughout the activity of classifying fruits and vegetables by one criterion, i.e., they use their observational skills (using different senses) to determine the criteria for classification. Further, students are acquainted with a simplified schematic dichotomous identification key for organisms from students' local environment (they have to identify snail, spider, earthworm, etc. with the key). They also create their own simplified schematic dichotomous identification key for plants in the context of making herbarium. With created identification key the plants included in an herbarium can be determined. At the tutorials pre-service primary school teachers also learn to use simplified textual dichotomous identification key for determining ground animals (Bajd, 1998), and winter branches of woody species (Bajd, 1997). On the seacoast, they also use a simplified identification key for marine snails and shells (Bajd, 2012). Throughout curricula described above, pre-service primary school teachers learn to use and create simplified dichotomous identification keys for children in primary school and understand the importance of including identification keys in primary school science subjects' curricula.

Pre-service biology teachers learn to use and create simplified dichotomous keys at the Didactics of Biology from their first year of study. They learn about the variety of identification keys. The theoretical, empirical, schematic, graphic, textual, dichotomous, polytomous, digital, paper-based (book) identification keys are presented. They experience using digital dichotomous identification key for woody plants and also designing digital dichotomous identification key for common animals. Pre-service biology teachers also learn to use scientific biological identification keys at the Systematic Botany, the Zoology of Invertebrates, the Slovenian flora and fauna. Besides the use of identification keys and adaptation of them for teaching and learning biology subjects, students also learn about the diversity of species.

Procedure and Instruments

Eight tasks for butterfly species identification were presented on a computer screen in the form of text and images. Each task was presented with one identification step: two different written descriptions (A and B), an illustration and a photograph of an organism (for example see Fig. 8.1). Illustrations and photographs were taken from existing identification keys (Polak, 2009; Tolman, 2008). Students had to identify which description (A or B) correctly describes the organism on the images. On four out of eight tasks illustrations were placed on the left side and photographs on the right side of the slide. The remaining four slides had the opposite pattern.

To detect students' identification process an eye-tracker device was used. In present study the screen-based Tobii Pro X2-30 eye-tracker apparatus with Tobii Studio Enterprise for recordings and analyses of students' eye movement when using identification key was used. Gaze data were captured at 30 Hz with an accuracy of 0.4 degrees of visual angle at distances ranging between 40 and 90 cm. With an eye-tracker device eye movements can be detected, such as fixations of the gaze to the specific area of the computer screen during a specific activity. Also, students'

Fig. 8.1 An example of a student focusing on illustration, placed on the left side when identifying the butterfly species (represented with a heat map)

visual attention toward different elements of the task on the computer screen, the total amount of time (total fixation duration, TFD) and number of fixations (fixation count, FC) spent in particular areas of interest (AOI) was measured (Tsai et al., 2012; Havanki & VandenPlas, 2014).

Also, a short questionnaire was used to gather information about students' experiences with identification keys and butterflies (questions are presented in Table 8.2).

Data Analysis

To determine students' visual attention toward different elements of the slides while identifying species, we focused on the total amount of time (total fixation duration - TFD) spent in particular areas of interest (AOI). The tasks displayed on the computer screen were divided into several AOIs with regard to the placement of the images/texts investigated. Fixations refer to maintaining the visual gaze on a certain location (Fig. 8.1), and fast eye movements from one location to another are called saccades (Fig. 8.2). The identification of saccades or fixations is based on the motion of gaze. When both the velocity and acceleration thresholds (in our case: 30 degrees per second and 8,000 degrees per second squared) are exceeded, a saccade begins; otherwise, the sample is labeled as a fixation. Besides, how often participants used, in addition to text, just an illustration or a photograph was analysed. The limit of TFD in AOI was set at half a second (this was overall the lowest total time on AOI for illustration and photograph). If those participants spent less than half a second

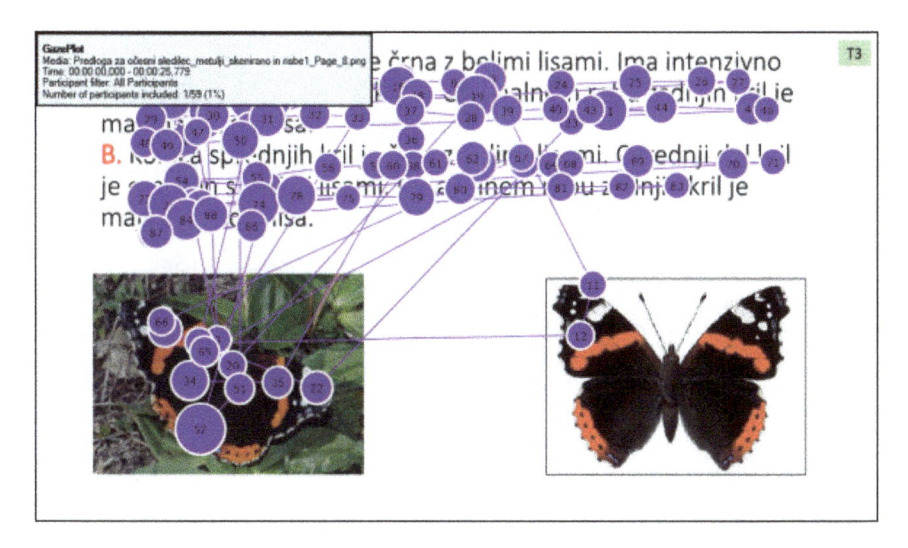

Fig. 8.2 An example of a student focusing on the photograph of the organism, placed on the left side, when identifying the butterfly species (represented with gaze plot). Through the representation of the gaze plot, the many integrations between images and text could be observed

on AOI illustration or photograph, we considered that the illustration or photograph was not used in the identification process.

Data entry and analysis were conducted using the Statistical Package for the Social Sciences (IBM SPSS 22.0). Basic descriptive statistics for numerical variables (mean, standard deviation, and frequency) were employed. The inferential statistical methods used were the Pearson product-moment correlation and the Student's *t*-test. In addition, effect sizes Cohen's *d* was calculated. Cohen's *d* statistic is a common measure to estimate effect size for independent samples *t*-tests (Cohen, 1988).

Results

Results show that 41.4% of pre-service teachers solved correctly all eight tasks, 27.6% seven, 22.4% six, 6.9% five, and 1.7% four tasks. Table 8.1 shows students' success in the identification of butterflies. On average they identified seven ($SD = 1.04$) butterfly species correctly.

Students were asked about their experiences with butterflies and identification keys (Table 8.2). A great amount of them (79.3%) said that they like observing butterflies in nature, but only 6.9% of students have sampled and determined butterflies so far. Most of the students (84.5%) mistakenly thought that butterfly species *Lycaena virgaureae* has yellow-colored wings. Also, only 5.2% of students would recognize butterfly species clouded yellow *Colias croceus*. Students also reported about their knowledge on the biodiversity of butterflies. Only one of the participants

Table 8.1 Correctly and incorrectly solved tasks

Tasks (*species*)	Task solved correctly		Task solved incorrectly		Total	
	f	f%	f	f%	f	f%
Task 1 (*Anthocharis cardamines*)	49	84.5	9	15.5	58	100.0
Task 2 (*Inachis io*)	52	89.7	6	10.3	58	100.0
Task 3 (*Vanessa atalanta*)	41	70.7	17	29.3	58	100.0
Task 4 (*Maniola jurtina*)	56	96.6	2	3.4	58	100.0
Task 5 (*Lycaena phlaeas*)	52	89.7	6	10.3	58	100.0
Task 6 (*Vanessa cardui*)	51	87.9	7	12.1	58	100.0
Task 7 (*Polygonia c-album*)	58	100.0	0	0.0	58	100.0
Task 8 (*Pararge aegeria*)	47	81.0	11	19.0	58	100.0

Table 8.2 Self-reported experiences with keys and butterflies

Self-reported experiences with identification keys and butterflies	Experienced		Not experienced	
	f	f%	f	f%
I like observing butterflies in the meadow.	46	79.3	12	20.7
I could recognize a butterfly species yellow clouded *Colias croceus*.	3	5.2	55	94.8
I have sampled and determined butterflies more than three times.	4	6.9	54	93.1
Butterfly species *Lycaena virgaureae* has yellow wings.	9	15.5	49	84.5

assessed her knowledge about butterfly species diversity as good (1.7%), 37.9% as average, 43.1% as fair, and 17.2% as poor.

They spent on average 18.19 s ($SD = 6.95$) for each task (tasks 1–8) presented on separate slides. When identifying butterflies, they devoted on average 26.17% of the total fixation duration (TFD) time to images and the rest of the time on written descriptions of butterfly species. They used on average 13.86% ($SD = 4.33$) of TFD on illustrations and 12.31% ($SD = 4.18$) of TFD on photographs of butterflies. The difference is statistically significant ($t = 1.966$, $df = 114$, $p = .050$). The effect size for this analysis ($d = .364$) was found to exceed Cohen's (1988) convention for a small effect ($d = .20$). In most cases students used at the same time photographs and illustrations (83.3%) in the identification process. Details are presented in Table 8.3.

Results in Table 8.4 show that the number of correctly identified butterfly species is not significantly correlated with self-reported prior experiences with identification keys and butterflies ($r = -.063, p = .637$). The number of correctly identified butterfly species is also not significantly correlated with average TFD for identifying butterflies ($r = -.236$, $p = .075$) neither with the average percent of TFD on illustrations ($r = -.056$, $p = .677$) and photographs ($r = -.204$, $p = .125$). It is also seen that

Table 8.3 Number (f, $f\%$) of times TFD in AOI < 0.5 s for illustrations and photographs

		Illustration		Photograph		Total	
Task	N	f	$f\%$	f	$f\%$	f	$f\%$
1	57	4	6.9	5	8.6	9	15.8
2	58	10	17.2	2	3.4	12	20.7
3	58	5	8.6	0	0.0	5	8.6
4	58	3	5.2	9	15.5	12	20.7
5	58	2	3.4	4	6.9	6	10.3
6	58	0	0.0	6	10.3	6	10.3
7	58	6	10.3	7	12.1	13	22.4
8	58	1	1.7	3	5.2	4	6.9
Total	463	31	6.7	36	7.8	77	16.6

Table 8.4 Correlations between the number of correctly solved tasks, self-reported experiences with identification keys and butterflies, and eye-tracking results

	(1)	(2)	(3)	(4)	(5)	(6)
Number of correctly solved tasks (1)	1					
Self-reported experiences with identification keys and butterflies (2)	−.063	1				
Average TFD for each task (3)	−.236	.083	1			
Average TFD% on images (4)	−.144	−.040	.340**	1		
Average TFD% on illustrations (5)	−.056	−.124	.283*	.894***	1	
Average TDF on photographs (6)	−.204	.056	.325*	.887***	.586***	1

Note *$p < .05$, **$p < .01$, ***$p < .001$

average TFD for identifying butterflies is not significantly correlated with students' experiences with identification keys and butterflies ($r = .083, p = .533$). The same is true for the percent of TFD on illustrations ($r = −.124, p = .353$) and photographs ($r = −.056, p = .674$) (Table 8.4).

Student's *t*-test indicated that pre-service primary school teachers had less prior experiences with identification keys and butterflies than pre-service biology teachers ($M = 2.75, SD = .95; M = 3.85, SD = .88$), $t(56) = 4.513, p < .001, d = 1.201$ (Fig. 8.3). The effect size for this analysis was found to exceed Cohen's (1988) convention for a large effect ($d = .80$). However, there are no statistically significant differences in a number of correctly solved tasks between pre-service primary school teachers and pre-service biology teachers ($M = 7.03, SD = 0.97; M = 6.96, SD = 1.14$), $t(56) = .251, p = .803, d = .066$. Pre-service biology teachers spent on average less TFD in identification process than pre-service primary school teachers ($M = 17.75, SD = 6.14; M = 18.53, SD = 7.62$), $t(56) = .425, p = .672, d = .113$. An average percent of TFD spent for observing images (illustrations and photographs) of butterfly species is smaller among pre-service biology teachers ($M = 25.22, SD =$

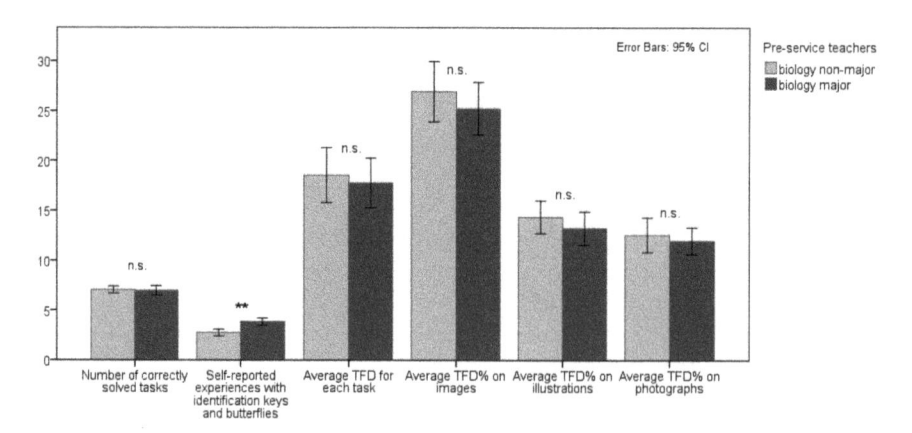

Fig. 8.3 Differences in results between biology major and non-major pre-service teachers (*Note* **p < .001; n.s.—not a statistically significant difference)

6.53) than among pre-service primary school teachers ($M = 26.93$, $SD = 8.36$), $t(56) = .854$, $p = .397$, $d = .228$. Cohen's d value above 0.2 suggest small effect (Cohen, 1988). Consequently, pre-service biology teachers average percent of TFD spent for observing illustrations of butterfly species is also smaller among pre-service biology teachers ($M = 13.24$, $SD = 4.11$) than among pre-service primary school teachers ($M = 14.36$, $SD = 4.49$), $t(56) = .980$, $p = .331$, $d = .260$. And pre-service biology teachers average percent of TFD spent for observing illustrations of butterfly species is smaller among pre-service biology teachers ($M = 11.98$, $SD = 3.32$) than among pre-service primary school teachers ($M = 12.57$, $SD = 4.81$), $t(54.68) = .553$, $p = .582$, $d = .143$.

Discussion

Pre-service primary school and biology teachers report having very little prior experiences with identifying keys for butterflies. The use of identification keys in teaching and learning biology is a great opportunity to develop students' scientific skills of observation and classification (Randler & Bogner, 2006), familiarize them with biodiversity, biological terminology, and the use of identification keys (Bajd, 2016). Identification keys can be simplified to make it more fitting for educational purposes. The main goal of the present research was to observe the determination of butterfly species with a simplified dichotomous key containing illustrations, photographs, and written descriptions, and to explore what type of visual representations (illustrations or photographs) pre-service teachers prefer to use and are more effective in the identification process. The findings of the present research suggest that pre-service biology and primary school teachers spent on average one-quarter of a total fixation duration on images (illustration and/or photograph of a butterfly) while identifying species of butterflies. Students used more illustrations than photographs for the determination of butterfly species, but the size of the effect is small. A possible explanation of the result is that students disliked photographic backgrounds because they distract them from the subject (Leggett & Kirchoff, 2011). Leggett and Kirchoff (2011) recommend that backgrounds should be of the standard color, which is also a common practice in scientific identification keys (e.g., Johnson, 2004; Svensson, 2010). A finding of the present research suggests that students rely more on illustration prepared by professional biological illustrators in comparison to a photograph even though later depicts butterfly more realistically. Drawings that are prepared by a trained botanical illustrator show specimens typical of the species and focus on and emphasize their specific features (Meicenheimer, 2007, 2009) which help learners identify organisms.

Before the study, all participating pre-service primary school teachers learned how to use and develop simplified dichotomous identification keys and besides that, pre-service biology teachers learned to use scientific identification keys in botany and zoology. As expected, pre-service biology teachers reported having more experiences with identification key and butterflies than pre-service primary school teachers, this had no significant effect on success in the identification process. Those students that reported having more experiences with identification keys and butterflies were not more successful in identification in comparison to less experienced students. These results can be explained in two ways. First, with the simplicity of examples presented in the tasks (as seen from Table 8.1), where a large majority of participants correctly determined the species. This is also one of the limitations of the present study, therefore, in future studies, complex examples and also scientific identification keys should be used, to make sure that tasks are sufficiently difficult for learners to differentiate them more. Secondly, Eberbach and Crowley (2009) explained that specific scientific skills of observation are developed within a disciplinary framework. Meaning, that being able to observe and identify plants does not help you much within disciplines such as entomology or ornithology.

Studying students' eye movements and total fixation durations in AOIs gave us information on their process of identification. Namely, the average total fixation duration time was slightly shorter, but not significantly, among students who provided more frequently correct answers. The same conclusion can be made for their total fixation duration on illustrations and photographs. Pre-service biology teachers used less percentage of time to observe images, probably due to better knowledge of the biological terminology used in the texts or higher confidence in a determination. Most students looked both images—photographs and illustrations (for example see Fig. 8.2) in the identification process which speaks in favor of Wisniewski's (2002) recommendation that taxon written descriptions should be supported with multiple images to help students visualize the presentative features of an organism. Furthermore, it is interesting finding that the number of students who used only illustration or photograph was lower within tasks that were most correctly answered. This is something to be further explored in future studies.

Conclusion

The present research provides some useful information about the determination of species with identification keys. Pre-service teachers reported having very little prior experiences with identification keys for butterflies, but still, they were very successful in the determination of eight butterfly species with provided simplified identification key. This speaks in favor of using identification keys for learners to autonomously observe and identify taxon. Eye-tracking technology proved to be useful in the present research, but it is hardly a standalone research method. Analysis of the eye movements showed that students preferred illustrations over photographs of butterflies in the identification process, but the effect size is small and needs to be further tested. Furthermore, students mostly looked at both images (photographs and illustrations) in the identification process, therefore, multiple images are recommendable in identification keys to help learners to observe the presentative features of a taxon. Due to relatively small numbers of incorrectly solved tasks, it is difficult to make solid conclusions on the effect of visual attention toward different elements of identification keys. The findings of the present study also contribute to the support of those who see identification keys as the way to enable school students to learn about biodiversity independently of a teacher's help.

References

Anđić, B., Cvijetićanin, S., Maričić, M., & Stešević, D. (2019). The contribution of dichotomous keys to the quality of biological-botanical knowledge of eighth grade students. *Journal of Biological Education, 53*(3), 310–326.
Bajd, B. (1997). *Moje prve zimske vejice*. DZS: Ljubljana.
Bajd, B. (1998). *Moje prve drobne živali tal*. DZS: Ljubljana.
Bajd, B. (2012). *Moji prvi morski polži in školjke*. Založba Hart: Ljubljana.
Bajd, B. (2016). Jednostavni biološki ključevi. *Educatio biologiae: časopis edukacije biologije, 2*(1), 91–99.
Bybee, R. W. (2011). Scientific and engineering practices in K-12 classrooms: Understanding a framework for K-12 science education. *Science and Children, 49*(4), 10–16.
Bromham, L., & Oprandi, P. (2006). Evolution online: Using a virtual learning environment to develop active learning in undergraduates. *Journal of Biological Education, 41*(1), 21–25.
Cohen, J. (1988). *Statistical power analysis for the behavioral sciences* (2nd ed.). Hillsdale, NJ: Erlbaum.
Dallwitz, M. J., Paine, T. A., & Zurcher, E. J. (2002). Interactive identification using the internet. Towards a global biological information infrastructure—Challenges, opportunities, synergies, and the role of entomology. *European Environment Agency Technical Report, 70*(3), 23–33.
Duschl, R. A. (2000). Making the nature of science explicit. In R. Millar, J. Leach, & J. Osborne (Eds.), *Improving science education: The contribution of research* (pp. 187–206). Philadelphia: Open University Press.
Eberbach, C., & Crowley, K. (2009). From everyday to scientific observation: How children learn to observe the biologist's world. *Review of Educational Research, 79*(1), 39–68.
Eurydice. (2019). *Slovenia overview*. Available at: https://eacea.ec.europa.eu/national-policies/eurydice/content/slovenia_en.

Ferk Savec, V., Hrast, Š., Devetak, I., & Torkar, G. (2016). Beyond the use of an explanatory key accompanying submicroscopic representations. *Acta Chimica Slovenica, 63*(4), 864–873.

Garcia Moreno-Esteva, E., Kervinen, A., Hannula, M. S., & Uitto, A. (2020). Scanning signatures: A graph theoretical model to represent visual scanning processes and a proof of concept study in biology education. *Education Sciences, 10*(5), 141.

Havanki, K. L., & VandenPlas, J. R. (2014). Eye tracking methodology for chemistry education research. In D. M. Bunce & R. S. Cole (Eds.), *Tools of chemistry education research* (pp. 191–218). Washington, DC: American Chemical Society.

Johnson, O. (2004). *Collins tree guide.* London: HarperCollins Publishers.

Johnston, J. (2009). What does the skill of observation look like in young children? *International Journal of Science Education, 31*(18), 2511–2525.

Johnston, J. (2011). *Early explorations in science.* Maidenhead: Open University Press.

Johnston, J. (2013). *Emergent science: Teaching science from birth to 8.* London: Routledge.

Klemm, J., & Neuhaus, B. J. (2017). The role of involvement and emotional well-being for preschool children's scientific observation competency in biology. *International Journal of Science Education, 39*(7), 863–876.

Laganis, J., Prosen, K., & Torkar, G. (2017). Classroom versus outdoor biology education using a woody species identification digital dichotomous key. *Natural Sciences Education, 46*(1), 1–9.

Lai, M.-L., Tsai, M.-J., Yang, F.-Y., Hsu, C.-Y., Liu, T.-C., Lee, S. W.-Y., Lee, M. H., Chiou, G. L., Liang, J. C., & Tsai, C.-C. (2013). A review of using eye-tracking technology in exploring learning from 2000 to 2012. *Educational Research Review, 10,* 90–115.

Lederman, N. G. (2018). Nature of scientific knowledge and scientific inquiry in biology teaching. In K. Kampourakis & M. J. Reiss (Eds.), *Teaching biology in schools* (pp. 216–235). London: Routledge.

Leggett, R., & Kirchoff, B. K. (2011). *Image use in field guides and identification keys: Review and recommendations.* AoB Plants. http://www.ncbi.nlm.nih.gov/pmc/articles/PMC3077818. Accessed 20 January 2019.

Meicenheimer, R. D. (2007). *Miami University dendrology expert system.* Oxford, OH: Miami University.

Meicenheimer, R. D. (2009). *Miami University dendrology expert system.* Oxford, OH: Miami University.

NAAEE (North American Association for Environmental Education). (2019). *Guidelines for excellence: K–12 environmental education.* Washington, DC: NAAEE.

NRC (National Research Council). (2012). *A framework for K-12 science education: Practices, crosscutting concepts, and core ideas.* Washington, DC: National Academies Press.

OECD (Organisation for Education Co-operation and Development). (2018). *The future of education and skills: Education 2030: The future we want* (Working Paper). Paris: OECD.

Pavlin, J., Glažar, S. A., Slapničar, M., & Devetak, I. (2019). The impact of students' educational background, interest in learning, formal reasoning and visualisation abilities on gas context-based exercises achievements with submicro-animations. *Chemistry Education Research and Practice, 20*(3), 633–649.

Polak, S. (2009). *Metulji Notranjske in Primorske.* Notranjski muzej, Postojna: Notranjski regijski park.

Randler, C. (2008). Teaching species identification—A prerequisite for learning biodiversity and understanding ecology. *Eurasia Journal of Mathematics, Science & Technology Education, 4*(3), 223–231.

Randler, C., & Bogner, F. X. (2006). Cognitive achievements in identification skills. *Journal of Biological Education, 40*(4), 161–165.

Randler, C., & Zehender, I. (2006). Effectiveness of reptile species identification—A comparison of dichotomous key with an identification book. *Eurasia Journal of Mathematics, Science and Technology Education, 2*(3), 55–65.

Rayner, K. (1998). Eye movements in reading and information processing: 20 years of research. *Psychological Bulletin, 124*(3), 372–422.

Rayner, K. (2009). The 35th Sir Frederick Bartlett Lecture: Eye movements and attention in reading, scene perception, and visual search. *The Quarterly Journal of Experimental Psychology, 62*(8), 1457–1506.

Silva, H., Pinho, R., Lopes, L., Nogueira, A. J., & Silveira, P. (2011). Illustrated plant identification keys: An interactive tool to learn botany. *Computers & Education, 56*(4), 969–973.

Stagg, B. C., Donkin, M. E., & Smith, A. M. (2015). Bryophytes for beginners: The usability of a printed dichotomous key versus a multi-access computer-based key for bryophyte identification. *Journal of Biological Education, 49*(3), 274–287.

Svensson, L. (2010). *Collins bird guide*. London: HarperCollins.

Tolman, T. (2008). *Collins butterfly guide*. London: HarperCollins.

Tomkins, S. P., & Tunnicliffe, S. D. (2015). Naming the living world: From the infant's perception of animacy to a child's species concept. In *Darwin-inspired learning* (pp. 147–163). Brill Sense.

Tomkins, S., & Tunnicliffe, S. D. (2007). Nature tables: Stimulating children's. *Journal of Biological Education, 41*(4), 150–155.

Torkar, G., Veldin, M., Glažar, S. A., & Podlesek, A. (2018). Why do plants wilt? Investigating students' understanding of water balance in plants with external representations at the macroscopic and submicroscopic levels. *Eurasia Journal of Mathematic, Science and Technology Education, 14*(6), 2265–2276.

Tsai, M. J., Hou, H. T., Lai, M. L., Liu, W. L., & Yang, F. Y. (2012). Visual attention for solving multiple-choice science problem: An eye-tacking analysis. *Computer Education, 58*, 375–385.

Wisniewski, E. J. (2002). Concepts and categorization. In D. Medin (Ed.), *Stevens' handbook of experimental psychology* (pp. 467–531). New York: Wiley.

Tanja Gregorčič Ph.D. student, is a teaching assistant for biology education at the University of Ljubljana, Faculty of Education, Slovenia. Her research focuses on the field of students' conceptions about animals, system thinking of natural complex system such as the human body system and on implementing new technologies (augmented reality, eye-tracking) in biology teaching and learning.

Gregor Torkar, Ph.D. is an Associate Professor for biological education at the Faculty of Education, University of Ljubljana, Slovenia. His research field in general is biology and environmental education within the primary, secondary and undergraduate level of education. His current research focus is ecology, evolution and conservational education. He is using eye-tracking technology in his research for evaluating textbooks and science education in general. He is involved in several national and international research projects on science education and nature conservation.

Chapter 9
Case Processing in the Development of Expertise in Life Sciences-What Can Eye Movements Reveal?

Ilona Södervik and Henna Vilppu

Introduction

The quality of life science experts' work contributes strongly to the well-being of human society and the whole globe on numerous levels; hence, fostering the development of expertise is a goal of utmost importance in life sciences at universities. Future experts of life sciences need adaptive and flexible reasoning skills in solving remarkably complex, multidisciplinary and still unpredictable problems, such as pandemics of severe infections, antibiotic resistance, biodiversity loss and climate change, that will require innovative ways of reasoning, as well as the ability to use knowledge and skills adaptively in unforeseen and sometimes even adverse contexts. In addition, frequent changes in current work environments as well as a rapidly changing society call for experts who possess the required domain expertise, can quickly overcome changes, and are able to update their competencies (Bohle Carbonell et al., 2014). To achieve this, future experts need reasoning skills that exceed conventional and traditional ways of thinking, i.e. adaptive expertise (Hatano & Inagaki, 1986). However, universities are often criticized for not producing graduates with sufficient adaptability or innovativeness, although their graduates typically succeed well in familiar tasks (Gube & Lajoie, 2020).

The challenge is that, according to current understanding, expertise is mostly domain-specific, meaning that it hardly transfers to novel tasks or other domains

I. Södervik (✉)
Centre for University Teaching and Learning,
University of Helsinki, Helsinki, Finland
e-mail: ilona.sodervik@helsinki.fi

I. Södervik · H. Vilppu
Department of Teacher Education, University
of Turku, Turku, Finland

(Bertram et al., 2013). Therefore, supporting the development of expertise at universities requires meaningful instructional operations that support the development of reasoning skills relevant to a particular discipline. Currently, it has been suggested that the development of expertise requires domain-specific processes at different levels: (a) conceptual understanding; (b) knowledge integration and (c) learning the links between theoretical knowledge and practice (Boshuizen & Schmidt, 2018). During the first stage of expertise development, students acquire a large amount of concepts that are relevant to a particular discipline and link them in a semantic knowledge network. Gradually, more concepts are added to the network and refined, and more and better connections are made between the concepts that activate frequently. Repeated activation of connections results in knowledge integration, the formation of so called macro-concepts (for example biodiversity, evolution or photosynthesis), in which knowledge networks are organized so that large amounts of lower-level conceptual details are clustered under higher-order concepts. Knowledge integration enables experts to, for example, effectively retrieve large amounts of information from their knowledge network, because of direct links that can be made between the first and last concepts in a certain line of reasoning, skipping some intermediate details (Boshuizen & Schmidt, 2018). This type of processing clears up cognitive space, allowing experts to exceed typical cognitive restrictions, such as a very limited working-memory capacity (Boshuizen & Schmidt, 2018).

At the further stage of knowledge integration, automaticity of learned tasks and routines starts to play a role, which allows learners not to be overwhelmed by the continual processing of previously learned material (Bransford et al., 2000). In medicine, for example, it has been well established that this phase in the development of expertise leads the students not actively using much basic biological knowledge while reasoning, but operating more actively with macro-concepts and generating certain 'fast tracks' for reasoning, i.e. scripts (Boshuizen & Schmidt, 2018). However, the use of 'fast tracks' and routines is a two-edged sword: on the one hand, it makes routine cognitive processing more effective, enabling us to exceed the limits of our cognitive capacity, but on the other hand, it may impede adaptability (Ericsson, 1996, 1998; Weisberg, 2006). For example, under certain conditions, cognitive biases, such as a tendency to view situations or problems as simpler than they really are—leading to misconceptions and inferior performance—have been detected (Feltovich et al., 1997). Thus, these tendencies ought to be taken into account in designing instruction at universities, because unless they are actively resisted, the education system may continue to produce graduates who possess expert knowledge, but cannot reliably access or apply it innovatively in novel situations (Gube & Lajoie, 2020; Hatano & Oura, 2003; Sternberg, 2003).

The findings mentioned above have led to the study of qualitatively different types of expertise: 'routine' and 'adaptive' expertise (Hatano & Inagaki, 1986). Research has noted that while routine experts continue improving their fluency and efficacy over time, adaptive experts possess superior abstract and theoretical conceptual understanding, as well as flexible access to their interconnected knowledge networks, allowing them to respond to novel situations more effectively (Bohle Carbonell et al., 2014; Schwartz et al., 2005). Thus, adaptive expertise is considered a fundamentally different conception of professionalism instead of 'the next step' after routine expertise. However, despite broad agreement that adaptive expertise is a worthy goal of university education (Hammerness et al., 2005) relatively little is currently known about the adaptive expertise capabilities of university students, nor about how to develop adaptive expertise within university education related to learning of life sciences. Understanding the distinctions between processing of routine and non-routine problem-solving may shed light on the development of adaptive expertise. Hence, the purpose of this chapter is to examine and compare findings from our previous studies related to the development of expertise in the life sciences, including medicine. In these studies, an eye-tracking method was used to investigate the processing of text-based cases among actors with different levels of expertise.

Cases in Supporting the Development of Adaptive Expertise in Learning of Life Sciences

Based on previous research, expertise is largely domain-specific, meaning that experts in a specific domain do not develop problem-solving skills that can be effectively applied across domains. Instead, knowledge and the associated skills to use the knowledge develop simultaneously and interdependently (Boshuizen & Schmidt, 2018). Therefore, the use of authentic, discipline-specific case tasks can effectively support learning, especially in the early stages of education, when learners have to perform reasoning related to conceptual knowledge, and when real hands-on problems can still be overwhelming (see e.g. Boshuizen & Schmidt, 2018; Boshuizen et al., 2020).

Case tasks can be defined as descriptions of specific events or problems that are drawn from the real world of professional practice (Ramaekers et al., 2011). Furthermore, case tasks should require activation and meaningful linking of learners' prior knowledge so that new knowledge can be effectively connected to existing knowledge structures (Boshuizen & Schmidt, 1992). Solving them should require mental activities and processes similar to those used in real work life (Brown et al., 1989). In real-life situations, for example, not all the required information is typically available at the beginning of the problem-solving situation, but becomes available step by step, requiring the evaluation of information during the action. This process relates to a script-verification process in which the expert attempts to determine whether any of the activated scripts adequately fits the findings, until all available information is received (see e.g. Charlin et al., 2007). Even more importantly, in real settings experts must address complex and multifaceted cases, and should therefore include contingencies, complexities and dilemmas requiring differentiating of relevant substance and aspects from less relevant noise. Effective case processing and knowledge restructuring are key concepts of expertise development (see Boshuizen et al., 2020), and thus learners' knowledge structures must become organized in a way that enables the effective processing of information.

Most of the research related to cognitive adaptation during expertise development has been conducted in medical domains (see e.g. Boshuizen & Schmidt, 1992; de Bruin et al., 2005; Feltovich & Barrows, 1984; Kuipers & Kassirer, 1984; Patel et al., 1989; Schmidt & Boshuizen, 1993; Schmidt & Rikers, 2007). However, based on a recent extensive review related to the theory of knowledge restructuring through case processing, similar cognitive processes and transitions on the path to expertise also seem to be relevant across domains (Boshuizen et al., 2020). Therefore, although also most of the studies related to the role of learning by cases during the development of expertise have been conducted in the medical domain (including the two example studies presented in this chapter), the findings can be somewhat generalized to other scientific fields too. There is a long history of using various problem-analysis methods in medical instruction, but since knowledge structuring through case processing takes place in all domains, some of these features could well be adapted for instruction in other disciplines too.

Over the last decades, classroom practices in medicine, and several other disciplines at universities, have increasingly evolved from content-centred traditional lectures, where students often listen passively to the teacher, towards learning-centred environments that facilitate students' active and personal knowledge construction (Vilppu et al., 2019). Furthermore, various pedagogical approaches have been developed that make operating with real-life problems, dilemmas or questions the core of the learning situation. Such specific instructional approaches include problem-based learning (PBL) (see Barrows & Tamblyn, 1980), case-based learning (CBL) and inquiry learning (IL), which are qualitatively different approaches with unique features and principles, but the shared characteristic of operating with authentic problems that aim to bridge the gap between theoretical knowledge and real hands-on problems (about PBL and CBL in the context of life sciences, see e.g. Allchin & Allen, 2017). Utilizing case-based texts for learning has been particularly popular in several areas of professional education, such as medicine, business, law and engineering (Boshuizen et al., 2020; Williams, 1992).

Although previous studies have provided interesting insights into the reasoning of cases (see e.g. Boshuizen et al., 2012), research focusing on the processes by which participants use the case description text while coming to a solution is scarce. Eye tracking offers a suitable method for investigating these processes, since there is a close connection between the direction of human gaze and the focus of attention (regarding the widely accepted eye—mind hypothesis, see Just & Carpenter, 1980).

Eye Movements in Investigating Professional Development in Life Sciences

Eye-tracking provides interesting insights into the development of expertise in various contexts. The area of visual expertise in particular has been widely studied, using static visual stimuli, such as gross anatomical images (Zumwalt et al., 2015), microscopic images (Jaarsma et al., 2014), radiology images (van der Gijp et al., 2017) or graphical data (Harsh et al., 2019), just to mention a few examples. The processing of dynamic visual stimuli has increasingly been studied as well, such as with fish locomotion patterns (Jarodzka et al., 2010) and patient video cases (Jarodzka et al., 2012). Eye-tracking research has shown that attention allocation is often influenced by expertise (Reingold & Sheridan, 2011). In comprehension of visualizations, experts exhibit shorter fixation durations, more fixations on task-relevant areas and fewer fixations on task-redundant areas compared to non-experts (Gegenfurtner et al., 2011).

Despite the large number of studies concerning expertise in comprehension of visualizations, eye-tracking studies using domain-specific, relevant texts as stimuli are scarce. However, processing various texts, such as journal articles, records, prescriptions and product descriptions, is an essential task for life science experts. This encouraged us to focus on written cases in our studies. In their future work, experts need to be able to effectively differentiate the important substance from competing noise when operating with complex written material. What is known from reading research is that the typical eye movement pattern in reading is for the reader to make a sequence of left to right eye movements from one word to another, such that most words are fixated on at least once (Kaakinen & Hyönä, 2019). Because of the close link between where the eyes are gazing and what the mind is engaged with (eye-mind hypothesis, see Just & Carpenter, 1980), readers' eye fixation patterns can be used to investigate the various ongoing mental processes of reading. Previous research has demonstrated, for example, that longer fixations might reflect difficulties in processing (Kaakinen & Hyönä, 2019; Rayner, 1998; Rayner & Slattery, 2009). Moreover, skilled readers' fixations are briefer than those of less skilled readers, indicating that fixation duration is a successful predictor of reading comprehension (Underwood et al., 1990). Additionally, highly important sentences have been found to attract greater visual attention than those that are less important (Hyönä & Niemi, 1990). Thus, attraction of visual attention might be a sign of (high) experienced relevancy to the reader.

Although research utilizing eye tracking to examine expertise in processing text cases is scarce, there are several studies concerning case processing among participants with different levels of expertise. According to previous research literature, novices' and experts' processing of information differ remarkably, regardless of discipline (Chi et al., 1981), and experts' knowledge structures have several advantages over those of novices. One of the main underlying mechanisms of the expertise development process is the increasing sophistication of cognitive schemas, which means that experts are able to identify, store and retrieve large meaningful chunks of domain-specific information (Kalyuga et al., 2012). Experts tend to seek information meaningful to the problem at hand, whereas novices are easily sidetracked towards superficial and often irrelevant material (Etringer et al., 1995; Södervik et al., 2017).

In this chapter, we present results from two example studies that use eye-tracking to investigate expertise development in the context of life sciences. In the first study (Study 1), medical students' and residents' processing of two written patient cases, routine and non-routine, was investigated via eye movements, stimulated recall interviews, and written tasks. In the second study (Study 2), medical students' processing of a non-routine patient case text was investigated using eye movements and written tasks to explore whether there were differences among the students' processes. Successful solving of these case tasks required understanding and greater or less adaptation of basic biological background knowledge. Therefore, students' biomedical knowledge, particularly that related to their understanding of anatomy and physiology of human cardiovascular system, was measured and compared with their success in case tasks utilizing a longitudinal design.

Study 1: Examining the Effect of the Level of Expertise on Case Processing

The first study example investigates how the level of expertise influences the processing and solving of patient cases in cardiovascular medicine (Vilppu et al., 2017). Relative novices, third-year medical students ($n = 39$) and more experienced residents ($n = 13$) read two patient cases of different difficulty levels. The first, routine, patient case concerned cardiac failure, and represented a typical textbook example of the condition. The second, non-routine, patient case about pulmonary embolus was more demanding, since it did not illustrate a prototypical manifestation of the disease (see e.g. Charlin et al., 2007) and thus required greater adaptivity. Solving both cases required an understanding of the pathophysiology underlying these conditions, as well as the ability to adapt basic biological background knowledge concerning the central cardiovascular system, a topic that was familiar to the students from their previous studies. Both cases were structured to depict a patient encounter in a health care centre, and thus they were divided into three phases: anamnesis (i.e. medical history of the patient), status and examination results from laboratory tests. All the information in the patient case texts was provided in written form (no images), and specific terminology was not used. Additionally, it was not necessary to remember information such as reference values or details of the case, since the text also included some interpretation of the results.

Both patient cases included semantically different sentences: key sentences that were essential to solving the case, supplementary sentences that complemented the key sentences and helped to rule out incorrect diagnoses, and irrelevant sentences that were unimportant or contained misleading information concerning the patient case. The case texts were divided into three pages (anamnesis, status and examination results), and the participants were not to go back and forth, but to proceed in the given order. After each textual slide, a question slide followed, in which the participants were asked to note the most essential symptoms/findings, and provide a (working) diagnosis. By dividing the text reading into three phases, we sought to optimize the timing of information and limit the cognitive load, which has been a problem in case-based teaching where all the available information is given at once (Kester et al.,

2001; Kirschner, 2002). During the text reading, the participants' eye movements were recorded. After the second case, a stimulated recall interview was conducted, in which the eye-tracking data was reviewed with the participants to obtain explanations for issues of interests, such as longer fixations. The purpose of the stimulated recall interview and written tasks between the text slides was to supplement and explain the observed eye movement events, since examining complex cognitive processes requires complementary measures to eye tracking (see Hyönä 2010).

The data analysis consisted of digitizing and scoring the diagnoses, and analyzing the eye-tracking metrics: total visit duration per slide (Vilppu et al., 2017) and total dwell time in sentence-by-sentence analysis (Södervik et al., 2017). Each slide, each key sentence and each irrelevant sentence was defined as an area of interest (AOI). Supplementary sentences were excluded from the analyses, since we were more interested about the division of visual attention between the key and irrelevant sentences, and based on earlier research (Hyönä & Niemi, 1990), hypothesized that key sentences would receive more visual attention.

The results indicated that the residents, being more experienced actors, were highly efficient case solvers. Their expertise was shown in both the accuracy of the diagnoses and remarkably shorter processing times compared to the students in both cases (Vilppu et al., 2017). From the viewpoint of knowledge integration (e.g. Boshuizen & Schmidt, 2018), the residents' superiority can be explained by their use of macro-concepts that enable the effective retrieval of large amounts of information, and thus faster problem-solving compared to students. On the other hand, students' clinical processing is slower since they must consciously activate their biomedical knowledge, which is more time-consuming compared to more experienced physicians who have access to ready-made structures (Schmidt & Boshuizen, 1993). However, most of the students (90%) also reached the correct diagnosis in the first, routine, case, but only under half (44%) in the second, non-routine, case. We will now take a closer look at the analyses of the latter case to see what differs in the processing of participants with differing expertise levels.

The residents were already able to diagnose the non-routine case correctly after reading the first page, anamnesis (Vilppu et al., 2017; Södervik et al., 2017). We suggest that this indicates the early identification of relevant hypotheses, which is a typical feature of expert behaviour in medicine (e.g. Charlin et al., 2007). We believe that some features of the text in the first page triggered script activation, which guided residents' efficient problem-solving right from the beginning (Boshuizen & Schmidt, 2018). A closer look at the sentence-by-sentence inspection confirmed this suggestion: the residents read the first key sentence of the case ('The patient is recuperating from knee surgery') relatively longer than the students, although residents were generally remarkably faster readers (see Södervik et al., 2017). It seems that this first key sentence, and particularly the macro-concept of 'knee surgery', may have activated script(s) in residents' knowledge networks, a finding that was also supported by stimulated recall interviews (see Södervik et al., 2017).

An interesting finding in comparing the students' and the residents' processing was the different processing patterns they demonstrated: the residents' processing time decreased after the first slide (i.e. after reaching the correct solution), whereas all students', regardless of their success in the task, showed the opposite pattern by increasing reading times towards the end of the case (Vilppu et al., 2017). However, the residents and the students who succeeded better in the case task focused more on irrelevant than relevant sentences on the second (status) page (Södervik et al., 2017). This might be due to residents' and better-succeeding students' critical awareness of the fact that sticking to the first hypothesis could be fatal: physicians are taught to systematically test their hypotheses in a script-verification process, which aims to determine whether (any of) the activated script(s) adequately fits the clinical findings until all information is received (see e.g. Charlin et al., 2007). It might be that residents were efficiently checking for excluding criteria concerning their initial diagnosis in the following slides, whereas students were continuing a more indiscriminate search for information.

Study 2: Students' Processing of a Non-routine Case and Its Relationship to the Level of Their Basic Biological Knowledge

In our second study example, we focused on comparing the processing of a non-routine case task between the students who gave a correct solution ($n = 15$, 45%) and students who were unable to reach the correct answer ($n = 18$, 55%) (Södervik et al., 2017). In this examination, the case task was the same as the second, more difficult and non-prototypical case described in the first study example. In addition, the materials and methods were the same, but supplemented with measurements concerning the level of biomedical knowledge (entrance exam scores, written assignments during first and second study years).

Overall, the students who supplied a correct diagnosis read the case faster than the other group. This supports the previously reported finding that overall reading time correlates positively with experienced text difficulty (e.g. Rayner, 1998). However, the difference was statistically significant only in the last slide, which the students with incorrect diagnoses read longer. Those students also reported more irrelevant aspects in the written, open-ended question concerning the most essential symptoms and findings after the last slide, whereas the students who diagnosed correctly reported a higher number of relevant aspects after reading the last slide (Södervik et al., 2017). Thus, the successful students had a greater capacity to distinguish between relevant and irrelevant information. When the student groups' development of biomedical knowledge was compared, we yielded some interesting findings: a total of 11/16 (69%) of those students who held misconceptions related to basic anatomy and physiology of human cardiovascular system in their 1st or 2nd study year were not able to solve the case successfully in their 3rd study year. In contrast, of those who had a scientific model of basic biology in the preceding study years, a total of 9/15 (60%) solved the case correctly (Södervik et al., 2019).

Thus, the quality of biomedical knowledge seems, to at least some extent, to be related to success in sophisticated case tasks. The result is in line with earlier findings according to which basic science or biomedical knowledge provides a foundation for clinical knowledge (Kaufman et al., 2008; Woods, 2007). This highlights the importance of basic biological background knowledge as a cornerstone of adaptive expertise. Moreover, revisiting the basic sciences in the clinical phase of medical school has proven advantageous in integrating biomedical science into clinical practice (Spencer et al., 2008). According to Spencer et al. (2008), senior medical students seem better able to appreciate the relevance of basic science concepts to clinical medicine after having spent time on clinical wards, and they often wish they had paid more attention during the first years of basic science courses. Further, as expertise develops in a cumulative manner (Ericsson, 2016), the initial gap between students with weaker and stronger biological background knowledge might even become wider during their studies if the problems cannot be tackled via instruction.

Educational and Methodological Implications for Higher Education

Over the last decades, several studies have aimed to explicate how university students acquire a high level of competence on their way to achieving expertise in different fields. During this journey, the students need to develop adequate knowledge structures, i.e. to obtain large amounts of conceptual knowledge and organize this knowledge to be meaningfully accessible and usable in real-life problem-solving situations. Therefore, teachers as well as learning researchers have begun to focus on adaptive expertise as an important cognitive capacity to understand and promote in an increasingly complex, knowledge-intensive, and fast-changing world (Bransford et al., 2000). Boshuizen and Schmidt (2018, p. 61) highlight that, during the early phases of the development of expertise, at a stage when knowledge accretion and validation take place, 'students should be given ample opportunity to test the knowledge they have acquired for its consistency and connectedness, to correct concepts and their connections and to fill the gaps they have detected'. This process benefits from various learning activities that simulate the reasoning processes required later. Learning by cases could provide a beneficial opportunity to practice using theoretical knowledge and solving authentic-like problems even in the early phases of studies (Boshuizen & Schmidt, 2018). To tap into this phenomenon, this chapter presented results from two earlier studies in which processing of text-based case tasks were investigated using eye-tracking methodology in a life science context.

Eye-tracking data revealed interesting aspects of participants' reasoning processes that can have implications for improving higher education. Firstly, eye movements of more experienced actors showed that script activation could be detected from the eye movements as relatively longer visit durations to those particular text parts. This notion was supported by the stimulated recall interviews after the case had been read, during which several participants of the experienced group explained that this (medical operation) was a critical point, where 'they could get the details to fit together' for script activation. The finding has both methodological and pedagogical implications: firstly, it proves that eye tracking is an excellent methodology to study the reasoning processes of text-based cases. Secondly, it supports earlier findings, according to which supporting the students in developing macro-concepts and scripts relevant to the particular discipline, would be of utmost importance in higher education (Boshuizen & Schmidt, 2008; Boshuizen et al., 2020). However, based on our findings, training of routine tasks may well foster students' efficiency in problem-solving, but does not necessarily prepare them to become flexible problem-solvers who are ready to deal with unexpected and non-routine situations.

Thus, complexity, structure and difficulty level should vary when designing learning activities based on cases that are to support the development of adaptive expertise. Bohle Carbonell and colleagues' (2014) review of adaptive expertise studies noted that training activities that: (a) stimulate learners to confront novel situations and new tasks; (b) allow learners to make errors and get feedback (it is important that a link is made between the errors and the knowledge to be learned) and (c) allow learners to try out different solutions support the creation of a flexible knowledge base associated with adaptive expertise. Thus, cases as instructional methods to promote active knowledge building should be designed to support students to become aware of their prior knowledge and reveal to learners the outcomes of their choices, to help the learners' self-regulation (Södervik et al., 2019; Vilppu et al., 2013). When students' processing starts to operate with macro-concepts instead of a large number of single details, that again enhances their self-regulation during processing, since monitoring of reasoning on integrated concepts in a network requires less control than monitoring of reasoning on detailed concepts (Boshuizen & Schmidt, 2018). The learning activities described above provide individuals with challenges that go beyond their current level of reliable performance—ideally in a learning context that allows immediate feedback and gradual refinement through repetition and intentional improvement (Ericsson, 2014).

Based on previous research, the structure of the scientific as well as the practical knowledge available and taught in the different domains plays a crucial role in the development of macro-concepts and scripts. In their review, Boshuizen and colleagues (2020) noted that theoretical knowledge lays the foundation for developing on macro-concepts and scripts. Our results confirmed this, because script activation seemed to play a role in non-routine case processing (a finding that could be detected from the eye movements of residents), and additionally the level and quality of students' basic biological knowledge was related to students' success in non-routine case tasks. These findings lead to the pedagogical conclusion that fostering forming of macro-concepts and scripts, and using them in non-routine problem-solving situations, is an aim of utmost importance in higher education. Based on the studies synthesized in this chapter and several other previous studies, university students would benefit from frequent exposure to authentic case tasks that align theory and practice.

Conclusions

Eye-tracking method shows great potential in assessing growing expert performance with written texts as stimuli. Our studies showed that the more experienced actors required less time and fewer fixations to produce more accurate answers compared to the students. In addition, the students appeared to demonstrate different reading patterns compared to more experienced actors. For the latter group, making a decision regarding the solution decreased their reading time of the following text, whereas the students increased their reading time towards the end of the case.

The most important finding from our studies is related to the processing of the non-routine case task, which revealed that script activation may be detectable from eye movements, considering that more experienced actors focused longer on the sentence including a relevant macro-concept and solved the case correctly based on the information provided in this part of the text. Although experienced actors made a correct working hypothesis at the beginning of the text reading process, both they as well as better-succeeding students focused even more on irrelevant text parts after the first working hypothesis. This was interpreted to indicate that the script-verification process is relevant in adaptive problem-solving, where sticking to the first working hypothesis without all the information available might lead to a false solution. Finally, since students' level and quality of basic biological knowledge was related to their success in the case task, where the basic biological knowledge had to be applied, it is necessary to design learning activities that support students to bridge basic science with authentic problem-solving reasoning. Repeated processing of domain-relevant cases thus seems to be important in supporting the development of adaptive expertise in life sciences. This notion should be taken seriously even though study programmes always struggle with allocation of time to theory and practice.

To conclude, facilitating the development of adaptive expertise is a vital aim in life sciences in higher education. Human society unquestionably needs experts who are able to use their knowledge structures flexibly and adaptively to protect well-being on Earth, even when unforeseen global catastrophes and crisis threaten it. Continuous changes mean that we do not know today the specific set of skills and knowledge that will be necessary for future experts to succeed and thrive in the decades to come, but we do know that it is imperative for life science experts to be able to use their knowledge and skills adaptively to face the challenges of rapidly changing requirements. Recent examples, such as the coronavirus (Covid-19) pandemic and climate change, have shown that preparing for the unexpected is a crucial skill that future experts will need to solve such wicked problems. Methods that enable investigating learning online at the processing level, such as eye tracking, have the potential to reveal important insights into learning via different-level cases, as well as the obstacles that students experience at various stages of development towards expertise. These findings should have implications on instruction in higher education in a rapidly changing world, where adaptability is an increasingly important skill for future experts.

References

Allchin, D., & Allen, D. (2017). Problem- and case-based learning in science: An introduction to distinctions, values, and outcomes. *CBE—Life Sciences Education, 12*(3).

Barrows, H., & Tamblyn, R. (1980). *Problem-based learning: An Approach to Medical Education.* New York: Springer.

Bertram, R., Helle, L., Kaakinen, J., & Svedström, E. (2013). The effect of expertise on eye movements in medical image perception. *PLoS ONE, 8*, e66169.

Bohle Carbonell, K., Stelmeijer, R. E., Könings, K. D., Segers, M., & van Merriënboer, J. J. G. (2014). How experts deal with novel situations: A review of adaptive expertise. *Educational Research Review, 12,* 14–29.

Boshuizen, H. P. A., Gruber, H., & Strasser, J. (2020). Knowledge restructuring through case processing: The key to generalise expertise development theory across domains? *Educational Research Review, 29,* Article 100310.

Boshuizen, H. P. A., & Schmidt, H. G. (1992). On the role of biomedical knowledge in clinical reasoning by experts, intermediates and novices. *Cognitive Science, 16*(2), 153–184.

Boshuizen, H. P. A., & Schmidt, H. G. (2008). The development of clinical reasoning expertise: Implications for teaching. In J. Higgs, M. Jones, S. Loftus, & N. Christensen (Eds.), *Clinical reasoning in the health professions* (3rd comp. rev. ed.). Oxford: Butterworth-Heinemann/Elsevier.

Boshuizen, H. P. A., & Schmidt, H. G. (2018). The development of clinical reasoning expertise. In J. Higgs, G. Jensen, S. Loftus, & N. Christensen (Eds.), *Clinical reasoning in the health professions* (4th ed., pp. 57–65). Edinburgh: Elsevier.

Boshuizen, H. P. A., van de Wiel, M. W. J., & Schmidt, H. G. (2012). What and how advanced medical students learn from reasoning through multiple cases. *Instructional Science, 40,* 755–768.

Bransford, J. D., Brown, A. L., & Cocking, R. (2000). *How people learn.* Washington, DC: National Academy Press.

Brown, J. S., Collins, A., & Duguid, P. (1989). Situated cognition and the culture of learning. *Educational Researcher, 18*(1), 32–41.

de Bruin, A. B. H., Schmidt, H. G., & Rikers, R. M. J. P. (2005). The role of basic science knowledge and clinical knowledge in diagnostic reasoning: A structural equation modelling approach. *Academic Medicine, 80*(8), 765–773.

Charlin, B., Boshuizen, H. P. A., Custers, E. J., & Feltovich, P. J. (2007). Scripts and clinical reasoning. *Medical Education, 41,* 1178–1184.

Chi, M. T. H., Feltovich, P. J., & Glaser, R. (1981). Categorization and representation of physics problems by experts and novices. *Cognitive Science, 5,* 121–152.

Ericsson, K. A. (1996). The acquisition of expert performance: An introduction to some of the issues. In K. A. Ericsson (Ed.), *The road to excellence: The acquisition of expert performance in the arts, and sciences, sports and games* (pp. 1–50). Mahwah, NJ: Erlbaum.

Ericsson, K. A. (1998). The scientific study of expert levels of performance: General Implications for optimal learning and creativity. *High Ability Studies, 9,* 75–100.

Ericsson, K. A. (2014). Adaptive expertise and cognitive readiness: A perspective from the expert-performance approach. In H. F. O'Neil, R. S. Perez, & E. L. Baker (Eds.), *Teaching and measuring cognitive readiness* (pp. 179–197). New York: Springer.

Ericsson, K. A. (2016). *Peak: Secrets from the new science of expertise.* Boston: Houghton Mifflin Harcourt.

Etringer, B. D., Hillerbrand, E., & Claiborn, C. D. (1995). The transition from novice to expert counselor. *Counselor Education & Supervision, 35,* 4–17.

Feltovich, P. J., & Barrows, H. S. (1984). Issues of generality in medical problem solving. In H. G. Schmidt & M. L. de Volder (Eds.), *Tutorials in problem-based learning.* Van Gorcum Assen: Maastricht.

Feltovich, P. J., Spiro, R., & Coulson, R. (1997). Issues of expert flexibility in contexts characterized by complexity and change. In P. J. Feltovich, K. M. Ford, & R. R. Hoffman (Eds.), *Expertise in context: Human and machine* (pp. 125–146). Menlo Park, CA: MIT Press.

Gegenfurtner, A., Lehtinen, E., & Säljö, R. (2011). Expertise differences in the comprehension of visualizations: A meta-analysis of eye-tracking research in professional domains. *Educational Psychology Review, 23,* 523–552.

Gube, M., & Lajoie, S. (2020). Adaptive expertise and creative thinking: A synthetic review and implications for practice. *Thinking Skills and Creativity, 35,* 1–14.

Hammerness, K., Darling-Hammond, L., & Bransford, J. (2005). How teachers learn and develop. In L. Darling-Hammond & J. Bransford (Eds.), *Preparing teachers for a changing world* (pp. 358–389). San Francisco: Jossey-Bass.

Harsh, J. A., Campillo, M., Murray, C., Myers, C., Nguyen, J., & Maltese, A. V. (2019). "Seeing" data like an expert: An eye-tracking study using graphical data representations. *CBE Life Science Education, 18*(3), ar 32.

Hatano, G., & Inagaki, K. (1986). Two courses of expertise. In H. Stevenson, H. Azama, & K. Hakuta (Eds.), *Child development and education in Japan* (pp. 262–272). New York: Freeman.

Hatano, G., & Oura, Y. (2003). Commentary: Reconceptualizing school learning using insight from expertise research. *Educational Researcher, 32*(8), 26–29.

Hyönä, J. (2010). The use of eye movements in the study of multimedia learning. *Learning and Instruction, 20*(2), 172–176.

Hyönä, J., & Niemi, P. (1990). Eye movements in repeated reading of a text. *Acta Psychologica, 73,* 259–280.

Jaarsma, T., Jarodzka, H., Nap, M., van Merrienboer, J. J., & Boshuizen, H. P. (2014). Expertise under the microscope: Processing histopathological slides. *Medical Education, 48*(3), 292–300.

Jarodzka, H., Schreiter, K., Gerjerts, P., & van Gog, T. (2010). In the eye of the beholder: How expert and novices interpret dynamic stimuli. *Learning and Instruction, 20*(2), 146–154.

Jarodzka, H., Balslev, T., Holmqvist, K., Nyström, M., Scheiter, K., Gerjerts, P., & Eika, B. (2012). Conveying clinical reasoning based on visual observation via eye movement modelling examples. *Instructional Science, 40,* 813–827.

Just, M. A., & Carpenter, P. A. (1980). A theory of reading: From eye fixations to comprehension. *Psychological Review, 87*(4), 329–354.

Kaakinen, J., & Hyönä, J. (2019). Eye movements during reading. In C. Klein & U. Ettinger (Eds.), *Eye movement research: An introduction to its scientific foundations and applications* (pp. 239–274). Cham: Springer.

Kalyuga, S., Rikers, R., & Paas, F. (2012). Educational implications of expertise reversal effects in learning and performance of complex cognitive and sensorimotor skills. *Educational Psychology Review, 24,* 313–337.

Kaufman, D. R., Keselman, A., & Patel, V. L. (2008). Changing conceptions in medicine and health. In S. Vosniadou (Ed.), *International handbook of research on conceptual change*. New York: Routledge.

Kester, L., Kirschner, P. A., van Merrienboer, J. J. G., & Baumer, A. (2001). Just-in-time information presentation and the acquisition of complex cognitive skills. *Computers in Human Behavior, 17*(4), 373–391.

Kirschner, P. A. (2002). Cognitive load theory: Implications of cognitive load theory on the design of learning. *Learning and Instruction, 12*(1), 1–10.

Kuipers, B., & Kassirer, J. P. (1984). Causal reasoning in medicine: Analysis of protocol. *Cognitive Science, 8,* 363–385.

Patel, V. L., Evans, D. A., & Groen, G. J. (1989). Biomedical knowledge and clinical reasoning. In D. A. Evans & V. L. Patel (Eds.), *Cognitive science in medicine: Biomedical modelling*. Cambridge, MA: The MIT Press.

Ramaekers, S., van Keulen, J., Kremer, W., Pilot, A., & van Beukelen, P. (2011). Effective teaching in case-based education: Patterns in teacher behavior and their impact on the students' clinical problem solving and learning. *International Journal of Teaching and Learning in Higher Education, 23*(3), 303–313.

Rayner, K. (1998). Eye movements in reading and information processing: 20 years of research. *Psychological Bulletin, 124*, 372–422.

Rayner, K., & Slattery, T. J. (2009). Eye movements and moment-to-moment comprehension processes in reading. In R. K. Wagner, C. Schatschneideder, & C. Phythian-Sence (Eds.), *Beyond decoding: The behavioral and biological foundations of reading comprehension* (1st ed., pp. 27–45). New York, NY: The Guilford Press.

Reingold, E. M., & Sheridan, H. (2011). Eye movements and visual expertise in chess and medicine. In S. P. Liversedge, I. D. Gilchrist, & S. Everling (Eds.), *Oxford handbook on eye movements* (1st ed., pp. 528–550). Oxford, UK: Oxford University Press.

Schmidt, H. G., & Boshuizen, H. P. A. (1993). On acquiring expertise in medicine. *Educational Psychology Review, 5*, 205–221.

Schmidt, H. G., & Rikers, R. M. J. P. (2007). How expertise develops in medicine: Knowledge encapsulation and illness script formation. *Medical Education, 41*, 1133–1139.

Schwartz, D. L., Bransford, J. D., & Sears, D. (2005). Innovation and efficiency in learning and transfer. In J. P. Mestre (Ed.), *Transfer of learning from a modern multidisciplinary perspective* (pp. 1–51). CT: Information Age Publishing Inc.

Spencer, A. L., Brosenitsch, T., Levine, A. S., & Kanter, S. L. (2008). Back to the basic sciences: An innovative approach to teaching senior medical students how best to integrate basic science and clinical medicine. *Academic Medicine, 83*, 662–669.

Sternberg, R. J. (2003). What is an "expert student"? *Educational Researcher, 32*(8), 5–9.

Södervik, I., Mikkilä-Erdmann, M., & Chi, M. T. H. (2019). Conceptual challenges in medicine during professional development. *International Journal of Educational Research, 98*, 159–170.

Södervik, I., Vilppu, H., Österholm, E., & Mikkilä-Erdmann, M. (2017). Medical students' biomedical and clinical knowledge: combining longitudinal design, eye tracking and comparison with residents' performance. *Learning and Instruction, 52*, 139–147.

Underwood, G., Hubbard, A., & Wilkinson, H. (1990). Eye fixations predict reading comprehension: The relationship between reading skill, reading speed, and visual inspection. *Language and Speech, 33*, 69–81.

van der Gijp, A., Ravesloot, C. J., Jarodzka, H., van der Schaaf, M. F., van der Schaaf, I. C., van Schaik, J. P. J., et al. (2017). How visual search relates to visual diagnostic performance: A narrative systematic review of eye-tracking research in radiology. *Advances in Health Sciences Education, 22*, 765–787.

Vilppu, H., Mikkilä-Erdmann, M., & Ahopelto, I. (2013). The role of regulation strategies in understanding science text among university students. *Scandinavian Journal of Educational Research, 57*(3), 246–262.

Vilppu, H., Mikkilä-Erdmann, M., Södervik, I., & Österholm-Matikainen, E. (2017). Exploring eye movements of experienced and novice readers of medical texts concerning the central cardiovascular system in making a diagnosis. *Anatomical Sciences Education, 10*(1), 23–33.

Vilppu, H., Södervik, I., Postareff, L., & Murtonen, M. (2019). The effect of short online pedagogical training on university teachers' interpretations of teaching-learning situations. *Instructional Science, 47*, 679–709.

Weisberg, R. W. (2006). Expertise and reason in creative thinking: Evidence from case studies and the laboratory. In K. A. Ericsson, N. Charness, P. J. Feltovich, & R. R. Hoffman (Eds.), *The Cambridge handbook of expertise and expert performance* (pp. 457–470). New York, NY: Cambridge University Press.

Williams, S. M. (1992). Putting case-based instruction into context. *The Journal of The Learning Sciences, 2*(4), 367–427.

Woods, N. N. (2007). Science is fundamental: The role of biomedical knowledge in clinical reasoning. *Medical Education, 41*, 1173–1177.

Zumwalt, A. C., Iyer, A., Ghebremichael, A., Frustace, B. S., & Flannery, S. (2015). Gaze patterns of gross anatomy students change with classroom learning. *Anatomical Sciences Education, 8,* 230–241.

Ilona Södervik, Ph.D. works as a senior lecturer at the Centre for University Learning and Teaching. Her major research interest concerns the development of expertise in learning and teaching of life sciences. She currently leads a multidisciplinary research project Cultivating Expertise in Learning of Life Sciences (CELLS) at the University of Helsinki, Finland. In the project, various research methods, such as eye-tracking and mixed reality environments are utilised in investigating the relation of conceptual understanding, problem-solving, and learning of hands-on skills during professional development among university life science students and teachers.

Henna Vilppu, Ph.D. works as a University Research Fellow at the UTUPEDA Centre for University Pedagogy, Department of Teacher Education, and Centre for Research on Learning and Instruction, University of Turku, Finland. Her research interests are in higher education learning and teaching, especially in regulation of learning, conceptions of teaching, and pedagogical development. She has been involved in research projects in the field of medical education and university pedagogy, which have utilised eye-tracking in the research on learning science and teachers' professional vision.

Chapter 10
Analysis of Aspects of Visual Attention When Solving Multiple-Choice Science Problems

Miroslawa Sajka and Roman Rosiek

Introduction

Since the dawn of time, scientists have been trying to explore the cognitive processes of humans. The measurement of eye fixations is one of the methods used in this context, as it can provide reliable and sensitive insights into otherwise unavailable cognitive processes (e.g. Sosnowski et al., 1993). Fixation duration is an accurate measurement of task difficulty, but also a measure of the *flow of attention* (Susac et al., 2014). Examining visual attention not only shows where and how one's gaze is directed, but also constitutes a basis for further analysis of problem solving, reasoning, attention, and mental images (Just & Carpenter, 1976; Benjamin et al., 1984; Zelinsky & Sheinberg, 1995; Ball et al., 2003; Yoon & Narayanan, 2004; Madsen et al., 2012, 2013).

That is why research with the use of eyetracking methodology for educational purposes started as early as the 2000s, with a huge rise in popularity during the next 2 decades. The publication of this book also proves the usefulness of this method and confirms its widespread usage around the world.

A survey on using eye-tracking technology for exploring learning was provided e.g. by Lai et al. (2013) on the basis of the papers included in the Social Science Citation Index from 2000 to 2012, which consisted in total of 81 papers describing 113 studies carried out up until that point. The authors have distinguished the following seven thematic areas on studying eye movements for educational purposes: (1)

M. Sajka (✉)
Institute of Mathematics, Pedagogical University of Krakow, Krakow, Poland
e-mail: miroslawa.sajka@up.krakow.pl

R. Rosiek
Institute of Physics, Pedagogical University of Krakow, Krakow, Poland

© Springer Nature Switzerland AG 2021
I. Devetak and S. A. Glažar (eds.), *Applying Bio-Measurements Methodologies in Science Education Research*,
https://doi.org/10.1007/978-3-030-71535-9_10

Patterns of information processing, (2) Effects of instructional strategies, (3) Reexaminations of existing theories, (4) Individual differences, (5) Effects of learning strategies, (6) Social/Cultural effects, and (7) Decision-making patterns.

"Individual differences" were discussed in 7 works from this period, and "Effects of learning strategies" were analyzed in 9 works, with both themes being raised in 3 works, including the paper "Visual attention for solving multiple-choice science problems: An eye-tracking analysis" (Tsai et al., 2012), in which the responses of 6 male university students to a multiple-choice task with more than one correct answer were analyzed.

The problem was formulated using both graphics and words. The respondents were asked to analyze the data provided by the drawings; they were encoded with the letters A, B, C, D and listed under the statement: "Please select the image(s) inferring a landslide would occur (single/multiple selection) and justify your selection(s)" (see Fig. 10.1).

The answer could be selected by clicking the appropriate checkbox provided next to the letter. After completing work on the problem and selecting all the correct answers, the respondents were asked to confirm their answers by clicking the submit button located at the bottom of the screen. Each of the drawings included four relevant factors: temperature, rainfall, slope, and debris. Among them only one (the temperature) had no effect on the response and was specified by the authors as irrelevant. The

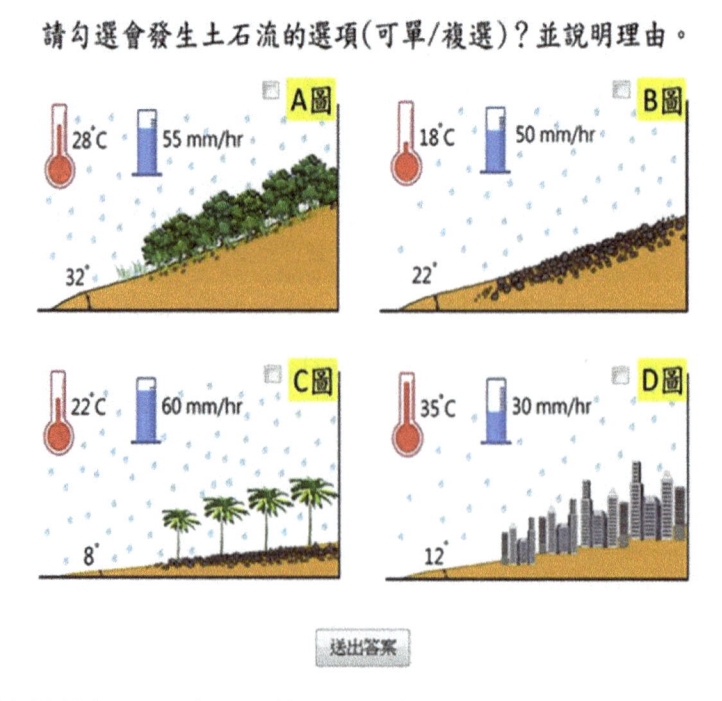

Fig. 10.1 Multiple-choice science problem analyzed by Tsai et al. (2012, p. 377)

task had two correct answers, among which answer B was the most adequate, since all of its factors were conducive to the formation of landslides, while the situation described in figure A, though less obvious, also caused the formation of landslides, therefore answer A was correct.

Tsai et al. (2012) posed the following four hypotheses to verify:

H1. Students, in general, spent more time inspecting their chosen options than rejected ones;
H2. Students, in general, spent more time inspecting relevant factors than irrelevant ones;
H3. Successful problem solvers inspected the options in a different pattern than unsuccessful problem solvers;
H4. Successful problem solvers inspected the factors in a different pattern than unsuccessful problem solvers.

The authors compared the cumulative fixation duration between the chosen and rejected options and between the relevant and irrelevant factors. Tsai et al. (2012, p. 375) confirm that they positively verified all the four hypotheses: *The results showed that, while solving an image-based multiple-choice science problem, students, in general, paid more attention to chosen options than rejected alternatives, and spent more time inspecting relevant factors than irrelevant ones.*

Aim of This Study

According to Tsai et al. (2012), limited research has been reported on how students solve image- or graphic-based problems containing multiple choices, which are frequently seen in science assessments. However, from that time many studies on this topic were conducted. Our study extends prior research on science problem solving in several aspects.

Firstly, our general aim is to understand how people at different levels of expertise solve mathematics and physics-related problems, and to examine how students inspect complex mathematical and physical graphics in the context of the problem-solving process. The research was designed to meet several specific goals. Chosen results and analyses thereof, including various points of view, were already partially published (e.g. Rosiek & Sajka, 2018, 2019; Sajka, 2017).

In this chapter, we will focus only on one group of goals, the aim of which was to mainly verify hypothesis H1 posed by Tsai et al. (2012), also in the context of hypothesis H3. However, hypothesis H3 has since been positively verified by numerous researchers in vastly different contexts. Our aim is to research the possible influence of strategy of solving the task on the results of H1.

Our research, however, differs from the one provided by Tsai et al. (2012) in methodology, numerous assumptions, and circumstances. Some of them are noted below.

1. The first group of obvious differences concerns the differences in culture and language between the participants from Taiwan and Poland. Both aspects can have a very meaningful influence on the research results. This was noted by our comparative research with Japan described in Rosiek et al. (2017). Of note were the different ways of reading coupled with fundamentally different approaches towards graphics. Polish humanists revealed basic troubles with interpreting schemes.
2. There is also a difference in the scope of the studied educational content as well as national curricula.
3. Tsai et al. (2012) positively verified hypotheses H1 and H3 for 6 people. In our study, we made use of a much greater group of 103 subjects. This allowed us to analyze the data in a quantitative way.
4. The respondents in the experiment of Tsai et al. (2012) were university students. Our study group is very diverse due to the age and the scientific experience of participants. It includes experts (academics with at least Ph.D. degrees in physics, mathematics, and computer science); Ph.D. students of physics, and different groups of university students: of physics, mathematics, computer science, and biology, as well as grade 11 high school students.
5. The final difference relates to the tasks used in the study. Although it is also a multiple-choice science problem, our research tasks concern physics and mathematics and mostly have only one correct answer. Moreover, they contain some additional difficulties that need to be overcome.

It is worth checking whether the conclusions reached by scientists from another continent are universal, or whether they are conditioned by other factors. We therefore wished to verify hypothesis H1 (also in the context of H3) and to examine whether the above differences between both experiments impact the results and conclusions, and whether they will enable us to formulate new theses, observations, or hypotheses. Therefore, in our opinion, the 5 areas of differences mentioned above constitute the methodological strength of the study, because if the general conclusion H1 is confirmed, it would mean that it is more stable and independent of omnifarious factors.

Summing up, we posed the following research questions:

Research questions

I. Do the respondents generally devote more time to analyze their chosen options, considered by them to be correct, than the rejected responses? (H1 posed by Tsai et al., 2012) What is the general trend?
II. Are there any significant differences in the general tendency for participants of various scientific expertise: experts, university students, and secondary school students?
III. Are there any factors which can influence or perturb the general tendency? Specifically—can the strategy of solving the task influence the tendency?

Method

Participants

The study involved 103 people at different levels of subject matter knowledge and mathematics/physics experience. Among them were four experts, namely: a full professor in physics and three academics with a Ph.D. degree: one in physics, one in physics & computer science, and one in mathematics; 9 Ph.D. students in physics, 2 physics students, 2 biology students, 55 computer science students, 7 math students, and 24 high school students, 11th grade (aged 17–18). The group included 73 men and 30 women.

The varied research sample and the difference in the number of people between groups of respondents is caused by the fact that the study participants were recruited as volunteers.

The subject matter knowledge of every participant was considered sufficient to solve the problems used in the study. The ability to interpret graphs of functions appears approx. at the level of grade 8. The students participating in the study learned the concept of a function and the use of its representation on mathematics lessons when they were 14–15 years old at the latest. The analysis of motion graphs is also a matter of physics education, and this sometimes happens even earlier in Polish education (which was the subject of analysis in: Sajka & Rosiek, 2019). Specifying the exact time of the introduction of function graphs is difficult because of the ever-changing national curriculum in Poland and the varied age of the participants, as the oldest expert was 50, and the youngest school student was 17.

During mathematics and physics classes, they also learn about the relationship between average velocity, displacement, and time, as well as the average acceleration for a constant velocity or constant acceleration motion.

The name and surname of each study participant is represented by a number.

Procedure

An SMI eyetracker was used to record the eye movements of the study participants. The data obtained in the experiment were processed by BeGaze software. All of the study participants passed the calibrations with less than 0.5° precision and were included in the eye-tracking experiment which involved solving a science problem. All respondents sat at a distance of 50 cm from a 24-inch monitor.

The duration of the experiment was fully adapted to the individual needs of participants and was not limited.

The participants were to choose an option by clicking the mouse in the appropriate field.

The participants' eye movement data, question responses, and mouse clicks were recorded by Experiment Center 3.1. Additionally, respondents were asked to orally confirm the selected answer.

Task

The respondents were to solve several problems in mathematics, physics, and computer science, presented as multiple choice tasks, concerning mainly the interpretation of a motion or other graphs. In this chapter, we mainly analyze the data in Problem 1. In another chapter of this book we analyze Problem 4, named the "*Stone problem.*" Figure 10.2 presents the formulation of Problem 1, which, after being translated from Polish, is as follows:

> "The motion graph illustrates the dependence of speed (v) in time (t) for two vehicles (I, II). Which of the following statements is incorrect?
> A. Vehicle (II) caught up with vehicle (I) after 10 min.
> B. In the timespan of 0–10 min, greater distance was traveled by vehicle (I).
> C. The speed of both vehicles was the same for $t = 10$ min.
> D. Vehicle (II) moved with greater acceleration.
> E. The distance traveled by vehicle (I) for ten minutes is twice the distance traveled by vehicle (II)."

Statement A is incorrect due to its assumption that the graphs represent trajectories. Statements B, C, and D are correct. Answers B and E concern the distance travelled by the vehicles. Statement B can be verified in an elementary way, based only on the analysis of speed values: vehicle (I) moves with a higher speed within the 0–10 min. timespan than vehicle (II), therefore its displacement is greater. Statement C concerns only the interpretation of values of the functions for the argument $t = 10$ min. Statement D essentially requires a basic understanding of acceleration, although the knowledge of monotonicity of linear functions would be sufficient.

Statement E is the most sophisticated, being true only at $t = 10$ min. This can be noticed by comparing the area of the figures—the rectangle and triangle bounded by

Fig. 10.2 Graph used for multiple-choice problem

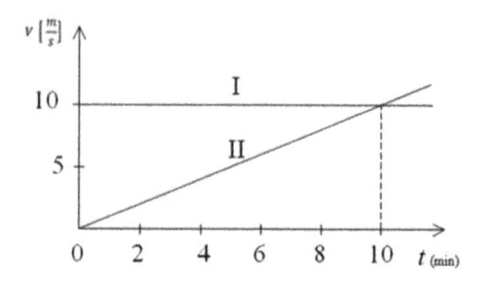

the graphs of the functions—and the *x*-axis (graphical interpretation of distance on the motion graph).

In the version for academics (without explicit information regarding $t = 10$), statement E is incorrect for all positive $t \neq 10$, thus the general statement is also incorrect. The aim of the difficulty of statement E was to make it a more challenging task for the academics. Other participants were asked to verify the statement for $t = 10$ min, therefore statement E for them was correct. However, all experts approached the task as if it was a multiple choice question with only one correct answer, and all of them indicated answer A.

Results and Discussion

Remarks on Data Analysis Methodology

We analyzed the consolidated data for all participants, as well as data separated by different levels of knowledge i.e. experts, academics (A, $N = 4$), university students (U, $N = 75$), secondary school students (S, $N = 24$), and individually.

First, we discuss the general results of the research in terms of the correctness of its subject matter knowledge.

Afterwards, visual attention is analyzed for the averaged data for all participants as well as in groups. Different areas of interest were defined in order to analyze the data; however, in this chapter, we focus mainly on the following 8 *Areas of Interest* (AOI):

AOI 1—wording of problem—first question (W);
AOI 2—graphical part of problem formulation (G);
AOI 3—answer A (A);
AOI 4—answer B (B);
AOI 5—answer C (C);
AOI 6—answer D (D);
AOI 7—answer E (E);
AOI 8—phrase: "is incorrect" (I).

The analysis covered the numerical data included in the defined AOI, i.e.: *dwell time, revisits, average fixation length*, and *average fixation count*. In this paper, we focus mainly on the analysis of options A–E.

Finally, individual analysis of selected solutions is presented.

Fig. 10.3 Results of the task in total

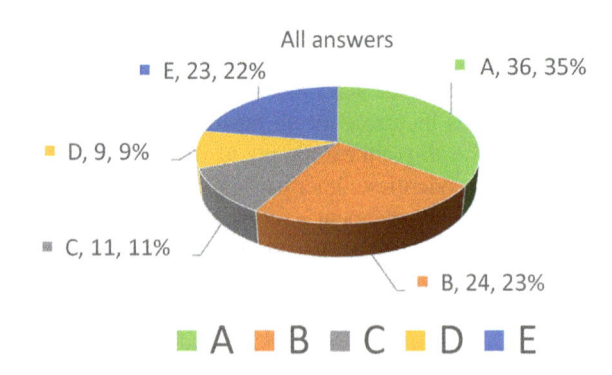

Table 10.1 Results of task for each group

Answers in groups	A	B	C	D	E
A (4)	4	0	0	0	0
U (75)	24 (32%)	19 (25%)	8 (11%)	9 (12%)	15 (20%)
S (24)	8 (33%)	5 (21%)	3 (13%)	0 (0%)	8 (33%)

General Results as Per Provided Answers

The current answer (the incorrect statement) was A, chosen only by 35% of respondents (36 people, including 4 experts), which constitutes, however, the majority of all answers. Each of the answers A, B, C, D, E was chosen by at least 9 people. The total distribution of answers is presented in Fig. 10.3.

Table 10.1 illustrates the results according to each group (A—academics, U—university students, S—secondary school students).

General Results for Visual Attention Measured by Dwell Time

The time devoted by the participants on task ranged from less than 20s (19,615 ms—high school student P43) to 2 min 40s (P87, student of computer science). The average time of solving this problem equaled slightly over 1 min (63,983.7 ms). Such a large disparity between individual total times of solving the problem was the reason to take into consideration percentage data for the purpose of comparison. Thus, the analysis covers mainly the *dwell time* [%] of every respondent for each of the area of interests (AOI) defined below.

The respondents' answers were very diverse. The *heat map* created for all participants (see Fig. 10.4) clearly shows that both the distribution of the selected options, (see red signs ♦ which registered mouse clicks), as well as the attention given to their analysis is surprisingly uniform.

10 Analysis of Aspects of Visual Attention ...

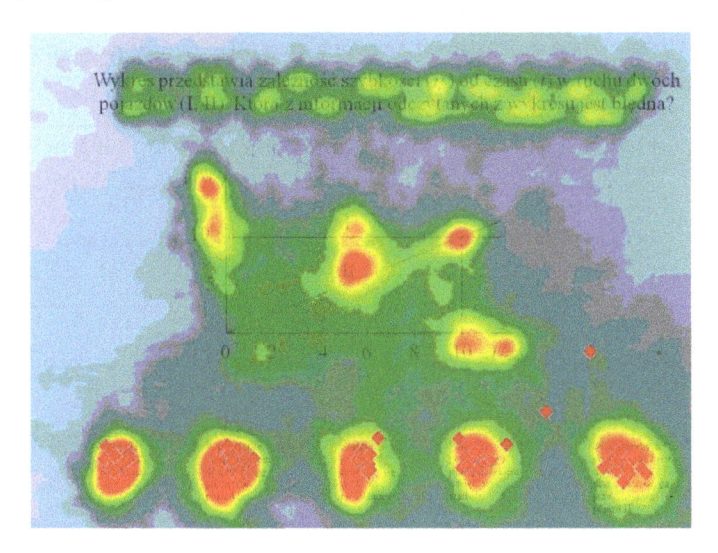

Fig. 10.4 Heat map for all study participants (103 people)

The participants' visual attention on the content of the task was more diverse. Concerning the wording of the problem, we notice that the respondents devoted most of their time and attention to two crucial formulations, in accordance with our expectations. The first one concerned the definition of the correspondence based on the phrase "speed (v) of time (t)," and the second concerned the incorrectness of the statement to be indicated: "is wrong." The utmost attention throughout the formulation of the task was dedicated by respondents to the symbol (v).

The participants dedicated, as might be expected, significantly more time and attention to the analysis of motion graphs i.e. the content of the task in its graphical form. The longest and most complete analysis was related to the description of both graphs (especially II), to the characteristic intersection point of the two graphs, and afterwards, to the descriptions of the axes, i.e. the symbols: v[*m/s*] and t[*min*] together with the number 10, and the second vehicle graph.

Figure 10.4 also illustrates that all the options: statements (A)–(E) have been thoroughly analyzed and the average of the participants' visual attention was quite homogeneously distributed between them.

Table 10.2 provides precise general average results of visual attention measured by *dwell time, revisits, average fixation length*, and *average fixation count* per AOIs of wording of the problem (W), provided graph (G), and options (A–E).

Analysis of the averaged data allows to notice that the utmost attention was given by the participants to options B and A. The longest dwell time is noted for answer B, where the total dwell time was 6008.7 ms, which meant 10.8% of time devoted on average by all participants to solve the problem. Similarly, this option was also the subject of the highest average number of revisits (4.6) and the average fixation count (21.1). Answer A was the second highest in terms of the participants' visual attention (5451.2 ms). They spent, on average, 10.6% of the time on this option, also

Table 10.2 Global average data of defined Areas of Interest (AOI) for 103 study participants

Average data (N = 103)	Wording	Graph	A*	B	C	D	E
Dwell time	8337.2 ms 16%	12497.3 ms 22.6%	5451.2 ms 10.6%	**6008.7 ms 10.8%**	3820.9 ms 6.9%	4059.5 ms 6.9%	4641.6 ms 7.8%
Revisits	3.6	12	3.7	4.6	3.4	3.3	2.9
Average fixation length	195.5 ms	206.3 ms	**306.7 ms**	254.4 ms	272.3 ms	278 ms	231.5 ms
Average fixation count	36.7	52.7	16.6	**21.1**	13.3	12.9	15.4

recording the highest average fixation length here (306.7 ms) in comparison to the other options.

Maximum Average Dwell Time Versus Chosen Answer—A General Tendency

The data of all respondents were grouped according to the type of the chosen answer. We calculated the average percentage of *dwell time* for each person from the groups for the AOIs of options A–E respectively. Table 10.1 shows the percentage value of the average *dwell time* for options A–E in relation to the total time of reading and solving the problem. For example, column (A) AOI 3 contains the collected data on the average *dwell time* for option A by all respondents who marked answers A, B, C, D, and E, respectively, as correct.

The values marked in bold represent the average percentage time of looking at the AOI of a chosen answer. We note that regardless of the choice, the respondents stayed the longest in the response field considered by them to be appropriate.

Maximum Dwell Time Versus Chosen Answer in Groups at Different Levels of Expertise

In the group of experts, who provided the correct answer A, as many as three out of four (P59, P104, P21) dedicated most of their time to analyzing the responses which they did not choose (respectively B, D, and E), rather than the selected answer, as presented in Table 10.4.

In the group of Ph.D. students who provided the correct answer A, one spent their maximum time on analyzing the discarded answer B, the second one dedicated equal time i.e., 16.5% of total work time, to analyze two answers: A and (not chosen) E.

10 Analysis of Aspects of Visual Attention … 195

Table 10.3 The average percentage *dwell time* for the AOI of options A–E according to the responders' choice

The chosen answer	(A) AOI3 Dwell time [%]	The chosen answer	(B) AOI4 Dwell time [%]	The chosen answer	(C) AOI5 Dwell time [%]	The chosen answer	(D) AOI6 Dwell time [%]	The chosen answer	(E) AOI7 Dwell time [%]
A	**17.36**	A	8.98	A	5.14	A	2.36	A	5.76
B	7.15	**B**	**16.00**	B	5.08	B	2.13	B	5.81
C	7.02	C	7.79	**C**	**17.82**	C	1.36	C	4.17
D	7.57	D	8.76	D	6.92	**D**	**8.22**	D	8.33
E	6.66	E	10.43	E	6.22	E	4.35	**E**	**14.67**

Table 10.4 Maximum *dwell time* of experts for answer AOIs

Experts $n = 4$	Number of people with the maximum dwell time				
Field AOI	A	B	C	D	E
Chosen answer A	1	1	0	1	1

In such situations, we consider two maxima. The other doctoral students spent most of their time analyzing their final answer (see Table 10.5).

In the group of all university students, 19 provided the correct answer A. Among them, only two computer science students analyzed a different field the longest (answer B). In the group of 47 students who provided incorrect answers, 14 of them analyzed a discarded answer the longest.

Detailed distributions of the maximum time in the group of university students are presented in Tables 10.6, 10.7, 10.8, and 10.9. Double maxima (indicated by

Table 10.5 Ph.D. students' maximum dwell time for answer fields

Ph.D. students in physics $n = 9$	Number of people with the maximum dwell time				
AOI	A	B	C	D	E
Chosen A	4	1	0	0	1
Chosen C	0	0	2	0	0
Chosen D	0	0	0	1	0
Chosen E	0	0	0	0	1

Table 10.6 Computer science students' maximum dwell time for different options

Computer science students n=55	Number of people with maximum dwell time				
AOI	A	B	C	D	E
Chose A (n=15)	17*	2*	0	0	0
Chose B (n=15)	3	11	0	0	1
Chose C (n=6)	1*	0	6*	0	0
Chose D (n=6)	0	0	0	5	1
Chose E (n=13)	0	1	1	1	10

Table 10.7 Mathematics students' maximum *dwell time* for answer fields

Mathematics students $n = 7$	Number of people with maximum dwell time				
AOI	A	B	C	D	E
Chose A (n=2)	2	0	0	0	0
Chose B (n=2)	0	2	0	0	0
Chose D (n=2)	1	0	0	0	1
Chose E (n=1)	0	0	1*	0	1*

10 Analysis of Aspects of Visual Attention ...

Table 10.8 Physics students' maximum dwell time for answer fields

Physics students n = 2	Number of people with maximum dwell time				
AOI	A	B	C	D	E
Chose A	2	0	0	0	0

Table 10.9 Biology students' maximum dwell time for answer fields

Biology students n = 2	Number of people with maximum dwell time				
AOI	A	B	C	D	E
Chose B	0	2	0	0	0

Table 10.10 High school students' maximum *dwell time* for answer fields

High school students n = 24	Number of people with maximum dwell time				
AOI	A	B	C	D	E
Chose A	8	0	0	0	0
Chose B	1	2	0	0	2
Chose C	0	0	1	1	1
Chose E	0	3	0	0	5

*) occurred for one mathematics student (see below paragraph in section Reversed tendency caused by precariousness) and for three computer science students.

In the group of 8 high school students who chose the correct answer A, all of them reached maximum *dwell time* for the chosen answer. Among the others, 8 people out of 16 analyzed a discarded response's AOI the longest (see Table 10.10).

Strategy of Eliminating Wrong Answers (Expert P62)—Inverted Maximum

Analyzing the work of expert P62, who is a researcher in computer science with a Ph.D. degree, we can observe a trend inverse to the general trend described in section General tendency on dependence between visual attention and chosen option based on average data. He devoted **minimum time** to analyze the chosen answer.

The duration of his work on the task was P62,928 ms, slightly below average. Figure 10.5 shows the *scan path* from his work on the task. Analysis of the real-time *scan path* from video allows us to note that his strategy of solving the problem consisted of constant and very thorough analysis of each proper statement and focusing attention on their faultless elimination.

The order of his visual attention is illustrated with the use of a sequence chart (Fig. 10.6), on which the subsequent examined AOIs were marked with colors and

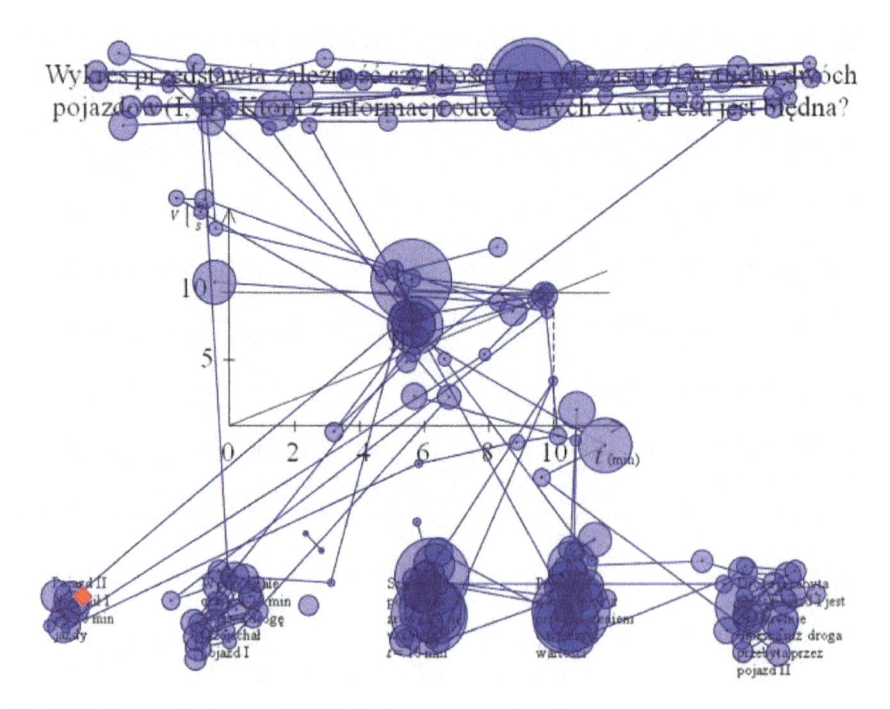

Fig. 10.5 *Scan path* of expert P62 (Ph.D. in computer science)

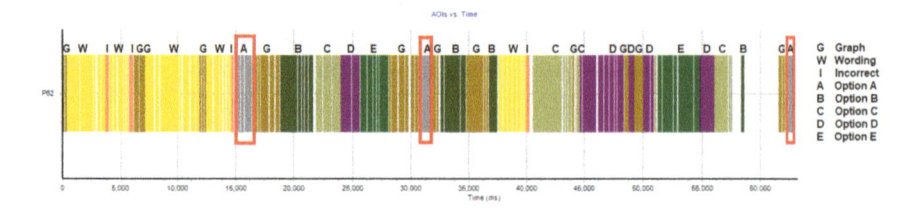

Fig. 10.6 Sequence chart of expert P62 (Ph.D. in computer science)

letters. The time of visual attention spent on the selected option A is highlighted with red frames.

The *scan path* and *sequence chart* strengthened by the numerical data collected in Table 10.11 prove that Expert P62 paid the lowest attention to the chosen answer A at only 1940.8 ms, which is about 3% of his whole visual attention on this task, with the lowest number of fixations (only 8), while other options were examined almost 4 times longer (e.g., option D approx. 12%, 7484.9 ms) and with more fixations (21, 22). His attention went back to the AOI of option A only twice, the last time only to quickly perform a mouse click.

It is worth mentioning that the maximum number of fixations was devoted to option B (26) which indicates his alertness while reading and analyzing the first correct statement.

Based on these eye-tracking data, we are entitled to claim that Expert P62 was sure about incorrectness of the option A immediately after reading it, and thoroughly examined the other options in order to eliminate them. At the same time, Expert P62 had no doubts or difficulties related to the non-standard question asked in Problem 1, which consisted of indicating the wrong answer, which is confirmed by the last rapid revisit on option A, only for the purpose of clicking the mouse.

In this case, the strategy of solving the problem coupled with a high level of expertise and motivation were the factors influencing, and even inverting the general tendency described in section General tendency on dependence between visual attention and chosen option based on average data.

Strategy of Eliminating Wrong Answers (Expert P21)—Cognitive Load While Making Decision

The expert P21 used the strategy of wrong-answer elimination several times (see the analysis of the *Stone Task* presented in Rosiek & Sajka, 2019).

Expert P21's method of analyzing the task and his strategy for solving the task was very similar to expert P062's visual attention (Figs. 10.7 and 10.8). P21 worked on the task about 2 s longer (64944 ms), and achieved a total of 193 fixations.

In this case, to diversify the analysis, we make a slightly more detailed, cinematic description of his visual attention, on the basis of different numerical data. Chronological work on the task was the following. Firstly, the expert read the task twice. The average fixation duration during the first reading of the wording of the problem was P215.8 ms. The longest fixation (no. 9) was noted on the symbol "v" after the

Table 10.11 Detailed data of visual attention of expert P62 while working on Problem 1 according to defined AOIs

P62 Chose A	Dwel time [ms]	Revisits	Fixation count	Dwell time [%]	Average fixation length [ms]
Wording	14633.7	3	63	23.3	194.9
Incorrect	930.4	2	3	1.5	180.9
Graph	10394.0	11	40	16.5	229.6
A	**1940.8**	2	8	3.1	256.1
B	6070.4	3	26	9.6	201.2
C	7218.8	5	22	**11.5**	308.9
D	7484.9	4	22	**11.9**	320.3
E	5824.3	1	21	9.3	248.4

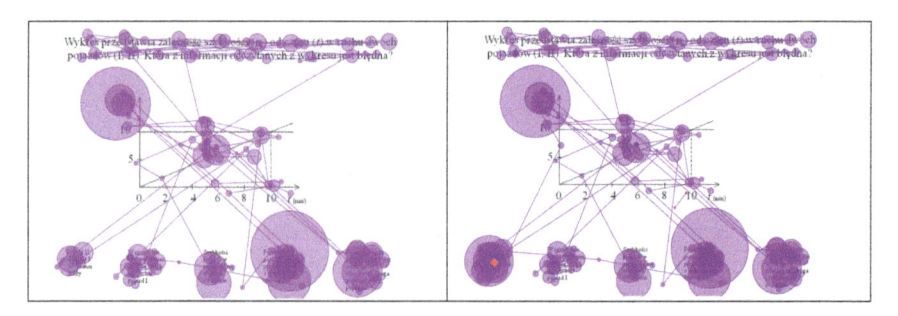

Fig. 10.7 Scan path of expert P21 before and after last reading of option A

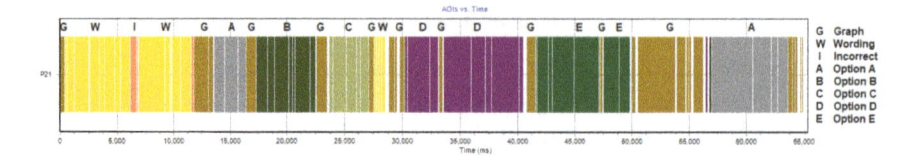

Fig. 10.8 Sequence chart of expert P21 (Ph.D. in computer science)

word "speed" (442 ms) followed immediately by fixation no. 10 (64 ms) on this symbol, two short fixations on the word "wrong," followed by re-reading the content of the task. In the second reading, longer fixations no. 31 and no. 32 (328 ms and 348 ms) happened on the description of the dependency, where the longest fixation was observed on the word "time" (up to 404 ms, no. 35). This indicates that the focus has been put on the description of the type of dependency presented in the task. Then, expert P21 paid attention to the description of the movement graphs (I and II), the value "10" on the v-axis, and the intersection point of the graphs. Next, he read option A, where only 8 fixations were observed, although two of them were very long, at P624ms and 504 ms. This was an essential part of the analysis of the whole problem, and it took the expert only about ¼ of the total work time (approx. 16 s).

Then, expert P21 proceeded to thoroughly analyze the remaining options. Option B was analyzed only once for about 5 s (18 fixations), and the analysis of subsequent options was interspersed with verification of their content on the graph. After reading option C, the expert returned to the wording of the problem by reading the description of the dependency in the graph. Non-standard long fixations were observed, no. 108 (1882 ms) and no. 131 (1038 ms) on options D, then on option E (no. 142, 1190 ms), and then on v-axis description no. 162 (1704 ms).

Such long fixations do not indicate any visual analysis of the image or text (Ober et al. 2009). Therefore, in order to find out the reason behind them, average relative changes in pupil diameter were calculated on fixations neighboring the long ones. The percentage changes of pupil diameter in relation to this average were calculated. For instance, the average relative pupil diameter change increased by 9% when perceiving

Table 10.12 Detailed data of visual attention of expert P21 while working on Problem 1 according to defined AOIs

P21 Chose A	Dwell time [ms]	Dwell time [%]	Revisits	Fixation count	Average fixation length [ms]
Wording	12034.8	18.5	1	48	215.9
Incorrect	858.4	1.3	1	5	155.3
Graph	12581.0	19.4	10	47	240.4
A	9319.7	14.4	1	23	385.3
B	5214.1	8	1	19	249.0
C	3459.4	5.3	1	9	361.3
D	**9803.9**	**25.1**	**2**	**26**	357.4
E	7711.0	11.9	1	20	363.5

the *y*-axis description. However, during a single fixation following the longest fixation no. 162 (1704 ms) in this area, the relative percentage change of pupil diameter was +12%. The increase in pupil diameter is interpreted as an indicator of cognitive load (see: Rosiek & Sajka, chapter 13 in this book).

To sum up, at the beginning, expert P21 analyzed the task similarly to expert P62. He also used the wrong-answer elimination strategy by consistently analyzing all options in turn, spending a short time on option A during the first reading (see Fig. 10.7, left), however the average length of fixation on option A was the longest, up to 385.3 [ms]. If he were to not have repeatedly read the entirety of option A at the end of his work on the problem—his work would be also an example of **a inverted tendency**. However, because of this, visual attention increased on option A, placing it at 2nd place in terms of *dwell time* and the *number of fixations* among all options (see Table 10.12).

The significant difference, however, lies at the end of his task-solving process. He returned to the option which he favored from the beginning, read it again and analyzed it thoroughly, spending the last 21 fixations (no. 173–193) on option A, before making the final decision. He wanted to make sure not to make any mistakes. During the first reading of option A, the relative pupil diameter change was negative and close to zero: −0, 3% while during the second, i.e., the last reading of option A, the relative change of pupil diameter was huge and positive: +15.1%, which is interpreted as an increase of cognitive load during the process of making a decision (see: Rosiek & Sajka, chapter *Task-evoked pupillary responses in context of exact science education* in this book). The expert's cognitive load during the decision-making process was the highest in comparison to his whole work on the task.

Cognitive Load Associated with Need to Indicate Incorrect Answer During Decision-Making Process (Expert P59)

In this section, we present the work of a third expert, P59. He worked on the task for 72,710 ms, reaching 232 fixations. Similarly to the previous analysis, we provide the scan path (Fig. 10.9), the AOI in time (Fig. 10.10), and overall data in Table 10.13.

Expert P59 only reads the content of the task once, very accurately, having longer fixations on the important symbols (v, t, I, II, "incorrect"), then analyzes the graph, looking at all of its important elements (axis descriptions, chart descriptions, point of intersection of graphs), then carefully analyzes option A, and after reading it again, analyzes the graph and its important elements, analyzes option B—returning for a moment to option A, and three times to the graph, then option C, briefly returning

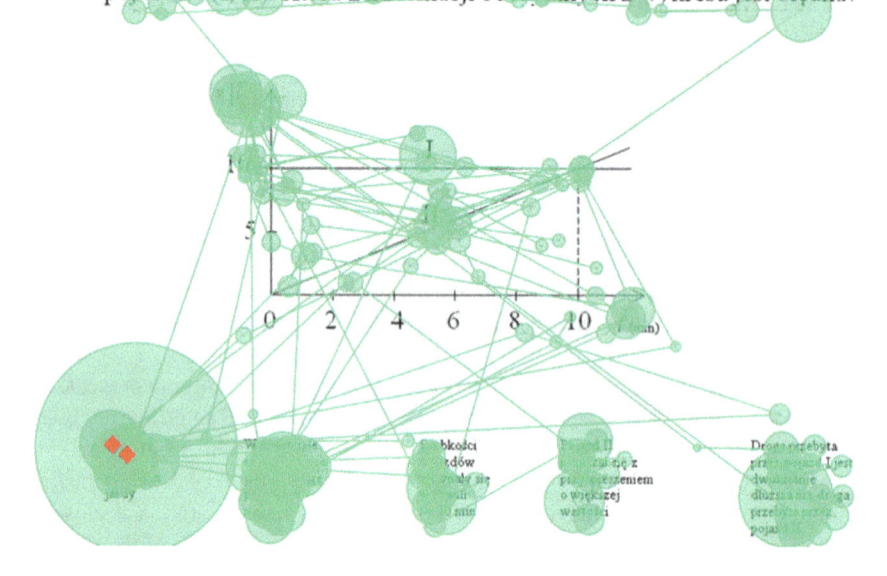

Fig. 10.9 Scan path of expert P59

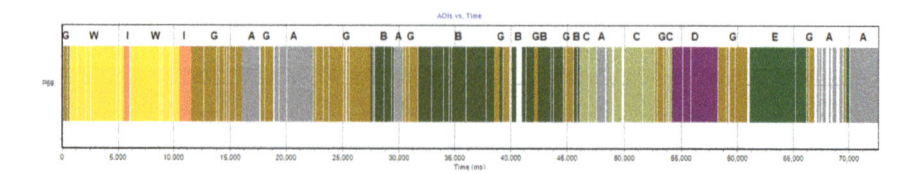

Fig. 10.10 Sequence chart of expert P59

Table 10.13 Detailed data of visual attention of expert P59 while working on Problem 1 according to defined AOIs

P21 Chose A	Dwell time [ms]	Dwell time [%]	Revisits	Fixation count	Average fixation length [ms]
Wording	10548.1	14.5	0	40	243.1
Incorrect	1408.6	1.9	2	4	345.6
Graph	18195.1	25	12	72	223.5
A	10412.1	14.3	6	28	364.6
B	**12554.8**	**17.3**	7	43	278.4
C	5292.0	7.3	3	20	259.1
D	3835.5	5.3	0	12	307.0
E	5140.0	7.1	1	15	334.9

again to option A and the graph, then option D, back to the graph after reading it, then to option E, before returning to option A and reading it again.

The work on the task is finished by a fixation on empty space and option E and one very long fixation no. 231 on option A (visible in Fig. 10.9, scan path, lasting up to 2 136 ms), which appears just before clicking the mouse and confirming the answer, and continues after the click.

As was already mentioned, such a long fixation does not indicate the process of analysis of the visual scene contained in the text or drawing where the fixation is placed (Ober), but about other processes—in this case, most likely, the expert analyzed the task, thought about the answer, and his sight went to an empty space, where the largest increase in pupil diameter was noted during fixation no. 228 and the relative percentage change in pupil diameter was +12%.

Expert P59 froze, still analyzing the task, and then double-clicked on option A. The relative change in average pupil diameter for the fixations present when reading option A for the last time and during the decision-making process (fixations no. 221–232) was positive +16%. The expert showed cognitive load at the time of making the decision.

The expert spent the most time on option B, i.e., the first correct option, and the relative pupil diameter change for the average data on option B is positive and equals 9%. Therefore, it can be concluded that cognitive load also increased during the analysis of answer B. This was probably caused by the necessity of indicating the incorrect answer and rejecting the correct statements.

Strategy of Singling Out Answer A and Then Verifying It (Expert P104)

Analyzing the work of expert P104, it can be hypothesized that he immediately selected option A as the answer. The fixations recorded on option A were twice as

Table 10.14 Detailed data of visual attention of expert P104 while working on Problem 1 according to defined AOIs

P104 Chose A	Dwell time [ms]	Dwell time [%]	Revisits	Fixation count	Average fixation time [ms]
Wording	5103.9	11.8	3	26	157.5
Incorrect	378.1	0.9	1	2	189.1
Graph	12937.0	30	12	41	285.5
A	**5314.0**	**12.3**	5	**16**	**312.4**
B	4591.8	10.6	4	**20**	204.5
C	2144.8	5	1	8	239.8
D	2739.1	6.3	2	9	284.3
E	3991.5	9.3	2	18	192.2

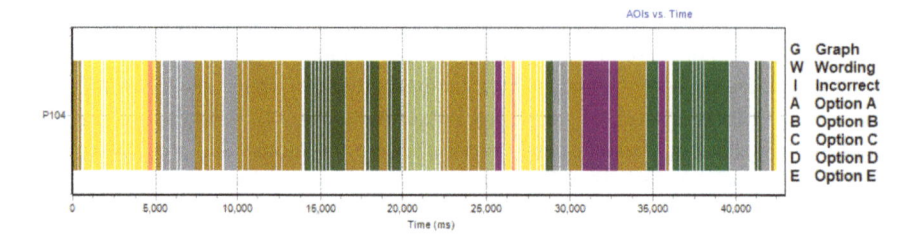

Fig. 10.11 *Sequence chart* of expert P104

long when reading the wording of the task (see Table 10.14). Less time was spent reading the other options, with negative relative percentage changes in pupil diameter (average data for B: −0.2%, C: −2.5%, D: −0.1%, E: 0.9%). This allows us to suppose that the other options were only skimmed through. Expert P105 returned to option A five times, also after analyzing the other options. Changes in pupil diameter were insignificant, but larger after the second and last return (relative percentage change +2.6%), i.e., at the time of making the decision, than during the first reading, where this change was negative (−4.9%). There was no disturbance due to the need to indicate the wrong answer. The focus of the expert was on option A from the beginning, and this option accumulated most of the maximum values presented in the table (Fig. 10.11).

Disrupted Tendency Caused by Precariousness

The second observation, which is to be revised, is the following: **a disrupted tendency was recorded in the case of the participants' precariousness of knowledge or choice**. For instance, one of the mathematics students, P99, who chose answer E as the correct one, had two maximal *dwell times* (equal to 14.8% of total time) for the fields of answers C and E. This person's *scan path* (see Fig. 10.12) shows that the distribution between the attention devoted to solve the problem and the total time was slightly higher than average, at 65,945 ms (Table 10.15).

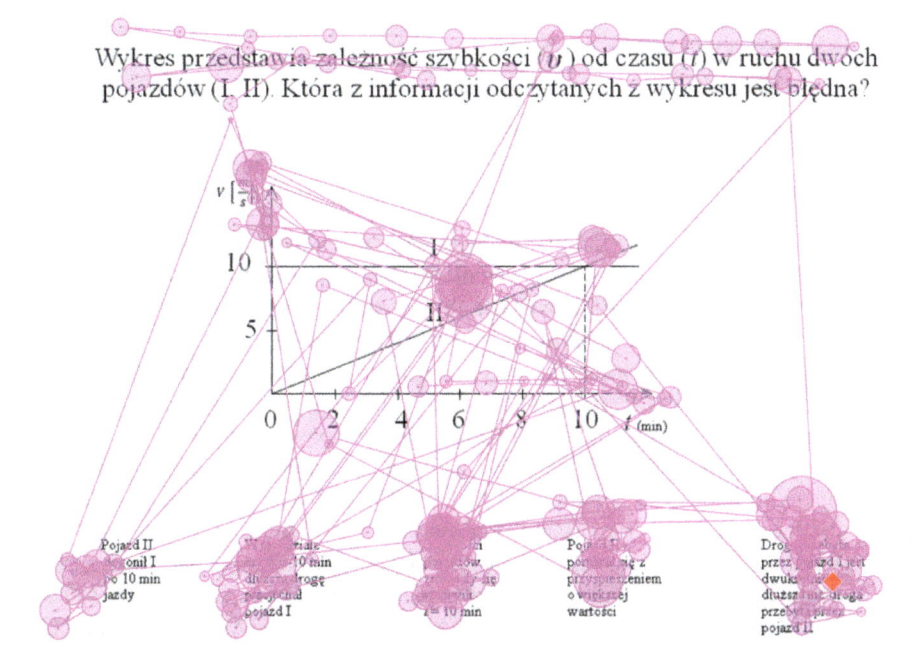

Fig. 10.12 Scan path of mathematics student P99, chose answer E

Table 10.15 Detailed data of visual attention of student P99 while working on Problem 1 according to defined AOIs

P99 Chose E	Dwell time [ms]	Dwell time [%]	Revisits	Fixation count	Average fixation time [ms]
Wording	8237.2	12.5	4	38	195.2
Incorrect	527.6	0.8	1	3	207.4
Graph	14310.2	21.7	18	59	229.8
A	3375.3	5.1	2	13	243.6
B	6254.4	9.5	8	27	215.7
C	**9791.8**	**14.85**	10	40	**234.5**
D	5806.2	8.8	7	23	**244.9**
E	9779.8	14.83	5	41	226.0

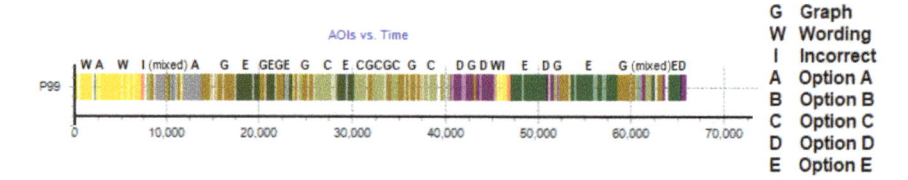

Fig. 10.13 AOI vs time for P99

Singling Out the Correct Answer Without Verification—Strong Conviction of Correctness (P43)

The third observation is the following: **the tendency of staying for the longest time at the field of the chosen answer was often observed among participants who were convinced of making the correct choice while expressing a lower level of precision, criticism, and motivation in comparison to the experts**. For instance, in the group of physics students, the correct answer is the one with the longest *dwell time*, and we observe the same for all the high school students who answered correctly. Their knowledge on the matter was already consolidated. Additionally, all school students, as well as mathematics, biology, and computer science university students, who pointed to answer B as the correct one, reached the maximal *dwell time* for that answer. It is important that answer B was the first correct statement—it seems that they chose the correct statement instead of the incorrect one. Such a case is illustrated by participant P43, and the scan path in Fig. 10.14 and Table 10.16. The mouse click at 18,381 ms was ahead of the sight movement into opinions C and D, so the other options were not analyzed nor seen at all. The student was sure about providing the correct answer.

Answering by Chance (P77)

Lastly, we present the solution of student P43 as a negative example. The student did not even read the content of the task, looked at option A and the graph only briefly, and then, not knowing the content of the task, marked option A (after about 7 s). It should be noted, however, that although the correct answer was selected and the largest visual attention on the chosen option A was noted—it does not yield any useful result and we cannot deduce anything about the subject matter knowledge of that person—only about his attitude and motivation toward solving the task (Fig. 10.15).

10 Analysis of Aspects of Visual Attention … 207

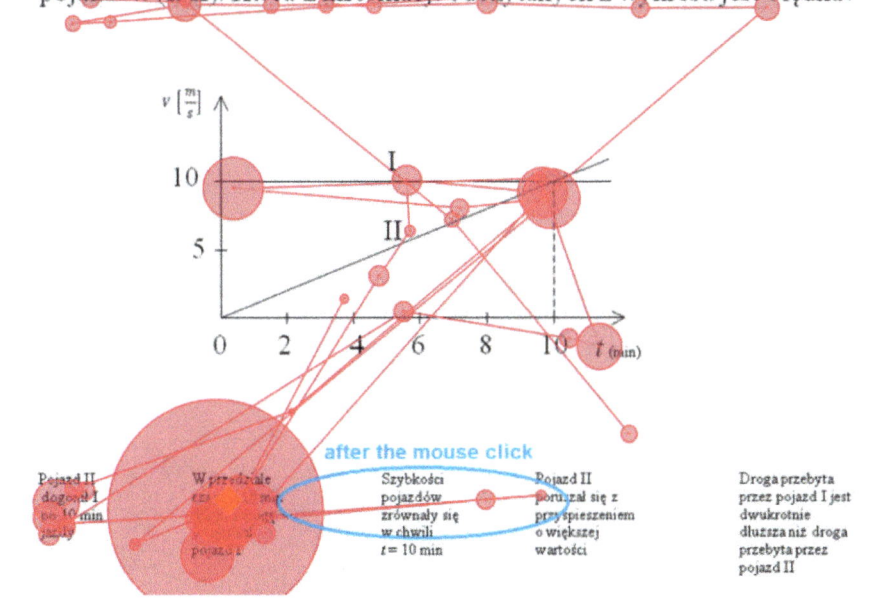

Fig. 10.14 *Scan path* of student P43

Table 10.16 Detailed data of visual attention of the participant P43 while working on Problem 1 according to defined AOIs

P143 Chose A	Dwell time [ms]	Dwell time [%]	Revisits	Fixation count	Average fixation time [ms]
Wording	4475.8	22.8	0	21	174.5
Incorrect	1070.4	5.5	1	5	192.1
Graph	3945.6	20.1	3	13	270.3
A	1788.7	9.1	2	8	212.1
B	7583	38.7	3	18	403.7
C	180.1	0.9	0	1	180.1
D	256.1	1.3	1	2	128.1
E	0	0	0	0	0

Wykres przedstawia zależność szybkości (v) od czasu (t) w ruchu dwóch pojazdów (I, II). Która z informacji odczytanych z wykresu jest błędna?

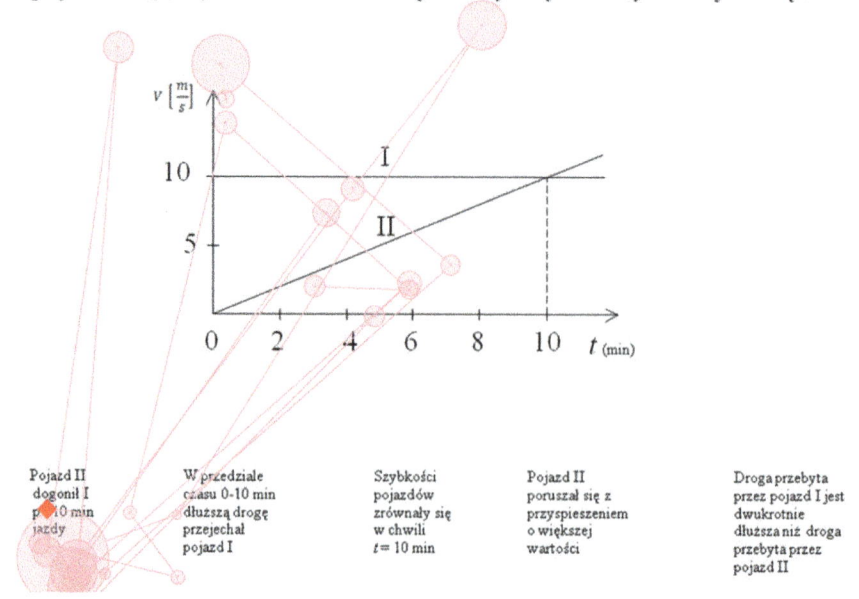

| Pojazd II dogonił I po 10 min jazdy | W przedziale czasu 0-10 min dłuższą drogę przejechał pojazd I | Szybkości pojazdów zrównały się w chwili $t = 10$ min | Pojazd II poruszał się z przyspieszeniem o większej wartości | Droga przebyta przez pojazd I jest dwukrotnie dłuższa niż droga przebyta przez pojazd II |

Fig. 10.15 *Scan path* of student P77

Conclusions

Remarks on General Answers

1. The problem formulation asked to indicate the incorrect statement. This constituted an added difficulty to the task as it required additional mental operations. The participants were to preserve the so-called *discipline and critical thinking* (Klakla, 2003).

   The existence of this difficulty can be confirmed by the knowledge of didactics and psychology, and it is also visible in the data obtained from the eye-tracking study. It has been shown in other studies that fixation duration is an accurate measurement of task difficulty, but also a measure of the *flow of attention*. (Susac et al., 2014). For our problem, therefore, *dwell time* of the selected answers describes both of these variables.

   Firstly, the highest average attention (dwell time, fixation count and revisits) of all respondents, regardless of the answer chosen, was directed to option B (see Table 10.2), which contained the first correct statement. Moreover, when we examine the answers of people who chose and marked the correct statement instead of the incorrect one, we observe this in most of these participants, as 24 out of 67 indicated answer B (see Fig. 10.16).

Incorrect answers

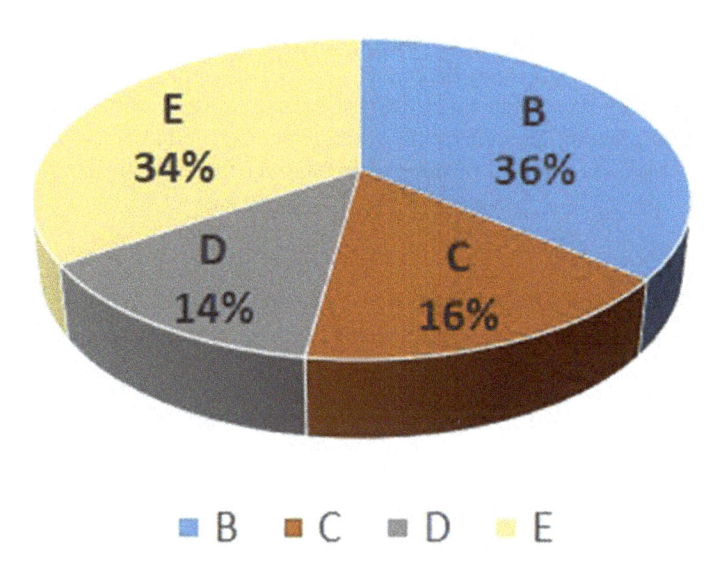

Fig. 10.16 Types of all incorrect answers given by 67 participants

Certainly, it is necessary to examine their work individually to find out how the necessity of indicating the incorrect answer affects their strategies of solving the problem. This is, however, a topic for further study. In this study, we proved the existence of an increase in cognitive load caused by the need to indicate the incorrect answer during the decision-making process on the basis of the visual attention of expert P59's work (see sections Cognitive load associated with need to indicate incorrect answer during decision-making process [Expert P59] and highlighting different strategies and approaches for solving tasks and their impact on compatibility, change, or inversion of trend described in General tendency on dependence between visual attention and chosen option based on average data).

2. Study participants spent significantly more time and attention analyzing the graphs of the given functions, which is the graphical part of the problem formulation (see Table 10.2). This confirms that the graph provided more complex information which needed to be processed.
3. It was also interesting (and mentioned in: Sajka, 2017) that none of academics questioned the correctness of Statement E, created especially for them. Here, we faced the limitation of pure eye-tracking methodology—mixed methods with interviews or written questionnaires may have exposed their reasoning. Thus, several months after the research, the experts were asked to solve the task again, without eyetracking. While solving the task, they confirmed answer A, read the other statements again and eliminated them, considering them to be correct.

When asked why they consider statement E to be correct, they answered that the displacement can be counted as the area under the graph. They indicated $t = 10$ min. After further questions they were surprised that "t=10 min" was not written explicitly in the statement. The previous statements for $t = 10$ min together with the graph and the dashed line indicating the point (10, 10) caused their certainty that statement E was also formulated for $t = 10$ min. That was their tacit assumption. One of them mentioned the routine of solving multiple choice tasks, usually with only one correct answer, therefore his inquisitive mind stopped paying attention after finding the correct answer A.

General Tendency on Dependence Between Visual Attention and Chosen Option Based on Average Data

The results of our experiment on a sample of 103 people, **based only on average data,** confirmed that subjects **generally** devote more time to analyze their final choices, which are considered by them to be correct, than to analyze discarded answers. The values presented in Table 10.3 show the general dependence, and also determine how large the differences are. Hypothesis H1 posed by Tsai et al. (2012) and confirmed by them for a sample of 6 people is thus confirmed by our research on the basis of analysis of the average *dwell time* for 103 participants. It is a surprising conclusion that the finding, drawn on such a small sample, is quite universal. Moreover, the general trend of the average *dwell time* turned out to be independent of differences in language and culture, of the scope and type of educational content, of the type of selected answers and their correctness, and, finally, of the type of multiple choice tasks—whether containing one, or multiple correct answers.

Significant Differences in General Tendency for Participants of Various Scientific Expertise: Experts, University Students, and Secondary School Students

Such a smooth and simple, general conclusion drawn on the basis of averages, as described in paragraph General tendency on dependence between visual attention and chosen option based on average data, **radically changes its power** when we analyze the data in groups and individually.

Based on the analysis of the results presented in Maximum dwell time versus chosen answer in groups at different levels of expertise, it can be stated that this tendency depends on the level of experience and knowledge of the respondents. Among the experts, *dwell time* for the selected correct answer was maximal only for one person.

In the group of students, the majority of them followed the trend, namely 77% of the group of university students and 66% of high school students. It is important to note that the high school students were systematically familiarized with solving such tasks and were a part of a "university class" with an extended mathematics and physics program, therefore their knowledge of the matter exceeded that of the university students, who did not have experience solving such tasks during their studies.

Summing up the average results in groups, it can also be noted that among university and high school students, those who chose answer A constituted the majority of people whose solutions followed the overall tendencies, for example, among 19 university students who answered A, only two students (of computer science) analyzed a different option the longest, i.e., answer B, which was the correct answer.

The individual analysis of data involving knowledge and experience in physics and mathematics provides other interesting conclusions.

We could distinguish various factors that disturb such a smooth and general tendency (General tendency on dependence between visual attention and chosen option based on average data). We have empirically proved the following: type of task-solving strategy, increase of cognitive load, precariousness, lack of self-confidence or certainty about the solution which was often associated with another factor such as a lower level of knowledge (resulting in incorrect answers), and the lack of motivation resulting from random answers. However, most of them have been distinguished based on the analysis of individual solutions.

Highlighting Different Strategies and Approaches for Solving Tasks and Their Impact on Compatibility, Change, or Inversion of Trend Described in Section General Tendency on Dependence Between Visual Attention and Chosen Option Based on Average Data

Our general aim is to understand how people at different levels of expertise solve mathematics and physics-related problems, to examine how students inspect complex, mathematical, and physical graphs in the context of the problem-solving process. The analysis of individual data allowed us to partially contribute to this issue and to formulate several new findings.

The group of experts, who were professionally active researchers, differed significantly from the other participants of the study. There is a meaningful difference concerning not only subject matter knowledge and experience in this group, but also the level of motivation to properly solve the task, the level of criticism of their own work in terms of error avoidance, and the level of responsibility for the result of their work.

The work of each of the experts has been analyzed, as in their case both subject matter knowledge and motivation did not raise any doubts, thus we can eliminate negative factors influencing their work, such as: lack of knowledge, lack of certainty, lack of motivation, or lack of concentration. This assumption allowed for analyzing their visual attention in terms of implemented strategy.

Such analysis of the experts' work allowed to distinguish several strategies and approaches to solve the task. Strategies 1–4 do not follow the trend described in section General tendency on dependence between visual attention and chosen option based on average data, while strategies 5–6 do.

1. *Strategy of eliminating wrong answers,* as described for expert P62 (section Strategy of eliminating wrong answers [expert P62]—inverted maximum), where the tendency was inverted. The expert devoted *minimum time* of visual attention for the chosen option. This was caused by an in-depth analysis of the rejected options coupled with the certainty regarding the correctness of option A.

2. *Strategy of eliminating wrong answers with cognitive load while making decision* described for expert P21 (section Strategy of eliminating wrong answers [expert P21]—cognitive load while making decision), where the tendency was inverted only until the last reading of the chosen option. We proved the increase in cognitive load while decision making (see section Strategy of eliminating wrong answers [expert P21]—cognitive load while making decision).

   We could pose the hypothesis, to be verified, that for people who use the *strategy of eliminating wrong answers* the tendency is either inverted, or at least disrupted, with the assumption of a high amount of motivation for the work and a high level of scientific expertise.

3. *Cognitive load associated with need to indicate incorrect answer during decision-making process* was diagnosed in the work of expert P59 (section Cognitive load associated with need to indicate incorrect answer during decision-making process [Expert P59]), where the expert only read the content of the task once, but very thoroughly, and had many revisits to the graphs (12) as well as options A (6) and B(7). Increased cognitive load was manifested by long fixations (up to 2136 ms) followed by an increased relative change in average pupil diameter (+16%) for the fixations during the decision-making process. The expert devoted the most of his visual attention to the first correct option B.

4. *Not preserving tendency due by precariousness* (e.g. section Reversed tendency caused by precariousness) consisted of the fact that the task analysis was chaotic (see chart in Fig. 10.13) and the subjects jumped to various elements of the task as well as various options, which manifested their uncertainty. In addition, the wording of the task and the analysis of the graphs was not as insightful and systematic.

5. *Strategy of singling out the correct answer and verifying it,* as described for expert P104 (see section Cognitive load associated with need to indicate incorrect answer during decision-making process [Expert P59]) was in line with the general trend (General tendency on dependence between visual attention and chosen option based on average data). School and university students frequently focused their attention on identifying the correct statement without attempting further analysis.

6. *Strategy of singling out the correct answer without verification—strong conviction of correctness* (see: section Singling out the correct answer without verification—strong conviction of correctness [P43]) was the most common reasoning in line with the general trend (General tendency on dependence between visual attention and chosen option based on average data), when providing both correct and incorrect answers. The participants who were underperforming in mathematics and physics did not always notice that they should have indicated the incorrect statement. This conclusion is manifested by the *dwell time* and the number and duration of *fixations* on the AOI 08 field: "information is incorrect" as well as by the analysis of the scan paths of individual participants (see e.g. P43).
7. *Answering by chance,* described on the basis of P43's work (see section Answering by chance [P77]), can only be used to stress the effectiveness of the eyetracking method to note the cases of blindly guessing the answer while solving multiple choice tests (very often the first from the left) and in the same time indicate the limitations of paper and pencil multiple choice which are in widespread use, including external Matura examinations.

Importance of Long Fixations

The selection and analysis of long fixations in a given area can constitute a useful methodology for indicating the thought processes of the examined people which cannot be examined using other tools.

The longest fixations, that is, those whose durations are longer than approx. 500 ms, can be precisely noted during the problem-solving process, and then analyzed. We know from the literature, i.e., Ober et al. (2009) and other sources, that these are fixations that are not related to the analysis of the visual scene and that information regarding the area of the visual scene where this fixation is taking place does not reach the brain when they happen, such fixations are not associated with image analysis. To define the role of long fixations (see: see section Strategy of eliminating wrong answers [expert P21]—cognitive load while making decision or Cognitive load associated with need to indicate incorrect answer during decision-making process [Expert P59], fixations lasting up to 1882 ms for expert P21, and up to 2136 ms for expert P59), we analyzed the percentage relative change in pupil diameter within the time and area accompanying the longest fixations. Relative changes in the pupil diameter during the fixations neighboring the longest one clearly show large increases in the relative value of pupil diameter changes. The hypotheses of neuronal efficiency and the hypothesis (e.g. Beatty, 1982) claim that the increase in pupil diameter is associated with either stress, negative emotions, or cognitive load. Stress and negative emotions should likely be ruled out in this case, as this happened at the end of the task-solving process and their duration was too short. Such short changes in the pupil diameter clearly indicate cognitive load during task content analysis and the decision-making process.

As it is assumed that during this time, a decision-making process takes place in the brain or the content of the task is being analyzed, it can be associated with the greatest cognitive load. This methodology can be helpful for identifying cognitive load while solving tasks.

References

Ball, L. J., Lucas, E. J., Miles, J. N. V., & Gale, A. G. (2003). Inspection times and the selection task: What do eye-movements reveal about relevance effects? *Quarterly Journal of Experimental Psychology, 56A,* 1053–1077.

Benjamin, L. T., Cavell, T. A., & Shallenberger, W. R. (1984). Staying with the initial answers on objective tests: Is it a myth? *Teaching of Psychology, 11,* 133–141.

Beatty, J. (1982). Task-evoked pupillary responses, processing load, and the structure of processing resources. *Psychological Bulletin, 91*(2), 276–292.

Just, M. A., & Carpenter, P. A. (1976). Eye fixations and cognitive processes. *Cognitive Psychology, 8,* 441–480.

Klakla, M. (2003). Dyscyplina i krytycyzm myślenia jako specyficzny rodzaj aktywności matematycznej. *Studia Matematyczne Akademii Świętokrzyskiej, 10,* 89–106.

Lai, M. L., Tsai, M. J., Yang, F. Y., Hsu, C. Y., Liu, T. C., Lee, S. W. Y., Lee, M. H., Chiou, G. L., Liang, J. C., & Tsai, C. C. (2013). A review of using eye-tracking technology in exploring learning from 2000 to 2012. *Educational Research Review, 10,* 90–115. https://doi.org/10.1016/j.edurev.2013.10.001.

Madsen, A. M., Larson, A. M., Loschky, L. C., & Rebello, N. S. (2012). Differences in visual attention between those who correctly and incorrectly answer physics problems. *Physical Review Special Topics-Physics Education Research, 8,* 010122-1–010122-13.

Madsen, A. M., Rouinfar, A., Larson, A. M., Loschky, L. C., & Rebello, N. S. (2013). Can short duration visual cues influence students' reasoning and eye movements in physics problems? *Physical Review Special Topics—Physics Education Research, 9,* 020104-1–020104-16.

Ober, J., Dylak, J., Gryncewicz, W., & Przedpelska-Ober, E. (2009). Sakkadometria – nowe możliwości oceny stanu czynnościowego ośrodkowego układu nerwowego. *Nauka, 4,* 109–135.

Rosiek, R., & Sajka, M. (2018). Eyetracking in research on physics education. In T. Greczyło & E. Dębowska (Eds.), *Key competences in physics teaching and learning selected contributions from the international conference GIREP EPEC 2015,* Wrocław Poland, 6–10 July 2015 (pp. 67–77). Cham: Springer International Publishing.

Rosiek, R., & Sajka, M. (2019). One task—many strategies of interpreting and reasons of decision making in the context of an eye-tracking research. In *AIP Conference Proceedings P2152,* 030028-1-030028-8. https://doi.org/10.1063/1.5124772.

Rosiek, R., Sajka, M., Ohno, E., Shimojo, A., & Wcisło, D. (2017). An excerpt from an eye-tracking comparative study between Poland and Japan with the use of Force Concept Inventory. *AIP Conference Proceedings, 1804*(1), 060003–1–060003–7. https://doi.org/10.1063/1.4974400.

Sajka, M. (2017, February 1–5). Visual attention while reading a multiple choice task by academics and students: A comparative eye-tracking approach. In T. Dooley & G. Gueudet (Eds.), *Proceedings of the tenth congress of the European society for research in mathematics education (CERME10).* Dublin, Ireland: DCU Institute of Education and ERME.

Sajka, M., & Rosiek, R. (2019). Struggling with physics and mathematics curricula based on the notion of function in the context of the educational reform in Poland. In *AIP Conference Proceedings,* P2152, 030029–1–030029–8. https://doi.org/10.1063/1.5124773.

Sosnowski, T., Zimmer, K., Zaborowski, P., & Krzymowska, A. (1993). *Metody psychofizjologiczne w badaniach psychologicznych: praca zbiorowa.* Warszawa: Wydawnictwo Naukowe PWN.

Susac, A., Bubic, A., Kaponja, J., Planinic, M., & Palmovic, M. (2014). Eye movements reveal students' strategies in simple equation solving. *International Journal of Science and Mathematics Education, 12,* 555–577.

Tsai, M.-J., Hou, H.-T., Lai, M.-L., Liu, W.-Y., & Yang, F.-Y. (2012). Visual attention for solving multiple-choice science problem: An eye-tracking analysis. *Computers and Education, 58,* 375–385.

Yoon, D., & Narayanan, N. H. (2004). Mental imagery in problem solving: An eye tracking study. In S. Spencer (Ed.), *Proceedings of the eye tracking research and applications symposium 2004* (pp. 77–83). NY: ACM Press.

Zelinsky, G., & Sheinberg, D. (1995). Why some search tasks take longer than others: Using eye movements to redefine reaction times. In J. M. Findlay, R. Walker, & R. W. Kentridge (Eds.), *Eye movement research: Mechanisms, processes and applications* (pp. 325–336). North-Holland: Elsevier.

Miroslawa Sajka, Ph.D. is an Assistant Professor and the head of Chair of Mathematics Education at the Institute of Mathematics, Faculty of Exact and Natural Sciences, Pedagogical University of Krakow, Poland. She is a contributing member of the Interdisciplinary Research Group of Cognitive Didactics at Pedagogical University of Krakow from 2014, runs research on using eye-tracking in mathematics education also from physics education point of view. Her research areas: in general mathematical reasoning, specifically understanding the notion of function; teacher cognition, education and knowledge in/for teaching; comparative studies; eyetracking and other psychophysiological methods in research on mathematics and science education. Dr. Sajka authored a scientific book (monograph), over 50 scientific papers on thinking, learning and teaching mathematics and several school resources and textbooks in mathematics for secondary school students. Dr. Sajka is Vice Editor-In-Chief of a scientific journal "Annales Universitatis Paedagogicae Cracoviensis. Studia ad Didacticam Mathematicae Pertinentia" in the field of mathematics education. She undertakes co-leaderships of working groups at international conferences on mathematics education (e.g "Comparative Studies in Mathematics Education" at CERME, "Knowledge in/for teaching at secondary level" at ICME) and she delivered national and international guest and plenary lectures at conferences.

Roman Rosiek, Ph.D. is an Associate Professor of Physics Education at the Pedagogical University, Division of Theoretical Physics and Didactics of Physics, Faculty of Exact and Natural Sciences, Krakow, Poland. His research and development focuses on issues related to the didactics of physics, in particular the education of physics and computer science teachers, understanding of concepts by students and teachers, and the use of new research and teaching methods. Since January 2014, he has been a member of the Interdisciplinary Cognitive Didactic Research Group, currently under his leadership. As part of this activity, he conducts research on the applications of eyetracking methodology, EEG and psychophysiological methods e.g.: HRV, EDA, Respiration for research in the field of cognitive science and didactics of physics. Habilitation subject: Psychophysiological Methods in Research on Didactics of Physics. The results of his scientific and research work have been published in over 50 publications. Among them there are publications indexed in the Web of Science and SCOPUS databases, chapters in monographs, a book. His works are quoted by many scientists. He delivered national and international guest lectures, plenary lectures at international conferences e.g. GIREP.

Chapter 11
The Impact of Students' Educational Background, Formal Reasoning, Visualisation Abilities, and Perception of Difficulty on Eye-Tracking Measures When Solving a Context-Based Problem with Submicroscopic Representation

Jerneja Pavlin and Miha Slapničar

Introduction

Science and technology continue to develop rapidly, which leads to the need for all-encompassing science education, starting in the early years (Lloyd et al., 1998; Millar, 2006). All students should benefit from the science education provided, which includes an understanding of the scientific dimension of phenomena and events, critical recognition of the possibilities and limitations of science, its role in society and its contribution to citizenship, as well as the development of critical thinking, oral communication, and writing skills (BSCS, 2008; ICSU, 2011; Vieira & Tenreiro-Vieira, 2014). In addition, Harlen (2010) suggested that science education should enable everyone to make informed choices and take appropriate action that will affect their well-being and the well-being of society and the environment. A school curriculum that develops an understanding of fundamental science concepts, ideas about the nature and limits of science, ethical reasoning, as well as the skills of argumentation, and opportunities for students to apply them in a range of novel contexts, is appropriate preparation for future work on social issues arising from the application of science (Lewis & Leach, 2006). The content that students learn in primary school is defined, usually consisting of living beings and their interactions with the environment, materials and their properties, and forces and their effects. However, no preparatory course can cover all the science content that teachers have to teach. The courses must also promote independent learning and enable in-service

---

J. Pavlin (✉) · M. Slapničar
Faculty of Education, University of Ljubljana, Ljubljana, Slovenia
e-mail: jerneja.pavlin@pef.uni-lj.si

© Springer Nature Switzerland AG 2021
I. Devetak and S. A. Glažar (eds.), *Applying Bio-Measurements Methodologies in Science Education Research*,
https://doi.org/10.1007/978-3-030-71535-9_11

teachers to recognise the further knowledge they need when confronted with new challenges in teaching.

Representations are basic tools that help students to take advantage of the quality of learning in various scientific disciplines such as mathematics, chemistry, and physics (Mozaffari et al., 2016). The term 'representation' has several interpretations; mostly, it only refers to concrete, external descriptions used by a problem solver. Some examples include pictures or sketches, physics-specific descriptions (e.g., free body diagrams, field line diagrams, ray diagrams, or energy bar diagrams), animations, concept maps, diagrams and equations or symbolic notation, and similar. Some researchers further distinguish between general and physics-specific representation (Docktor & Mestre, 2014).

However, the complexity of teaching and learning concepts of physics, and more often of concepts of chemistry, can be explained on three levels: the macroscopic, the submicroscopic, and the symbolic level, which can be imagined as the corners of a triangle in which no form of representation is superior to the others but rather complements them (Johnstone, 1982, 1991). The macroscopic level of chemical concepts is illustrated by the observation of chemical phenomena. The interpretation of the observations is explained by the interaction of particles at the submicroscopic level. To illustrate chemical concepts at the particle level, static or dynamic submicroscopic representations (SMRs) can be used (Devetak & Glažar, 2010). It is also important to note that, at lower stages of chemical education, SMRs can be represented as a particle even if a represented molecule is more complex, while SMRs illustrate the actual structure of molecules at higher stages of chemical education. At the symbolic level of concept representation, these SMRs could be translated into established symbols, such as symbols of the elements, formulae and equations, mathematical equations, and various graphical, schematic representations (Johnstone 2001; Levy & Wilinsky, 2009; Devetak, 2012; Taber, 2013). Students often have difficulties in understanding SMRs, and teaching about the world of particles is challenging, since particle theory is abstract. Therefore, the use of visualisation material is necessary for classroom presentation (Johnstone, 2001; Exerciseer & Dalton, 2006; Kautz et al., 2005; Lin et al., 2016; Cheng & Gilbert, 2017).

In addition, the use of SMRs plays a crucial role in the teaching of chemistry, as it can enable the visualisation of phenomena that cannot be directly observed due to the size of the entities in these processes (e.g., atoms, ions, molecules, and subatomic particles) (Phillips et al., 2010). Visualisation abilities might be considered as the students' ability to recognise and manipulate visual objects (Barnea, 2000; Gilbert, 2005). The study by Wu and Shah (2004) introduced the role of visualisation in chemistry teaching and learning and suggested that one of the main features of visualisation tools should be to provide multiple representations (e.g., static 2D or 3D SMRs, computer animations of particles, physical models, etc.) that should be implemented in the classroom in such a way that students can visualise the connections between representations and relevant concepts.

The relationship between the students' perception of task or problem difficulty and the actual difficulty, measured by the successful completion of a task, is low. Students are not in a position to predict which tasks are difficult. Nunan and Keobke

(1995) cited three factors: lack of familiarity with the nature of the task, confusion on the part of students over the purpose of the task, and cultural knowledge that might justify students' views on the difficulty of the task. In addition, Andrzejewska and Stolińska (2016) investigated the comparison of task difficulty using eye-tracking in combination with subjective and behavioural criteria. They found that there was no correlation between the activity of eye movement parameters, which are considered indicators of mental effort, and a student's opinion about the difficulty of the task. On the basis of the theoretical background of the paper, it can be concluded that the average fixation time can be considered as an index of the difficulty of the problem or task to be solved. However, an analysis of the data obtained made it possible to observe discrepancies in the categorisation of the difficulty of the tasks according to subjective and behavioural criteria. A significant and strong correlation was found between the difficulty of the task, determined by the percentage of correct answers, and the fixation parameters, although no such relationship was found with the blinking parameters.

Docktor and Mestre (2014) prepared the paper *Synthesis of discipline-based education research in physics* in which they present various aspects of research in physics education in detail. They mentioned that eye-tracking has attracted the attention of the science community, including the research community for physics education (Han et al., 2017; Yen & Yang, 2016) and that several studies have been conducted (Pavlin et al., 2019). Otherwise, in cognitive psychology, eye-tracking methods (tracking eye gazes by capturing or recording position and duration during cognitive tasks) are quite common and are used to study what people pay attention to without disturbing their thought processes. It is generally agreed that the place where the eyes are looking indicates what people are looking for, so by tracking eye fixations one can inconspicuously observe what is being looked for during a cognitive task and thereby draw conclusions about what the mind is processing (Hoffman & Rehder, 2010; Rayner, 1998; Stuart et al., 2019).

Problem-solving is of significant interest in physics education research. There are many different definitions of what problem-solving means (Maloney, 2011). In order to uncover the students' approach, problem-solving has also been studied using the eye-tracking method in recent years (Tai et al., 2006). Smith et al. (2010) investigated which aspects of problem solution students look at while studying the examples. The results show that while students spent a large amount of time reading conceptual, textual information in the solution, their ability to remember this information later was poor. The students' eye-gaze patterns also showed that they often jumped back and forth between textual and mathematical information when attempting to integrate these two sources of information. The fact that performance is affected by the representation format is confirmed by studies that show that some students give inconsistent answers to the same problem-solving question when presented in different representation formats (Kohl & Finkelstein, 2005; Meltzer, 2005).

Carmichael et al. (2010) and Madsen et al. (2012) have used eye-tracking to examine how expert and novice learners allocate visual attention to physics diagrams in cases in which information critical to answering a question was contained in the diagram. They found that experts spend more time looking at thematically relevant areas of diagrams than novice learners did. Similarly, Pavlin et al. (2019) showed that the total fixation duration is longer for animations containing examples of novel stimuli (e.g., air pump), but that the percentage of fixation duration spent by students of different age groups on the area of interest with correct animations is significantly different. These results suggest that students' age and, therefore, the number of years of chemistry studies turn novices into experts and influence the fixation duration in the area of interest with the correct animation.

Several other studies including eye-tracking were conducted regarding physics education, for example, investigations of the trouble-shooting of malfunctioning circuits were done by Van Gog et al. (2005); comprehending malfunctioning mechanical devices was the topic of the study carried by Graesser et al. (2005); Hegarty (1992) on how mechanical systems work; Kozhevnikov et al. (2007) studied the relationship between spatial visualisation and kinematics problem-solving abilities; Smith et al. (2010) explored what novices look at while studying example problem solutions in introductory physics, comparing the time spent on mathematical information (equations) to the textual or conceptual information, and similar. The study by Klein et al. (2020) showed that graphical representation can be beneficial for data processing and data comparison. In addition, graphical representation aids in visualising data and thus reduces the cognitive load on students when performing measurement data analysis, so students should be encouraged to use it (Susac et al., 2014).

Furthermore, eye-tracking can make unique contributions to the validation of concept inventories at the behavioural level without using interview or survey data. While simple measures of time (length of visit durations on the question or options) are well suited to differentiate between different confidence levels of test-takers, they do not distinguish between the correct from the incorrect performers for this type of question (Viiri et al., 2017).

As shown, the eye movements of individuals can be measured and, after careful consideration, used to interpret processes during solving the tasks, since the direction of the human gaze is closely related to the focus of attention as individuals are processing the observed visual information, as indicated by Just and Carpenter (1980) and Rayner (2009). An eye-tracker device can be used to measure eye movements, such as fixations of the gaze on a specific area of the computer screen during a certain activity and saccades (i.e., eye movements between fixations) (Havanki & VandenPlas, 2014). The total time (total fixation duration [TFD]) spent in specific areas of interest (AOI) can be measured to capture students' visual attention to different elements of the context-based problem on the computer screen, which is also known as the 'visual attention' or 'attention allocation' of the student (Tsai et al., 2012).

According to Hyönä et al. (2002), a longer total fixation duration (TFD) on an AOI could indicate the salience, meaning deeper and more complex information processing, difficulties in processing this area, or less efficiency in finding the information on the computer screen (Green et al., 2007). Research has shown that experts had shorter fixations on task-irrelevant information than novices did, and longer fixations on relevant information (compared to novices) (Gegenfurtner et al., 2011). However, to ensure a proper fixation allocation, the problem or task displayed on the computer screen must be divided into several wisely chosen AOIs according to the placement elements that are of interest with regard to the research problem (Ferk Savec et al., 2016).

Visit count is another frequently used key figure and can be informative when examining participant interest or ease of understanding. The measure used to count the number of visits provides information on the number of individual returns to an AOI and is an indicator of its attractiveness, usefulness, or lack of clarity (West et al., 2006). Mean pupil dilation is a measure for identifying cognitive load (Bassok, 1990). As reported by Beatty and Lucero-Wagoner (2000), it is a common eye-tracker measure. However, Just et al. (2003) report that mental effort is reflected in the enlargement of the pupil diameter and showing the difficulty of the task. Lang et al. (2020) report that the pupil is demonstrably enlarged when individuals are confronted with cognitively challenging tasks, proving that autonomic pupil response can be used as a marker of mental effort (Korbach et al., 2018). Eye-tracking technology makes it possible to precisely measure pupil dilation, resulting in the real-time measurement of mental effort during task processing. Compared to retrospective self-reports, this measure of mental effort has the added value of providing more spontaneous and unbiased answers. Karch (2018) notes that research has shown that pupil responses can be correlated with higher processing loads. If a stimulus is carefully designed and the data are carefully recorded, the task defined by the stimulus can be correlated with a change in processing load.

Accordingly, the purpose of this paper is to investigate the total fixation duration, the average pupil dilation, visit count of students of different age groups to the AOI with macro- and submicroscopic representations and the perception of the difficulty of context-based problems, and to identify correlations between eye-tracker measures and the level of logical thinking/visualisation ability.

Aims and Research Questions

The present research aims to study the eye-tracker measurements of Slovenian students at different levels of education (primary school, upper secondary school, and undergraduate education level), in a context-based problem with SMR on AOI with macroscopic and submicroscopic representation. The context-based problem used in this study covers only the macroscopic and submicroscopic levels of concept representation; the symbolic level was not used. Emphasis is also placed on examining the differences between specific groups of students, including perceived problem difficulty, level of formal reasoning, and visualisation abilities in eye-tracker measures on AIOs.

With regard to the research aims, the following research questions can be addressed:

RQ1: Does educational level influence eye-tracker measures of AOIs at the macro- and submicroscopic levels?
RQ2: Are there significant differences between students of different levels of perception of the difficulty of eye-tracker measures on AIOs with macro- and submicroscopic levels of representation?
RQ3: Is there a relationship between the level of thinking ability or visual ability and eye-tracker measures at the macroscopic and submicroscopic levels of representation?

Method

In this research, a quantitative research approach with descriptive and non-experimental methods was used.

Participants

Seventy-nine students of three different age groups (primary education, upper secondary education, and university education levels) took part in this study, which was conducted in their native language, Slovenian. The group of primary school students comprised 30 12.0-year olds ($IQR = 0.0$ years); 29 upper secondary school students aged 16.0 years ($IQR = 1.0$ years) were in the second group; the group of university students had 20 pre-service teachers for two subjects (chemistry and biology) aged 23.0 years ($IQR = 2.0$ years). The students came from the Ljubljana region and participated in the study on a voluntary basis. For the students of primary and upper secondary education, the consent of school authorities, teachers, and parents was obtained, in accordance with the judgement of the Ethics Committee for Pedagogy Research of the Faculty of Education of the University of Ljubljana. The students were selected from a mixed urban population. All participants had normal or corrected-to-normal vision, and all were competent readers. To ensure anonymity, each student was assigned a code.

Instruments

Various instruments were used to collect data to answer the research questions, e.g., a context-based problem, a test of logical thinking, a test of visualisation ability, and eye-tracking apparatus.

Context-Based Problem About the Process of Opening a Bottle of Mineral Water

The problem consists of the macroscopic, submicroscopic representation and the text. The specific context-based problem is one of 11 science problems studied in the research funded by the Slovenian Research Agency (ARRS) entitled 'Explaining Effective and Efficient Problem Solving of the Triplet Relationship in Science Concepts Representations' (J5-6814) (Pavlin et al., 2019). The SMRs were designed by science educators; according to the ideas they developed, they were completed by the computer expert with expertise in creating chemistry animations. There was no limit to the time in which the participants viewed the animations. If the participants needed more time to solve the problem, the animations started again. However, the participants could not control the animations. The text of the context-based problem was written in Slovenian (Fig. 11.1). The problem starts with a photo from everyday life (closed and open bottle of mineral water) and SMR of the process. In the part of the study with the eye-tracker, the focus is on macroscopic and submicroscopic representations. The screen image was divided into seven AOI, four of which were the focus of our interest: two photos (Photo 1—closed bottle, Photo 2—open bottle) and two parts of the animation (upper and lower part).

Pri odpiranju steklenice mineralne vode sprva opazimo meglico, ki hitro izgine. Dobro si oglej predstavitev na nivoju delcev.

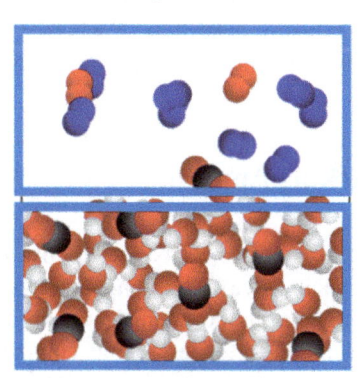

2. stopnja

Pojasni dogajanje ob odprtju steklenice mineralne vode. Pri razlagi upoštevaj obe stopnji (1 in 2) predstavitve.

Fig. 11.1 Screen image of the mineral water context-based problem with four labelled areas of interest with blue rectangles

Test of Logical Thinking

The Test of Logical Thinking (TOLT) is a multiple-choice paper-pencil test that assesses five skills of logical reasoning relevant for science teaching (Tobin & Capie, 1984). The test contains ten problems that require some consideration and the use of problem-solving strategies in different areas (i.e., controlling variables as well as proportional, correlational, probabilistic, and combinatorial reasoning). Participants were given a point for a correct answer and its explanation (in Exercises 1–8) and for the correct combinations and their correct number (in Exercises 9–10). These points were combined into a total score (maximum 10 points), which was used as the main result of the test (Devetak & Glažar, 2010).

Visualisation Ability Test

The pattern-based approach was used to assess students' visual processing skills (visualisation ability) with the application of the Pattern Comparison Test (PCT) from the Psychology Experiment Building Language (PEBL) test battery, a series of psychological tests for researchers and clinicians. In the PCT, there were 60 pairs of two grid patterns, 30 of which were the same and 30 different, displayed on the computer screen (Mueller & Piper, 2014). The participants individually had to compare the stimuli in pairs and respond as quickly as possible by pressing a specific key on a keyboard, regardless of whether the patterns were the same or different. Reaction time and the correctness of the answers were measured.

Eye-Tracking Apparatus

An eye-tracking device can be used to measure eye movements (i.e., fixations of the gaze on a specific area of the computer screen during a certain activity) and saccades (i.e., eye movements between fixations). To capture students' visual attention to different elements of the task on the computer screen, the eye-tracker measures the TFD, the average pupil size, the visit count to particular areas of interest (AOI) can be measured. This is also defined as the visual attention or attention allocation of the participant (Havanki & VandenPlas, 2014; Tsai et al., 2012). To ensure proper fixation allocation, the task displayed on the computer screen must be divided into several carefully and clearly divided AOIs, according to the placement elements that are of interest from the perspective of the research problem (Ferk Savec et al., 2016). The identification of saccades/fixations is based on the motion of gaze during each sample collected. If both velocity and acceleration thresholds (in our case: 30 degrees per second and 8000 degrees per second squared) are exceeded, a saccade begins; otherwise, the sample is labelled as a fixation. The screen-based eye-tracker apparatus EyeLink 1000 (35 mm lens, horizontal orientation) and associated software (Experiment Builder to prepare the experiment and connect to EyeLink; Data Viewer to collect the data and basic analysis) to record and analyse the students' eye movement when solving context-based problems was used. Data were collected at the right eye (monocular data collection that followed corneal reflection and pupil responses) at 500 Hz (Torkar et al., 2018).

Data Collection

TOLT was applied to the groups of participants in the standard environment one week before the eye-tracker study. The eye-tracker study was conducted from November 2016 to March 2017 at the Laboratory of the Department of Psychology at the Faculty of Arts, University of Ljubljana. Each participant completed the PCT study from the PEBL individually on the same day, but in a different room of the laboratory. After completing the PCT, participants entered the eye-tracker room. There was no time limit for the participants to solve eleven context-based problems; they needed about 30 min to complete them. Prior to the test, each participant was individually informed about the purpose of the study, the method used, and their role in the study. They sat about 60 cm from the screen (distance to the eyes) and had to place their head in a special headrest stand to ensure stability and gather the most optimal recordings. After the initial calibration and validation (through the nine-point algorithm), the participants solved all problems aloud: they gave oral answers (in the same order for all participants) while the experimenter wrote down their answers. To exclude distorting influences on pupil dilation, the lighting conditions were kept constant. The same procedure was performed with participants of each age group. The eye movements were measured with the eye-tracker. All data were collected in the Slovenian language (Pavlin et al., 2019).

Data Analysis

Participants' think-aloud responses were obtained during their solving of the context-based problem. To determine how students allocate attention to the macroscopic and submiscroscopic areas of interest, eye movement measures were obtained (recorded and analysed) with an EyeLink 1000 device and associated software. Experiment Builder software was used to prepare the experiment, and a connexion to the EyeLink Data Viewer was used to obtain the data for the basic analysis.

All data were collected in Excel and statistically processed in SPSS (Statistical Package for the Social Sciences). Basic descriptive statistics (median Mdn and an interquartile range IQR) of the numerical variables were determined. Spearman's Rank Order Correlation (rho, r) is used to calculate the strength of the relationship between two continuous variables. The Kruskal–Wallis nonparametric test was used to explain the relationship between eye-tracker measures on macroscopic and submicroscopic areas of interest among students of different groups and perceived problem difficulty due to the small sample size and non-normal distribution (Pallant, 2011). The Kruskal–Wallis test enables the comparison of results on a continuous variable for three or more groups. Post hoc Dunn tests are used to determine the differences between different pairs of groups (Pallant, 2011). Statistical hypotheses were tested at an alpha error rate of 5%. To describe whether the effects have a relevant magnitude, the effect size measure eta squared η^2 was used to describe the strength of a phenomenon. Cohen (1988) provided benchmarks to define small (.01), medium (.06), and large (.14) effects.

Results and Discussion

The results and discussions are presented according to the research questions.

Eye-Tracker Measures Among Students of Different Levels of Education

The first research question is related to the identification of eye-tracker measures of AOIs with macroscopic and submicroscopic representation level among students of different education level. Tables 11.1, 11.2, and 11.3 show whether the eye-tracker measures (TFD, VC, and APS) differ statistically significantly between students of different age groups.

The results in Table 11.1 show that students in all three educational levels spent more time on submicroscopic than on macroscopic representations. The TFDs on Photo 2 account for about half of the TFDs on Photo 1, which could indicate that familiar data represent stimuli that are processed more quickly (Topczewski et al., 2016).

However, there are statistically significant differences in TFDs between students of different age groups among AOIs with macroscopic representations (Photo 1 and Photo 2). The post hoc tests to test the pairwise comparisons were performed to show which group differs statistically from which group. Primary school students spent more time on AOIs with macroscopic representations than upper secondary school students or university students did. However, there are no statistically significant differences in TFDs on 3D dynamic SMR (divided into upper and lower parts) between students of different educational levels. This confirms the fact that primary school students spend more time observing AOIs with macroscopic representation than students of other educational levels and are more familiar with macroscopic imaging than with submicroscopic ones. As educational levels increase, students are less and less focused on the macroscopic level when solving a new context-based

Table 11.1 TFD on different AOIs in context-based problem related to a context-based problem of the processes occurring during the opening of a bottle of mineral water

AOI	Primary school students (Group 1)		Upper secondary school students (Group 2)		University students (Group 3)		Kruskal–Wallis test		
	Mdn [s]	IQR [s]	Mdn [s]	IQR [s]	Mdn [s]	IQR [s]	χ^2	p	η^2
Photo 1	8.894	7.449	3.124	3.889	2.186	3.654	22.975	.000[1]	.276
Photo 2	4.894	6.969	1.726	3.631	1.080	1.970	21.734	.000[2]	.260
3D SMR upper part	18.182	16.166	13.034	9.985	12.810	18.192	5.625	.060	–
3D SMR lower part	27.352	22.896	16.640	15.963	18.592	13.162	3.005	.223	–

Results of post hoc Dunn's tests:
[1]Group 1–2: $p = .004$; group 2–3: $p = .288$; group 1–3: $p = .000$; 1 > 2, 2 = 3, 1 > 3
[2]Group 1–2: $p = .020$; group 2–3: $p = .096$; group 1–3: $p = .000$; 1 > 2, 2 = 3, 1 > 3

Table 11.2 VC on different AOIs in a context-based problem about the processes occurring during the opening of a bottle of mineral water

AOI	Primary school students (Group 1)		Upper secondary school students (Group 2)		University students (Group 3)		Kruskal–Wallis test		
	Mdn	IQR	Mdn	IQR	Mdn	IQR	χ^2	p	η^2
Photo 1	15.00	9.00	8.00	3.50	7.00	9.00	17.194	.000[1]	.200
Photo 2	11.00	9.50	4.00	5.00	4.00	5.00	15.441	.000[2]	.177
3D SMR upper part	21.00	16.50	12.00	9.50	16.00	11.00	9.645	.008[3]	.101
3D SMR lower part	31.00	18.00	23.00	8.00	24.00	19.00	7.536	.023[4]	.073

Results of post hoc Dunn's tests:
[1]Group 1–2: $p = .005$; group 2–3: $p = .944$; group 1–3: $p = .000$; 1 > 2, 2 = 3, 1 > 3
[2]Group 1–2: $p = .049$; group 2–3: $p = .283$; group 1–3: $p = .000$; 1 > 2, 2 = 3, 1 > 3
[3]Group 1–2: $p = .007$; group 2–3: $p = .173$; group 1–3: $p = 1.000$; 1 > 2, 2 = 3, 1 = 3
[4]Group 1–2: $p = .036$; group 2–3: $p = 1.000$; group 1–3: $p = .106$; 1 > 2, 2 = 3, 1 = 3

problem (Chittleborough, 2014). However, research has shown that experts spent less time on task-irrelevant information than novices do (Gegenfurtner et al., 2011). This is also evident from Table 11.1, since university students who could be labelled experts have shorter TFDs, on AOIs with macroscopic levels of representation, than novices—primary school students do (Slapničar et al., 2017; Slapničar et al., 2018).

Table 11.3 APS on different AOIs in context-based problem about the processes occurring during the opening of a bottle of mineral water

AOI	Primary school students (Group 1)		Upper secondary school students (Group 2)		University students (Group 3)		Kruskal–Wallis test		
	Mdn [10^5 a.u.]	IQR [10^5 a.u.]	Mdn [10^5 a.u.]	IQR [10^5 a.u.]	Mdn [10^5 a.u.]	IQR [10^5 a.u.]	χ^2	p	η^2
Photo 1	1.78	.87	1.54	.77	1.04	1.41	10.285	.006[1]	.109
Photo 2	1.60	.76	1.42	1.70	.14	1.30	13.894	.001[2]	.157
3D SMR upper part	1.78	.59	1.51	1.02	1.35	.83	4.610	.100	–
3D SMR lower part	1.74	.64	1.67	.92	1.42	.79	2.601	.272	–

Results of post hoc Dunn's tests:
[1] Group 1–2: $p = .000$; group 2–3: $p = .064$; group 1–3: $p = .005$; $1 = 2, 2 = 3, 1 > 3$
[2] Group 1–2: $p = .310$; group 2–3: $p = .060$; group 1–3: $p = .001$; $1 = 2, 2 = 3, 1 > 3$

Table 11.2 presents VC on four different AOIs. Repeated visits to an area of interest (AOI) may indicate characteristics that the observer considers important or interesting (West et al., 2006). It is obvious that statistically significant differences in VC occur between students of different stages of education. AOIs with macroscopic representation were more frequently revisited by primary school students than by students of upper secondary school and university. The differences are statistically significant. However, VCs on AOIs with submicroscopic representation among students of different educational levels are also statistically significant. Differences occur between primary and upper secondary education, while there are no statistically significant differences between primary and university students and between upper secondary and university students.

Table 11.3 shows that there are differences between students at different levels of education in APS on AOIs with macroscopic representations. The APS decreases with increasing age of the students. The results suggest that mental effort decreases with the levels of education due to pupil diameter if the results of the study by Just et al. (2003) are taken into account. Furthermore, pupil dilation could be used to identify mental effort during problem processing (Korbach et al., 2018). The results show that there are statistically significant differences in APS at the macroscopic level. However, there are no statistically significant differences in APS at submicroscopic levels among students at different levels of education. It could be interpreted that the SMR represents a similar cognitive load for students of all ages.

Eye-Tracker Measures Among Students' Perceived Task Difficulty

The second research question relates to the identification of differences between students of different levels of their perception of context-based problem difficulty in eye-tracker measures on AIOs with macro- and submicroscopic representation levels. According to Nunan and Keobke (1995), one of the three factors is the lack of familiarity with the type of task, which could justify students' perception of task difficulty as moderate on average (*Mdn*=3.00, *IQR*=2.00).

Nevertheless, Tables 11.4, 11.5, and 11.6 show whether the eye-tracker measures (TFD, VC, and APS) differ statistically significantly among students who self-assess the difficulty of the context-based problem differently. The results show that there are no statistically significant differences in the eye-tracker measures among the students who assessed the context-based problems at the given difficulty level. However, the context-based problem including the process of opening the bottle of mineral water is new to the students, as it has never occurred before in the existing curricula and textbooks (Bačnik et al., 2009, 2011; Balon et al., 2011, Planinšič et al., 2009; Skvarč et al., 2011; Verovnik et al., 2011). Therefore, it might turn out that the self-assessment of a context-based problem according to the specifics of the participants is complex, and it is difficult to conclude anything if the deeper analysis is missing.

Viiri et al. (2017) report that simple measures of time (visit durations on the question or options) are well suited to distinguish between different confidence levels of test participants. However, this was not at the forefront of the present study. Furthermore, Andrzejewska and Stolińska (2016) found that there was no correlation between the activity of eye movement parameters, which are considered indicators of mental effort, and the opinion of a student about the difficulty of the task, as confirmed in the presented study.

Eye-Tracker Measures Among Students of Different Levels of Logical Thinking and Visualisation Abilities

The third research question relates to the relationship between the level of logical reasoning or visual ability and eye-tracker measures at the macroscopic and submicroscopic levels of representation. The relationship between eye-tracker measures on AOIs with macroscopic and submicroscopic representations and first TOLT and then PCT was investigated using the Pearson product-moment correlation coefficient (Tables 11.7 and 11.8).

There is a medium, negative correlation between VC on macroscopic representations and TOLT between TFD on macroscopic representations and TOLT, which means that the student with greater formal thinking skills spends less time on AOIs with macroscopic representations and returns less often to the AOIs mentioned. There is a medium, negative correlation between APS on submicroscopic representation

Table 11.4 TFD on different AOIs in a context-based problem related to the process of opening a bottle of mineral water

AOI	Very easy (N=1) Mdn [s]	IQR [s]	Easy (N=20) Mdn [s]	IQR [s]	Moderate (N=34) Mdn [s]	IQR [s]	Difficult (N=18) Mdn [s]	IQR [s]	Very difficult (N=6) Mdn [s]	IQR [s]	Kruskal–Wallis test χ^2	p	η^2
Photo 1	2.620	–	4.196	6.650	6.532	6.634	7.118	7.933	4.359	8.066	2.977	.562	–
Photo 2	1.026	–	1.738	3.814	4.680	7.460	2.710	3.465	2.948	3.578	.750	.945	–
3D SMR upper part	9.830	–	12.794	8.056	18.182	15.948	14.550	13.387	10.440	15.370	8.892	.064	–
3D SMR lower part	12.218	–	16.690	16.962	25.356	20.828	24.728	15.887	15.448	37.144	3.979	.409	–

Table 11.5 VC on different AOIs in context-based problem related to the process of opening of a bottle of mineral water

AOI	Very easy (N=1) Mdn	IQR	Easy (N=20) Mdn	IQR	Moderate (N=34) Mdn	IQR	Difficult (N=18) Mdn	IQR	Very difficult (N=6) Mdn	IQR	Kruskal–Wallis test χ^2	p	η^2
Photo 1	9.00	–	9.00	6.00	12.00	13.00	10.00	9.00	10.00	15.50	1.106	.893	–
Photo 2	2.00	–	5.00	6.00	9.00	10.00	6.00	6.50	8.50	8.75	1.580	.812	–
3D SMR upper part	10.00	–	15.00	8.00	21.00	18.00	15.00	15.00	18.00	14.50	7.913	.095	–
3D SMR lower part	16.00	–	23.00	12.00	27.00	12.00	28.00	16.50	28.50	20.00	4.277	.370	–

Table 11.6 APS on different AOIs in context-based problem related to the process of opening of a bottle of mineral water

AOI	Very easy (N=1) Mdn [10^5 a.u.]	Very easy (N=1) IQR [10^5 a.u.]	Easy (N=20) Mdn [10^5 a.u.]	Easy (N=20) IQR [10^5 a.u.]	Moderate (N=34) Mdn [10^5 a.u.]	Moderate (N=34) IQR [10^5 a.u.]	Difficult (N=18) Mdn [10^5 a.u.]	Difficult (N=18) IQR [10^5 a.u.]	Very difficult (N=6) Mdn [10^5 a.u.]	Very difficult (N=6) IQR [10^5 a.u.]	Kruskal–Wallis test χ^2	Kruskal–Wallis test p	Kruskal–Wallis test η^2
Photo 1	1.71	–	1.41	1.19	1.33	1.03	1.71	.66	1.64	1.18	1.106	.893	–
Photo 2	.18	–	1.60	1.43	1.20	1.51	1.34	1.65	1.67	1.17	1.580	.812	–
3D SMR upper part	1.61	–	1.67	.87	1.47	.89	1.70	.69	1.94	1.10	7.913	.095	–
3D SMR lower part	1.71	–	1.54	.89	1.47	.80	1.76	.59	1.99	.71	4.277	.370	–

Table 11.7 Spearman's rho correlations between TFD, VC, or APS in AOIs with macroscopic representations for a context-based problem related to the process of opening a bottle of mineral water and TOLT or PCT ($N=79$)

	VC Photo 1		VC Photo 2		APS Photo 1		APS Photo 2		TFD Photo 1		TFD Photo 2	
	r	p	r	p	r	p	r	p	r	p	r	p
TOLT	−.341	.003	−.345	.002	−.192	.097	−.121	.324	−.362	.129	−.334	.003
PCT	.079	.490	.005	.964	−.014	.903	.147	.221	.001	.258	−.019	.866

Table 11.8 Spearman's rho correlations between TFD, VC, or APS in AOIs with submicroscopic representations for a context-based problem related to the process of opening a bottle of mineral water and TOLT or PCT ($N=79$)

	VC 3D SMR upper part		VC 3D SMR lower part		APS 3D SMR upper part		APS 3D SMR lower part		TFD 3D SMR upper part		TFD 3D SMR lower part	
	r	p	r	p	r	p	r	p	r	p	r	p
TOLT	−.066	.570	−.135	.246	−.312	.006	−.147	.206	−.026	.823	.076	.515
PCT	.119	.296	.124	.277	.147	.199	−.015	.898	.061	.591	.029	.802

and TOLT. It could be interpreted that this representation represents a lower cognitive load for students with higher intellectual capacity, considering that it has been shown that the pupil becomes larger when individuals are confronted with cognitively challenging tasks, which proves that the autonomic pupil response can be used as a marker for mental effort (Korbach et al., 2018; Lang et al., 2020).

However, the ability to visualise refers to the recognition and manipulation of visual objects by the students (Barnea, 2000). The results from Tables 11.7 and 11.8 show that there are no correlations between PCT and eye-tracker measures of AOIs with both macroscopic and submicroscopic representations, which means that the visualisation ability identified by the PCT (i.e., a test of visual processing and pattern recognition) did not come to the fore in solving a new context-based problem with SMR (Phillips et al., 2010). However, the study by Pavlin et al. (2019) also shows that visualisation abilities do not play a significant role in choosing the correct one from among the three animations. It may, therefore, seem logical that visualisation ability is not reflected in the degree of cognitive load identified by pupil dilation.

Conclusions

This paper aimed to investigate and explain the eye-tracker measures (total fixation duration, average pupil dilation, visit count) of students of different age groups at the AOIs with macroscopic and submicroscopic representation and context-based problem difficulty perception and to identify correlations between eye-tracker measures and the level of logical thinking as well as visualisation ability.

When we conclude that educational level affects the decreasing time spend on the macroscopic representation, we find that older students process familiar stimuli faster, while the influence on submicroscopic stimuli is not significant. The number of visits to the macroscopic representations as well as to the submicroscopic representations varies among students of different age groups. In addition, in the upper part of the animation, which was crucial in answering the question, differences in pupil dilation, which identifies cognitive load, are found in students of different age groups.

However, the students found the context-based problem to be from very easy to very difficult. The self-assigned difficulty level of the context-based problem is not reflected in eye-tracker measures for AOIs with macroscopic and submicroscopic representations.

Visualisation abilities are not significant when observing, returning to, and processing the information on AOIs with macroscopic and submicroscopic representations over time. Furthermore, the results of the test of logical thinking correlate with the time spent and the number of returns on AOIs with macroscopic representations. Its results correlate with pupil dilation, which represents the effort of students solving problems.

Limitations of This Research

The main limitation of this research is the limited number of characteristics of the participants that allow for in-depth analysis. A second limitation results from the measurement of pupil dilation; despite ample evidence that pupil dilation is associated with complex cognitive processing, there may be other factors that affect pupil diameter. While physical factors, such as light conditions and screen distance, were kept constant, we did not consider other variables on the participant side that might affect pupil dilation, such as emotional arousal and autonomic activation (Bradley et al., 2008; Lang et al., 2020). Furthermore, the triple nature of science concepts should be integrated into the context-based problem where appropriate; the problem discussed in this research was not designed to require the symbolic level. The screen image should be developed so that eye-tracking measurements are clearly defined. An additional limitation of this study is the breadth of students' knowledge.

Implications for the Educational Process

The research suggests that more emphasis should be placed on teaching context-based problems of varying difficulty levels and that submicroscopic representations should be presented as simply as possible so that they do not require too much cognitive effort to the student. When designing teaching materials, the teacher must ensure that a sufficient number of carefully selected visualisation elements are included to arouse the interest of the student.

Further Research Guidelines

One possible area for future research would be to examine how students' self-esteem and visualisation skills are reflected in their problem-solving abilities. In addition, the number of repeat visits provides information on how often a participant returned his gaze to a specific location defined by an AOI. The participant may be attracted to a certain AOI because it is attractive, because it is confusing, or even because it is frustrating. From the eye-tracking measures, it is not evident how someone felt while looking at something; this issue should be further investigated. Further work is needed to determine whether the symbolic level of science concepts influences context-based problem-solving from an information processing perspective. The question of the role of teachers in the presentation of particle animations in class about specific science concepts is an interesting question that could be usefully investigated in further research.

Acknowledgements This research was supported by the project 'Explaining Effective and Efficient Problem Solving of the Triplet Relationship in Science Concepts Representations' (J5-6814), financed by the Slovenian Research Agency (ARRS).

References

Andrzejewska, M., & Stolińska, A. (2016). Comparing the difficulty of tasks using eye tracking combined with subjective and behavioural criteria. *Journal of Eye Movement Research, 9*(3), 1–16.
Bačnik, A., Bukovec, N., Poberžnik, A., Požek Novak, T., Keuc Z., Popič, H., & Vrtačnik, M. (2009). *Učni načrt, Program srednja šola, Kemija: Gimnazija: klasična, strokovna gimnazija* [Curriculum, program of secondary school, chemistry: Gymnasium: Classical, professional gymnasium]. Ljubljana: National Education Institute Slovenia.
Bačnik, A., Bukovec, N., Vrtačnik, M., Poberžnik, A., Križaj, M., Stefanovik, V., Sotlar, K., Dražumerič, S., & Preskar, S. (2011). *Učni načrt, Program osnovna šola, Kemija* [Curriculum, program of primary school, chemistry]. Ljubljana: national education institute Slovenia.
Balon, A., Gostinčar Blagotinšek, A., Papotnik, A., Skribe Dimec, D., & Vodopivec, I. (2011). *Učni načrt, Program osnovna šola, Naravoslovje in tehnika* [Curriculum, program of primary school, science and technology]. Ljubljana: National Education Institute Slovenia.
Barnea, N. (2000). Teaching and learning about chemistry and modelling with a computer managed modelling system. In J. K. Gilbert & C. Boulter (Eds.), *Developing models in science education* (pp. 307–324). Dordrecht: Kluwer Academic Publishers.
Bassok, M. (1990). Transfer of domain-specific problem solving procedures. *Journal of Experimental Psychology: Learning, Memory, and Cognition, 16*(3), 522–533.
Beatty, J., & Lucero-Wagoner, B. (2000). The pupillary system. In *Handbook of Psychophysiology* (2nd ed., pp. 142–162). New York: Cambridge University Press.
Bradley, M. M., Miccoli, L., Escrig, M. A., & Lang, P. J. (2008). The pupil as a measure of emotional arousal and autonomic activation. *Psychophysiology, 45*(4), 602–607.
BSCS. (2008). *Scientists and science education*. Retrieved from http://science.education.nih.gov/SciEdNation.nsf/EducationToday1.html.
Carmichael, A., Larson, A., Gire, E., Loschky, L., & Rebello, N. S. (2010). How does visual attention differ between experts and novices on physics problems? In *AIP Physics Education Research Conference 2010* (pp. 93–96). Manhattan: Kansas State University.
Cheng, M. M. W., & Gilbert, J. K. (2017). Modelling students' visualisation of chemical reaction. *International Journal Science Education, 39*(9), 1173–1193.
Chittleborough, G. (2014). The development of theoretical frameworks for understanding the learning of chemistry. In I. Devetak & S. A. Glažar (Eds.), *Learning with understanding in the chemistry classroom* (pp. 25–40). London: Springer, Dordrecht Heidelberg New York.
Cohen, J. (1988). *Statistical power analysis for the behavioural sciences* (2nd ed.). Hillsdale, New York: Lawrence Erlbaum Associates.
Devetak, I. (2012). *Zagotavljanje kakovostnega znanja naravoslovja s pomočjo submikroreprezentacij, Analiza ključnih dejavnikov zagotavljanja kakovosti znanja v vzgojno-izobraževalnem sistemu* [The analysis of the key factors in ensuring the quality of knowledge in educational system]. Ljubljana: Faculty of education, University of Ljubljana.
Devetak, I., & Glažar, S. A. (2010). The influence of 16-year-old students' gender, mental abilities, and motivation on their reading and drawing submicro representations achievements. *International Journal of Science Education, 32*(12), 1561–1593.
Docktor, J. L., & Mestre, J. P. (2014). Synthesis of discipline-based education research in physics. *Physical Review Special Topics—Physics Education Research, 10*(2), 1–58.

Exerciseer, R., & Dalton, R. (2006). Research into practice: Visualisation of the molecular word using animations. *Chemistry Education Research and Practice, 7*(2), 141–159.

Ferk Savec, V., Hrast, Š., Devetak, I., & Torkar, G. (2016). Beyond the use of an explanatory key accompanying submicroscopic representations. *Acta Chimica Slovenica, 63*(4), 864–873.

Gegenfurtner, A., Lehtinen, E., & Saljo, R. (2011). Expertise differences in the comprehension of visualisations: A meta-analysis of the eye-tracking research in professional domains. *Educational Psychology Review, 23*(2), 523–552.

Gilbert, J. K. (2005). Visualisation: A metacognitive skill in science and science education. In J. K. Gilbert (Ed.), *Visualisation in science education* (pp. 9–27). Dordrecht: Kluwer Academic Press.

Graesser, A. C., Lu, S., Olde, B. S., Cooper-Pye, E., & Whitten, S. (2005). Question asking and eye tracking during cognitive disequilibrium: Comprehending illustrated texts on devices when the devices break down. *Memory and Cognition, 33*(7), 1235–1247.

Green, H. J., Lemaire, P., & Dufau, S. (2007). Eye movement correlates of younger and older adults' strategies addition. *Acta Psychologica, 125*(12), 257–278.

Han, J., Chen, L., Fu, Z., Fritchman, J., & Bao, L. (2017). Eye-tracking of visual attention in web-based assessment using the force concept inventory. *European Journal of Physics, 38*(4), 1–16.

Harlen, W. (Ed.). (2010). *Principles and big ideas of science education*. Hatfield, England: Association for Science Education.

Havanki, K. L., & VandenPlas, J. R. (2014). Eye tracking methodology for chemistry education research. In D. M. Bunce & R. S. Cole (Eds.), *Tools of chemistry education research* (pp. 191–218). Washington, DC: American Chemical Society.

Hegarty, M. (1992). The mechanics of comprehension and comprehension of mechanics. In K. Rayner (Ed.), *Eye movements and visual cognition* (pp. 428–443). New York: Springer-Verlag.

Hoffman, A. B., & Rehder, B. (2010). The costs of supervised classification: The effect of learning task on conceptual flexibility. *Journal of Experimental Psychology, 139*(2), 319–340.

Hyönä, J., Lorch, R. F., & Kaakinen, J. K. (2002). Individual differences in reading to summarise expository text: Evidence from eye fixation patterns. *Journal of Education Psychology, 94*(1), 44–55.

International Council for Science. (2011). *Report of the ICSU ad-hoc review panel on science*. Paris, France: Author. Retrieved from http://www.icsu.org/publications/reports-and-reviews/external-review-of-icsu.

Johnstone, A. H. (1982). Macro- and micro-chemistry. *School Science Review, 64*(227), 377–379.

Johnstone, A. H. (1991). Why is science difficult to learn? Things are seldom what they seem. *Journal of Computer Assisted Learning, 7*(2), 75–83.

Johnstone, A. H. (2001). Teaching of chemistry-logical or psychological? *Chemistry Education Research and Practice, 1*(1), 9–15.

Just, M. A., & Carpenter, P. A. (1980). A theory of reading: From eye fixations to comprehension. *Psychological Review, 87*(4), 329–354.

Just, M. A., Carpenter, P. A., & Miyake, A. (2003). Neuroindices of cognitive workload: Neuroimaging, pupillometric and event-related potential studies of brain work. *Theoretical Issues in Ergonomics Science, 4*(1), 56–88.

Karch, J. M. (2018). Beyond gaze data: Pupillometry as an additional data source in eye tracking. In J. R. VandenPlas, S. J. R. Hansen, & S. Cullipher (Eds.), *Eye tracking for the chemistry education researcher* (pp. 145–163). Washington, DC: American Chemical Society.

Kautz, C. H., Heron, P. R. L., Shaffer, P. S., & McDermott, L. C. (2005). Student understanding of the ideal gas law, Part II: A microscopic perspective. *American Journal of Physics, 73*(11), 1064–1071.

Klein, P., Lichtenberger, A., Küchemann, S., Becker, S., Kekule, M., Viiri, J., Baadte, C., Vaterlaus, A., & Kuhn, J. (2020). Visual attention while solving the test of understanding graphs in kinematics: An eye-tracking analysis. *European Journal of Physics, 41*(2), 1–17.

Kohl, P. B., & Finkelstein, N. D. (2005). Student representational competence and self-assessment when solving problems. *Physical Review Special Topics—Physics Education Research, 1*(1), 1–11.

Korbach, A., Brünken, R., & Park, B. (2018). Differentiating different types of cognitive load: A comparison of different measures. *Educational Psychology Review, 30*(4), 503–529.

Kozhevnikov, M., Motes, M. A., & Hegarty, M. (2007). Spatial visualisation in physics problem solving. *Cognitive Science, 31*(4), 549–579.

Lang, F., Kammerer, Y., Oschatz, K., Stürmer, K., & Gerjets, P. (2020). The role of beliefs regarding the uncertainty of knowledge and mental effort as indicated by pupil dilation in evaluating scientific controversies. *International Journal of Science Education, 42*(3), 350–371.

Levy, S. T., & Wilinsky, U. (2009). Crossing levels and representations: The connected chemistry (CC1) curriculum. *Journal of Science Education and Technology, 18*(3), 224–242.

Lewis, J., & Leach, J. (2006). Discussion of socio-scientific issues: The role of science knowledge. *International Journal of Science Education, 28*(11), 1267–1287.

Lin, Y. I., Son, J. Y., & Rudd, J. A. (2016). Asymmetric translation between multiple representations in chemistry. *International Journal of Science Education, 38*(4), 644–662.

Lloyd, J. K., Smith, R. G., Fay, C. L., Khang, G. N., Wah, L. L. K., & Sai, C. L. (1998). Subject knowledge for science teaching at primary level: A comparison of pre-service teachers in England and Singapore. *International Journal of Science Education, 20*(5), 521–532.

Madsen, A. M., Larson, A. M., Loschky, L. C., & Rebello, N. S. (2012). Differences in visual attention betweenthose who correctly and incorrectly answer physics problems. *Physical Review Special Topics—Physics Education Research, 8*(1), 1–14.

Maloney, D. (2011). An overview of physics education: Research on problem solving. *Research in Problem Solving, 2*(1), 1–31.

Meltzer, D. E. (2005). Relation between students' problem-solving performance and representational format. *American Journal of Physics, 73*(5), 463–478.

Millar, R. (2006). Twenty first century science: Insights from the design and implementation of a scientific literacy approach in school science. *International Journal of Science Education, 28*(13), 1499–1521.

Mozaffari, S., Klein, P., Bukhari, S. S., Kuhn, J., & Dengel, A. (2016). Entropy based transition analysis of eye movement on physics representational competence. In *Proceedings of the 2016 ACM International Joint Conference on Pervasive and Ubiquitous Computing: Adjunct* (pp. 1027–1034).

Mueller, S. T., & Piper, B. J. (2014). The psychology experiment building language (PEBL) and PEBL test battery. *Journal of Neuroscience Methods, 222*(14), 250–259.

Nunan, D., & Keobke, K. (1995). Task difficulty from the learner's perspective: Perceptions and reality. *Hong Kong Papers in Linguistics and Language Teaching, 18*(3), 1–12.

Pallant, J. (2011). *SPSS survival manual: A step by step guide to data analysis using SPSS* (4th ed.). Crows Nest: Allen & Unwin.

Pavlin, J., Glažar, S. A., Slapničar, M., & Devetak, I. (2019). The impact of students' educational background, interest in learning, formal reasoning and visualisation abilities on gas context-based exercises achievements with submicro-animations. *Chemistry Education Research and Practice, 20*(3), 633–649.

Phillips, L. M., Norris, S. P., & Macnab, J. S. (2010). *Visualisation in mathematics, reading and science education*. Dordrecht: Springer.

Planinšič, G., Belina, R., Kukman, I., & Cvahte, M. (2009). *Učni načrt, Program srednja šola, Fizika: gimnazija: klasična, strokovna gimnazija* [Curriculum, Program of secondary school, Physics: Gymnasium: Classical, professional gymnasium]. Ljubljana: National education institute Slovenia.

Rayner, K. (1998). Eye movements in reading and information processing: 20 years of research. *Psychological Bulletin Journal, 124*(3), 372–422.

Rayner, K. (2009). Eye movements and attention in reading, scene perception, and visual search. *Quarterly Journal of Experimental Psychology, 62*(8), 1457–1506.

Skvarč, M., Glažar, S. A., Marhl, M., Skribe Dimec, D., Zupan, A., Cvahte, M., Gričnik, K., Volčini, D., Sabolič, G., & Šorgo, A. (2011). *Učni načrt, Program osnovna šola, Naravoslovje* [Curriculum, program of primary school, science]. Ljubljana: National education institute Slovenia.

Slapničar, M., Devetak, I., Glažar, S. A., & Pavlin, J. (2017). Identification of the Understanding of the states of water and air among Slovenian students aged 12, 14 and 16 years through solving authentic exercises. *Journal of Baltic Science Education, 16*(3), 308–323.

Slapničar, M., Tompa, V., Glažar, S. A., & Devetak, I. (2018). Fourteen-year-old students' misconceptions regarding the sub-micro and symbolic levels of specific chemical concepts. *Journal of Baltic Science Education, 17*(4), 620–632.

Smith, D., Mestre, J. P., & Ross, B. H. (2010). Eye-gaze patterns as students study worked out examples in mechanics. *Physical Review Special Topics—Physics Education Research, 6*(1): 1–9.

Stuart, S., Hickey, A., Vitório, R., Welman, K. E., Foo, S., Keen, D., & Godfrey, A. (2019). Eye-tracker algorithms to detect saccades during static and dynamic tasks: A structured review. *Physiological Measurement, 40*(2), 1–26.

Susac, A., Bubic, A., Kaponja, J., Planinic, M., & Palmovic, M. (2014). Eye movement reveal students' strategies in simple equation solving. *International Journal of Science and Mathematics Education, 12*(3), 555–577.

Taber, K. S. (2013). Revisiting the chemistry triplet: Drawing upon the nature of chemical knowledge and the psychology of learning to inform chemistry education. *Chemistry Education Research and Practice, 14*(2), 156–168.

Tai, R. H., Loehr, J. F., & Brigham, F. J. (2006). An exploration of the use of eye-gaze tracking to study problem-solving on standardised science assessments. *International Journal of Research and Method in Education, 29*(3), 185–208.

Tobin, K., & Capie, W. (1984). The test of logical thinking. *International Journal of Science and Mathematics Education Southeast Asia, 7*(1), 5–9.

Topczewski, J. J., Topczewski, A. M., Tang, H., Kendhammer, L. K., & Pienta, N. J. (2016). NMR spectra through the eyes of a student: Eye tracking applied to NMR items. *Journal of Chemical Education, 94*(1), 29–37.

Torkar, G., Veldin, M., Glažar, S. A., & Podlesek, A. (2018). Why do plants wilt? Investigating students' understanding of water balance in plants with external representations at the macroscopic and submicroscopic levels. *Eurasia Journal of Mathematic Science Technology and Education, 14*(6), 2265–2276.

Tsai, M. J., Hou, H. T., Lai, M. L., Liu, W. L., & Yang, F. Y. (2012). Visual attention for solving multiple-choice science problem: An eye-tacking analysis. *Computer and Education, 58*(6), 375–385.

Van Gog, T., Paas, F., & Van Merriënboer, J. J. G. (2005). Uncovering expertise-related differences in trouble-shooting performance: Combining eye movement data and concurrent verbal protocol data. *Applied Cognitive Psychology, 19*(2), 205–221.

Verovnik, I., Bajc, J., Beznec, B., Božič, S., Brdar, U. V., Cvahte, M., Gerlič, I., & Munih, S. (2011). *Učni načrt, Program osnovna šola, Fizika* [Curriculum, program of primary school, physics]. Ljubljana: National education institute Slovenia.

Vieira, R. M., & Tenreiro-Vieira, C. (2014). Fostering scientific literacy and critical thinking in elementary science education. *International Journal of Science and Mathematics Education, 14*(4), 659–680.

Viiri, J., Kekule, M., Isoniemi, J., & Hautala, J. (2017). *Eye-tracking the effects of representation on students' problem solving approaches.* Paper presented at the Proceedings of the Annual FMSERA Symposium 2016 (pp. 88–98). Joensuu: FMSERA.

West, J. M., Haake, A. R., Rozanski, E. P., & Karn, K. S. (2006). *EyePatterns: Software for identifying patterns and similarities across fixation sequences.* Paper presented at the Proceedings of the 2006 Symposium on Eye Tracking Research and Applications (pp. 149–154). New York: ACM Press.

Wu, H., & Shah, P. (2004). Exploring visuospatial thinking in learning. *Science Education, 88*(3), 465–492.

Yen, M. H., & Yang, F. Y. (2016). Methodology and application of eye-tracking techniques in science education. In M. H. Chiu (Ed.), *Science education research and practices in Taiwan* (pp. 249–277). Springer: Singapore.

Jerneja Pavlin, Ph.D. is an Assistant Professor of Physics Education at University of Ljubljana, Faculty of Education, Slovenia. Her research focuses on the investigation of different approaches to physics and science teaching (e.g. using 3D sub-microscopic representations, didactic games, outdoor teaching and learning, peer instruction etc.), aspects of scientific literacy and the introduction of contemporary science into physics teaching, from the development and optimization of experiments in the field of modern materials to the evaluation of the developed of learning modules, etc. She was involved in several research projects in the field of science education. She is a member of GIREP (International Research Group on Physics Teaching) and ESERA (European Science Education Research Association). In total she has published about 150 different publications. Dr. Pavlin is a member of the editorial board of CEPS Journal.

Miha Slapničar Ph.D. student, is a teaching assistant of chemical education at University of Ljubljana, Faculty of Education, Department of Biology, Chemistry and Home Economics. His research focuses on students' redox reaction comprehension in connection with triple nature of chemical concepts using eye-tracking technology, experimental work in the field of general, inorganic and organic chemistry, spectroscopy aspects of the chemistry of natural compounds, and the chemical knowledge assessment.

Chapter 12
Students' Understanding of Diagrams in Different Contexts: Comparison of Eye Movements Between Physicists and Non-physicists Using Eye-Tracking

Pascal Klein, Stefan Küchemann, Ana Susac, Alpay Karabulut, Andreja Bubic, Maja Planinic, Marijan Palmovic, and Jochen Kuhn

Introduction

The understanding of graphs and their adequate handling plays an important role in physics and in the other STEM (Science, Technology, Engineering, and Mathematics) disciplines. Graphs serve to simplify the representation of complex relationships, and to facilitate the exchange of information between individuals as well as the development of conceptual knowledge in a domain (Curcio, 1987; Freedman & Shah, 2002; Pinker, 1990; Strobel et al., 2018). They are also important in everyday

---

P. Klein (✉)
Faculty of Physics, Physics Education Research, University of Göttingen, Friedrich-Hund-Platz 1, 37077 Göttingen, Germany
e-mail: pascal.klein@uni-goettingen.de

S. Küchemann · A. Karabulut · J. Kuhn
Department of Physics, Physics Education Research, Technische Universität Kaiserslautern, Erwin-Schrödinger-Str. 46, 67663 Kaiserslautern, Germany

A. Susac
Department of Applied Physics, Faculty of Electrical Engineering and Computing, University of Zagreb, Unska 3, 10000 Zagreb, Croatia

A. Bubic
Chair for Psychology, Faculty of Humanities and Social Sciences, University of Split, Sinjska 2, 21000 Split, Croatia

M. Planinic
Department of Physics, Faculty of Science, University of Zagreb, Bijenicka 32, 10000 Zagreb, Croatia

M. Palmovic
Laboratory for Psycholinguistic Research, Department of Speech and Language Pathology, University of Zagreb, Borongajska cesta 83h, 10000 Zagreb, Croatia

life, because information in newspapers, on the Internet and TV is often conveyed through graphs. Student understanding of graphs has been investigated in many studies in physics education, mathematics education, and educational psychology. For example, McDermott et al. (1987) reported about dominant student difficulties in connecting graphs to physical concepts: discriminating between the slope and height of a graph, interpreting changes in height and changes in slope, relating one type of graph to another, matching narrative information with relevant features of a graph, and interpreting the area under a graph. A few years later, Beichner (1994) developed the Test of Understanding of Graphs in Kinematics (TUG-K), which became one of the most widely used PER assessment instruments and which includes well-examined student difficulties in the context of graphs. This test was recently modified by Zavala et al. (2017). More recently, a number of studies investigated and compared university students' understanding of graphs in mathematics, physics, and other contexts (Christensen & Thompson, 2012; Planinic et al., 2012; Wemyss & van Kampen, 2013; Planinic et al., 2013; Ivanjek et al., 2016; Bollen et al., 2016; Ivanjek et al., 2017). While most studies explored student interpretation of graph slope, there are only a few studies on student understanding of area under a graph. Recent studies using parallel (isomorphic) problems in mathematics, physics, and other contexts have shown that parallel problems with added context (physics or other context) were more difficult than the corresponding mathematics problems (Planinic et al., 2012; Wemyss & van Kampen, 2013; Planinic et al., 2013; Ivanjek et al., 2016, 2017). This suggested student difficulties with transfer of knowledge between mathematics and physics (or other contexts). It was found that students solved questions on water level vs time graphs better than the corresponding questions on distance vs time graphs, although they had never encountered former graphs in the formal educational setting. The analysis of students' responses and the categorization of their strategies revealed that they used similar correct and incorrect strategies regardless of country (Ireland, Belgium, and Spain in Bollen et al., 2016; Croatia and Austria in Ivanjek et al., 2017) or the level of mathematical proficiency (algebra-based or calculus-based physics courses).

Many researchers investigated students' ability to transfer mathematical skills to a different context such as physics or chemistry. The results have shown that direct transfer rarely occurs, i.e. students rarely apply problem solving strategy learned in a particular context (mathematics) to another context (e.g. physics). Obviously, students' ability to transfer certain mathematical skills depends on their possession of the required mathematical knowledge. Some researchers reported that the main cause of students' difficulties with transfer is a lack of the mathematical skills to be transferred (e.g. Potgieter et al., 2008; Hoban et al., 2013). However, studies on student reasoning about graphs in different contexts have shown that students who successfully solve problems in (purely) mathematical context, often fail to solve corresponding problems in physics or other contexts (Planinic et al., 2012; Wemyss & van Kampen, 2013; Planinic et al., 2013; Ivanjek et al., 2016, 2017). Some authors, such as Bransford and Schwartz (1999), suggested departure from studies that are looking for direct transfer to a broader view on transfer which includes students' "preparation for future learning". This perspective assumes that students who do not directly apply prior knowledge in a new context, might still be able to adapt prior knowledge and use it as a support in new learning. Hammer et al. (2005) discussed a resources framework as basis for this broader view of transfer. When students learn a new idea, they activate existing resources in new combinations, and activation of the resources depends on the context and the provided scaffolding. Indeed, a previous study on student reasoning about graphs in different contexts reported some examples of transfer of knowledge in the sense of preparation for future learning, such as using dimensional analysis (acquired in physics) in solving problems in other contexts (Ivanjek et al., 2016).

All previous studies on the influence of context on performance were conducted on physics students, so we decided to additionally explore non-physics students' understanding of graphs in both physics and everyday contexts (Klein et al., 2019; Susac et al., 2018). For the everyday context we have chosen the finance context that all students are familiar with. In addition to students' scores, we used eye tracking in our studies to investigate where students allocate visual attention during problem solving. Eye tracking has proven to be a powerful method that complements previous research with a data source on visual attention, i.e. how students extract information from graphs. This yields process data of learners while they solve problems with graphs. The measurement of eye movements is an increasingly used method in the educational sciences. There are a number of eye-tracking studies on understanding of graphs (Kozhevnikov et al., 2007; Viiri et al., 2017; Kekule, 2014; Carpenter & Shah, 1998; Goldberg & Helfman, 2011). In some studies kinematics graphs were used to investigate graph comprehension (Kozhevnikov et al., 2007; Viiri et al., 2017; Kekule, 2014), otherwise different graphs with everyday contexts were employed (Carpenter & Shah, 1998; Goldberg & Helfman, 2011). In the context of understanding kinematics graphs, Madsen et al. (2012) have shown that students who answer a question correctly focus longer on specific relevant areas of a graph such as the axes. Their findings also suggest that previous experience with a topic can increase the focus on the important regions (Madsen et al., 2013). Conversely, it can be assumed that learning difficulties and misunderstandings in the use of graphs (e.g. point- interval confusion; Leinhardt et al., 1990), which are well studied and well known in the literature, are reflected in certain eye-movement patterns and attention distributions that are shifted to conceptually irrelevant areas. Kekule (2015) reported the first approaches in this respect, comparing the distribution of visual attention between students with best and worst performance while working on the TUG-K. Overall, although eye tracking was previously used in several studies in which participants were solving problems with graphs, usually only a small number of problems was used and/or a small number of students participated. Student understanding of important concepts related to graphs, such as graph slope and area under a graph, as well as the performance in different contexts were usually not in the focus of these studies. Therefore, we decided to use eye-tracking to investigate students' understanding of the slope of a graph and the area under a graph in different cohorts and in different contexts.

Research Questions

We aimed to answer the following research questions:

1. How do the physics and the non-physics students solve tasks associated with slopes and the area under a graph in the context of physics and in the context of finance?
2. Do the eye movements of the students reveal differences between finance and physics questions with respect to their performance?

Methods

The data analysed in this paper come from two studies conducted by the authors in 2018 and 2019 in Croatia and Germany (Susac et al., 2018; Klein et al., 2019). There are already two publications in which the results of the individual studies are presented separately and here the two data sources are merged. Both studies used identical materials and the same schedule, and subtle differences have already been mentioned in Klein et al. (2019). In particular, no comparison of the performance data was possible in the individual studies, as Susac et al. (2018) had made an adjustment of the score based on written protocols, which Klein et al. dispensed with. For the evaluation here, the raw performance data (i.e. without adjustment to the written statements of the students) were used to allow a direct comparison. Eye-tracking data from different systems were extracted and fed into a common database. In the methods section, the materials, the participants, the procedure, and the analysis methods are described again in order to give a new reader a holistic insight into the study.

Participants

The total data basis consists of 157 students which ranks this study among the larger studies in STEM disciplines using eye-tracking technology. The physics students that participated in the study were from the University of Kaiserslautern ($N = 29$), Germany, and the University of Zagreb ($N = 45$), Croatia. The non-physics students were economics students from the University of Mainz ($N = 40$) and psychology students from the University of Zagreb ($N = 45$).

In Germany, physics is taught as a compulsory subject for 4 years in the lower secondary level (grades 5–10, participant age 11–16). In kinematics the basic quantities (time, distance) are introduced and basic ideas are taught, such as the concept of velocity and acceleration and their (indirect) measurement. The explicit teaching of kinematics graphs begins in the physics courses at the beginning of upper secondary school (participants aged 16–17), where physics is not a compulsory subject. Most physics students (89%) chose the physics course at school, while only a minority of economics students (16%) did so. None of the students were confronted with kinematics graphs after school. Especially the physics students had not learned anything about kinematic graphs in the university courses, because the experiment took place in the first weeks of the students' first year of study. According to the faculty, the economics students did not encounter similar graphs in their finance courses as in this study. Obviously, the business students did not learn about kinematics graphs after school, since kinematics is not part of a business curriculum.

In Croatia, physics is a compulsory subject taught in the last two grades of all primary schools and during the four years of most grammar schools (gymnasia). Students are taught kinematics graphs at the age of 15 and 16 (last class of primary school and first year of grammar school). Psychology students were not confronted with kinematics graphs after high school, while physics students also learned about kinematics graphs in several university courses. Students of physics and psychology had not been exposed to graphs in terms of prices, money, etc. in their formal education. In Croatia the participants were prospective physics teachers whereas in Germany, the students chose the subjects physics and finance as scientific majors.

Materials

Eight multiple-choice test items were developed or modified from a previous study (Planinic et al., 2013). Four sets of isomorphic questions on graphs in the context of physics (kinematics) and finance related to the qualitative and quantitative understanding of the slope and area under a graph. Isomorphic questions required the same mathematical approach to kinematics graphs and graphs related to prices (we will call them finance graphs). The text of the question and the appearance of the graph were similar for the isomorphic test items to allow comparison of the effects of the two contexts, see Fig. 12.1. We prepared two versions of the test, with a balanced sequence of physics and finance questions. The isomorphic questions were never asked one after the other.

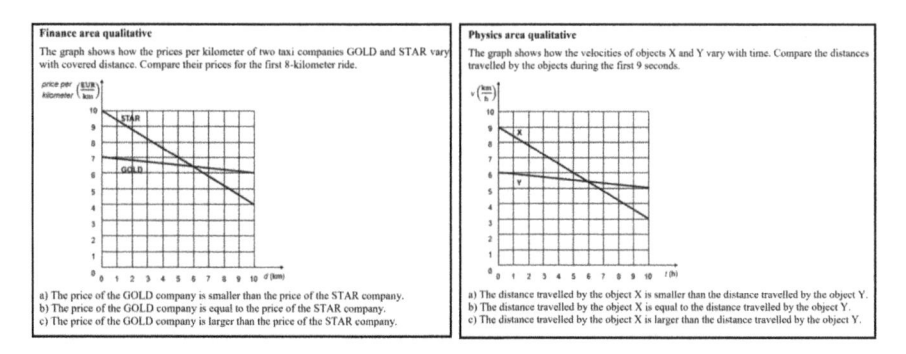

Fig. 12.1 An isomorphic pair of qualitative questions about the area under a curve

Apparatus

In Croatia, eye-movement data were recorded using a stationary eye-tracking system with a temporal resolution of 500 Hz and a spatial resolution of 0.25°–0.50° (SMI iView Hi-Speed system, Senso Motoric Instruments G.m.b.H.). The distance between the eyes and the monitor was 50 cm. In Germany, the eye movements were recorded with a Tobii X3-120 stationary eye-tracking system[1] which had an accuracy of less than 0.40 degree of visual angle (as reported by the manufacturer) and a sampling frequency of 120 Hz. The system allows a relatively high degree of freedom in terms of head movement (no chin rest was used). To detect fixations and saccades, an I-VT algorithm was adopted (Salvucci & Goldberg, 2000). A fixation can be defined as the state when the eye remains still over a period of time, while a saccade is the rapid motion of the eye from one fixation to another. Smaller eye movements that occur during fixations, such as tremors, drifts, and flicks are called microsaccades. Regarding both systems, an eye movement was classified as a saccade (i.e. in motion) if the acceleration of the eyes exceeded 8500 degrees/s2 and velocity exceeded 30 degrees/s.

Procedure

First, the participants were familiarized with the apparatus and the way to answer the questions (by pressing a key on the keyboard, and by choosing the answer using the mouse). The participants were asked to keep their head fixed during the measurements, so they could not use paper and pencil. After calibration, questions were presented to a participant one by one. The eight multiple-choice items were presented in a partially counterbalanced sequence (i.e. isomorphic questions were never presented one after another). Each slide contained the question, the diagram, and the answer options. By choosing the answer, the participant advanced to the next question. There was no time limit to answer the questions. The whole procedure, including preparation, eye-movement calibration and recording, lasted around 20 min.

Results

Students' Scores

Table 12.1 shows the performance data (raw test scores) for the German and Croatian physics students per item type (upper part of the table). With the exception of the qualitative slope questions, the physics students' performance decreases when

---

[1] More specifications can be found on the product website https://www.tobiipro.com.

12 Students' Understanding of Diagrams in Different Contexts ...

Table 12.1 Students' scores for each question

	Slope				Area			
	Qualitative		Quantitative		Qualitative		Quantitative	
	Physics	Finance	Physics	Finance	Physics	Finance	Physics	Finance
Physics students								
German (N = 27)	0.82	0.89	0.74	0.52	0.89	0.59	0.59	0.52
Croatia (N = 45)	0.87	0.84	0.91	0.71	0.80	0.67	0.69	0.44
Non-physics students								
German (N = 40)	0.83	0.83	0.48	0.20	0.55	0.38	0.15	0.28
Croatia (N = 45)	0.82	0.89	0.47	0.29	0.64	0.60	0.07	0.11

changing from the physics to the finance context. This observation is quantified in the following using repeated-measure ANOVAs with context as the main factor (physics vs. finance) and group as the between-subject factor (German vs. Croatian physics students).

For the quantitative slope questions, the analysis yields a significant main effect of the factor context ($F(1, 70) = 11.8, p < 0.001, \eta^2 = 0.15$) and a significant between-subject effect ($F(1, 70) = 4.9, p = 0.03, \eta^2 = 0.07$). The Croatian students solved the problems better than the German students, and the physics questions were easier than the finance questions. There was no interaction effect. For the qualitative area questions, the same analysis procedure was applied. We found a significant main effect of the factor context ($F(1, 70) = 10.5, p = 0.002, \eta^2 = 0.13$), without any group or interaction effects. The same holds for the quantitative area questions, i.e. there is a significant main effect of the factor context ($F(1,70) = 5.4, p = 0.02, \eta^2 = 0.07$), without other effects. In both cases, the physics problems were solved better than their isomorphic pairs with finance context. For the qualitative slope questions, there were no significant effects.

For the non-physics students (lower part of Table 12.1), the same analysis procedure was applied. The context of the question had a significant impact on the scores only for the quantitative slope question ($F(1, 83) = 9.6, p = 0.002, \eta^2 = 0.11$). Similar to the physics students, the non-physics students had more difficulties with the finance question compared to the physics questions.

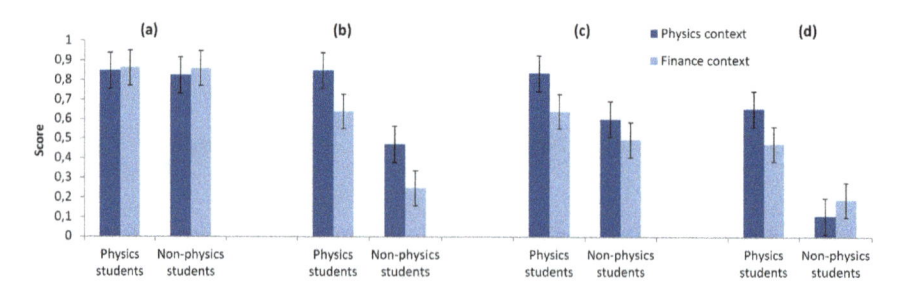

Fig. 12.2 Students' scores for the slope questions (**a** qualitative, **b** quantitative) and for the area questions (**c** qualitative, **d** quantitative). The error bars represent the standard error of the mean

Physics Vs. Non-physics Students

Since only marginal differences between the Croatian and German populations were found, the results were aggregated from the German and Croatian data for each type of question, see Fig. 12.2.

For the qualitative slope questions, a repeated-measure ANOVA with context as the main factor and study domain (physics vs. non-physics) revealed no main or interaction effects. For all other question types, main effects of context were found, which is in line with the results reported above. Additionally, there were significant differences between physics and non-physics students [slope quantitative: $F(1, 155) = 50.0, p < 0.001, \eta^2 = 0.24$; area qualitative: $F(1, 155) = 9.0, p = 0.002, \eta^2 = 0.06$; area quantitative: $F(1, 155) = 56.7, p < 0.001, \eta^2 = 0.27$]. For the quantitative area question, we also found an interaction effect between study domain and context ($F(1, 155) = 11.2, p < 0.001, \eta^2 = 0.07$). That means, physics students solved better the physics questions than the finance questions whereas non-physics students performed better on finance questions compared to physics questions. A post hoc t-test for this question type (area quantitative) revealed that the difference between non-physics students' scores on physics and finance questions is statistically different ($p = 0.03$), and also the difference between physics students' scores on physics and finance questions is statistically different ($p < 0.001$).

Dwell Times

From the eye-tracking data, we extracted total dwell times for each student and each item, see Fig. 12.3.

Regarding population effects, we observe that physics students and non-physics students spent similar time on the graphs; with one exception, that is the qualitative area question. For the qualitative area question, the physics students spent more time on the graphs (physics/finance context) compared to the non-physics students. Regarding context effects, we observe that physics students and non-physics students

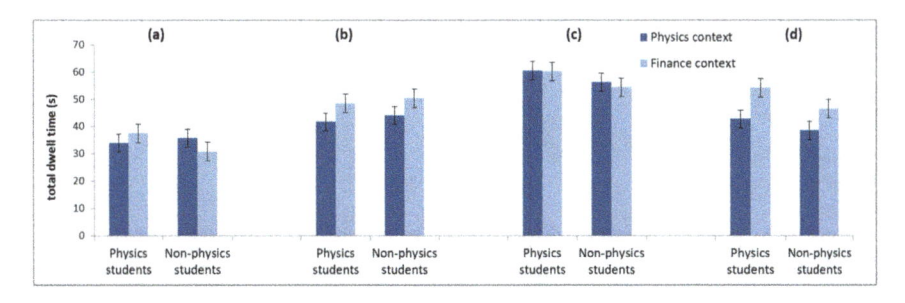

12 Students' Understanding of Diagrams in Different Contexts … 251

Fig. 12.3 Total dwell times for the slope questions (**a** qualitative, **b** quantitative) and for the area questions (**c** qualitative, **d** quantitative)

spent similar time on the qualitative questions in both physics and finance contexts. However, for the quantitative slope question, the students (both physics and non-physics) spent more time on the finance question than they spent on the physics question ($F(1, 155) = 6.13, p = 0.01, \eta^2 = 0.04$), and the same holds for the quantitative area question ($F(1, 155) = 13.6, p < 0.001, \eta^2 = 0.08$).

Attentional Distribution on Quantitative Area Question

In order to gain a better understanding for the decrease of the physics students' performance when switching from the physics to the finance context, we analyzed the visual attention distribution for one pair of isomorphic items in more detail. We chose the quantitative area question for this purpose because it was the most difficult question for the students, and technically, the graph had exactly identical dimensions in the German and the Croatian study which is crucial for applying the pattern analysis. Figure 12.4 shows the heatmaps for the physics students while solving the physics and finance question.

The heatmaps are presented for the German and Croatian physics students separately since different eye-tracking systems have been used in both studies. The quantitative area question requires to determine the area under the graph between the abscissa values "0" and "8".

Figure 12.5 shows the difference plot of a pattern analysis that was performed to the fixation count data. Data extraction was restricted to the graph region (excluding the question, the alternatives, and axis labels) in order to obtain a measurement of cognitive activity with the graph itself. The figure presents the differences in visual attention between both isomorphic items (physics context vs. finance context). If students spent more time to an AOI when the question was presented with physics context, the AOI is coloured green. Otherwise, e.g. when an AOI received more attention in the finance context, it is coloured red. Uncoloured AOIs reflect areas with no difference in visual attention between physics and finance context, occurring either if there was no visual attention at all or both the students allocated the same number of fixations when solving the items.

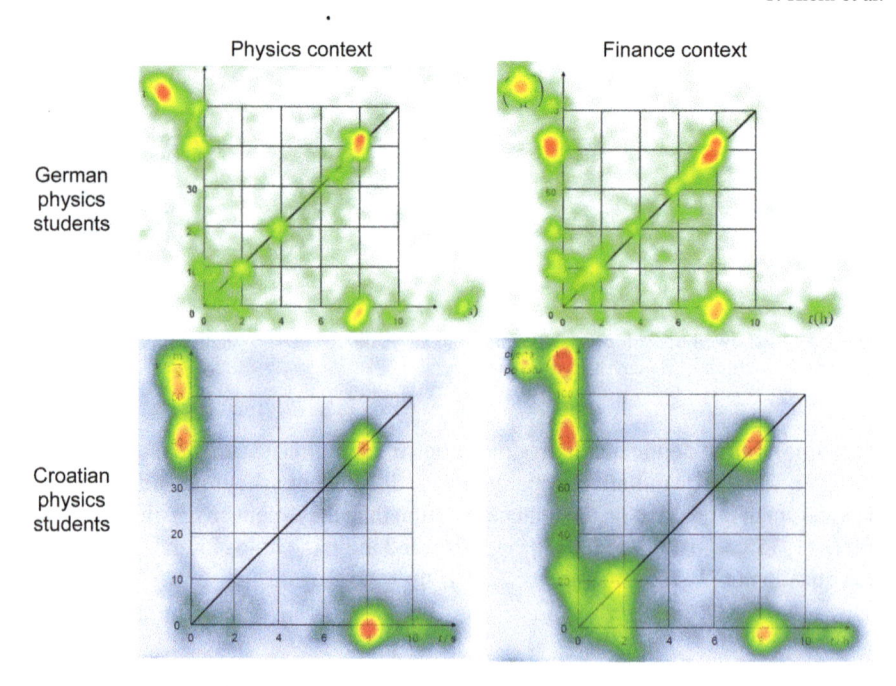

Fig. 12.4 Heatmaps of Germans and Croatian students solving the isomorphic pair of quantitative area tasks

Fig. 12.5 Analysis of visual attention and saccadic eye movements for the isomorphic pair of quantitative area questions (all data). The green (red) colour indicates areas that received more attention when the question was presented with the physics (finance) context

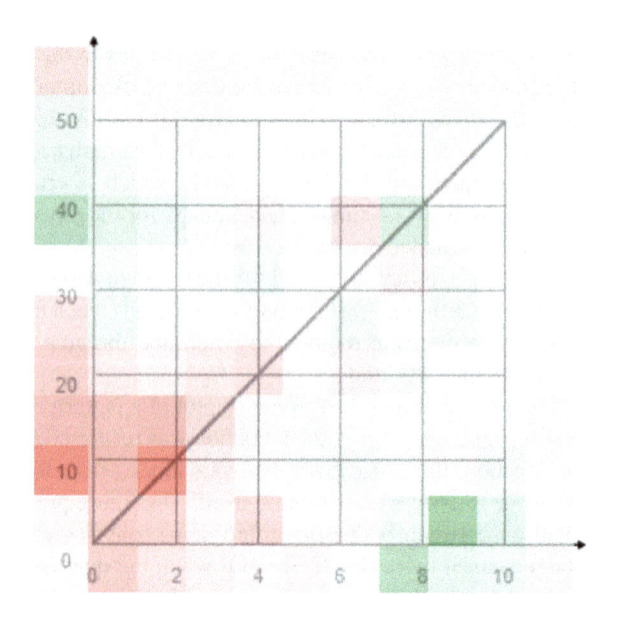

Attentional Distribution of Qualitative Slope Questions

Additionally, we analyzed the distributions of visual attention in the same manner as in as above for the isomorphic pair of qualitative slope items. According to Table 12.1 and Fig. 12.2a, most of the students solved this task successfully, indicating that they either transferred the mathematical procedures (slope concept) between both domains or had an intuitive idea about slope (meaning steepness) in both domains.

Figure 12.6 shows the attentional distribution of the Croatian and German physics students while solving the isomorphic pair of qualitative slope questions. The item pairs were not 100% isomorphic on the surface level, i.e. students were asked to compare the slopes of both graphs at the abscissa "3" for the physics question and at the abscissa "5" for the finance question, respectively. However, the mathematical procedure, the scaling of the axes, the location of the graphs, and their shapes were identical.

Due to the small difference between the framing of the question, i.e. comparing the slopes at different abscissa values, we cannot apply a pattern analysis that highlights the difference between both isomorphic questions conveniently as we did for the quantitative area question.

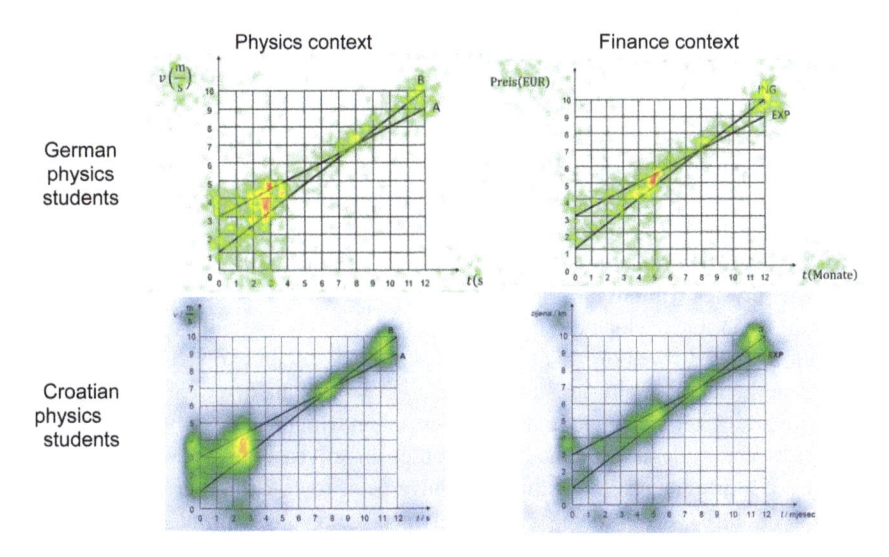

Fig. 12.6 Heatmaps of German and Croatian physics students while solving the isomorphic pair of qualitative slope questions. Note that students were asked to compare the slopes of both graphs at the abscissa "3" for the physics question and at the abscissa "5" for the finance question, respectively

Discussion

Discussion of Scores

A. Physics students achieved higher scores than non-physics students

Physics students performed significantly better overall than non-physics students. Apart from the qualitative questions about the slope of the graph, physics students had significantly higher scores in all combinations of question type and concept. This is true for the first-year physics students from Germany and the prospective fourth-year physics teacher students from Croatia. The combination of data from both studies also showed that psychology students and economics students both scored worse than the physics students, and both samples of non-physics students are comparable in terms of performance. This conclusion could not be drawn from the original data presented in Klein et al. (2019) and Susac et al. (2018) because the raw test results were not reported.

B. The slope of the graph seems to be a simpler concept for physics and non-physics students than the area under a graph

All students solved the questions about the slope of the graph better than the questions about the area under a graph. In particular, the qualitative questions about the slope of the graph were correctly solved by about 80% of the students in both contexts, which suggests that this idea is intuitive for physics and economics students (e.g. in connection with consumer or producer surplus). The non-physics students also had no difficulty with the qualitative question of the slope of the graphs in the physics context. This suggests that both the economics and psychology students were able to identify acceleration as a slope in velocity-time graph and growth rate of prices as a slope in the price-time graph.

C. Quantitative questions are more difficult than qualitative questions for first-year physics students

For the quantitative slope question, the analysis showed a significant main effect of the group factor, revealing that the Croatian physics students solved the questions better than the German physics students. In other words, the first-year students had more difficulties with the quantitative problems than with the qualitative ones, while the fourth-year students examined by Susac et al. solved both types of problems equally well. Similar results of this kind have been reported earlier for freshmen (Planinic et al., 2013), so the difference between the results is probably due to the differences between the two physics samples. As Susac et al (2018) emphasize, "studying physics improves [students'] ability to solve quantitative problems on graphs".

D. Physics students solved finance question better than non-physics students that indicates transfer

Physics students solved the physics questions better than the finance questions which is not surprising considering that they chose physics as their field of study. They also solved the finance questions very well and even better than the economics students. Since physics students have probably never been confronted with this kind of questions before, our results show that physics students seemed to be able to successfully transfer the mathematical strategies they developed at school in physics or mathematics to solve problems in a different context.

Discussion of Students' Visual Attention Using Dwell Time Analysis

The aggregated data set shows that both groups of students (physicists and non-physicists) spend more time on questions about the area under the graph than on questions about the graph slope. This proves that the area concept is more difficult for students than the slope concept, as longer viewing times are associated with a higher cognitive effort (Gegenfurtner et al., 2011). Furthermore, it was confirmed that solving quantitative questions about graph slope is more time-consuming than solving the qualitative question. Longer viewing times for quantitative questions on graph slope are usually due to longer viewing times of the axis tickets (Klein et al., 2019), which supports the idea that information extraction and processing contributes to the difference, and extends the previous results: Susac et al. (2018) explain the different viewing times by the fact that students have to extract more information from the graph when they have to make calculations.

The qualitative assessment of the area below the graph required the longest time across all items. It took even longer than the quantitative calculation of the area under the graph, which is an opposite trend of the viewing times when comparing the qualitative and quantitative questions of the slope. The result again agrees well with the literature and can be explained by the fact that, first, the area under a graph is not estimated as fast as the slope of the graph and second, these types of questions were likely to be new for both groups of students so maybe they needed more time to evaluate what to look for and where.

Analysis of Visual Attention Distribution on the Graphs: Success and Failure of Transfer

For the quantitative area question, the physics students solved the physics problem better than the finance problem. Since the graph in that question is linear and crosses the origin (compare Fig. 12.4), the area under the graph corresponds to a triangle and one possible correct solution strategy consists of the following steps

1. extracting the height of the graph at the abscissa "8"
2. multiplication of "8" and the height of the graph
3. division by 2.

Therefore, the relevant points of interest are the value "8" on the abscissa, the graph height, and the corresponding ordinate value. Additionally, students have to recognize that the graph passes the origin and that the graph is a straight line. As can be seen from Fig. 12.4, these points attract much attention in the physics context, especially for the Croatian students who solved this task correctly in 69% of cases (cf. Table 12.1). In the finance context, the Croatian physics students allocate more attention around the origin and the point (2/20); the performance drops to 44%. The heatmaps of the German students are similar, yet the difference and the drop of performance (59% --> 52%) are less pronounced. As can be seen, the units on the axis are also important areas of interest because they helped students to understand that they need to determine the area. In previous analyses it was shown that physics students spent more time on the finance axis, i.e. they needed more time to extract information from the axes in graphs with the context that was unfamiliar to them (Klein et al., 2019; Susac et al., 2018). The pattern analysis, i.e. the direct comparison of visual attention on the isomorphic pairs of question (Fig. 12.5), revealed that physics students allocate more attention to relevant regions when the question is posed in the physics context. They spent more time viewing the area around the origin in the finance version of the question and also spent more time viewing the y-axis values and x-axis values below that are smaller than the relevant numbers. It is possible that students were confused which procedure they had to apply for solving the question. Susac et al. (2018) reported that some calculated the slope instead of the area and the attention to smaller numbers possibly comes from this confusion. So possibly, some students tried to calculate the slope and when they saw it was not offered as the answer, they tried another strategy. Few students developed (incorrect) strategy to sum y-axis values for each of the eight hours. From the original protocol data of Susac et al. (2018), we found some examples that explained this issue. One student's explanation was "For the first hour he earns 10 kn, the second 20, the third 30, etc. So, for 8 h he earns $10 + 20 + 30 + 40 + 50 + 60 + 70 + 80 = 360$ (which I did wrong: on summation I got result 320)". Another student calculated the area by counting squares ("A total of 8 squares below the line, the area of each is 40 kn, in total earns 320."). The higher dwell times on the finance questions is also an indicator of unconfidence (Klein et al., 2019). When physics students encountered graphs in new context (finance) they spent some time to develop a strategy, i.e. to understand what they were supposed to do. For example, in quantitative questions about area under a graph, they had about 10 s longer total dwell time for finance question than for physics question (Fig. 12.3). The majority of physics students knew as a fact that covered distance corresponded to the area under the v vs t graph, so they calculated the area under a graph, or they used physics formulas. In the finance question, they could not rely on learned facts or formulas from physics, so they had to invent new strategies or modify the strategies learned in physics. Figure 12.4 indicates that physics students spent more time attending parts of the graph that are not relevant for the easiest solution (i.e. calculation of the area of triangle) in finance context than in physics context.

For the qualitative slope question, the transfer seemed to be achieved successfully. From Fig. 12.6, one can clearly observe similarities between the German and the Croatian physics students concerning both items (vertical comparison of the figures). The student allocated their attention at the lower part of the diagram ("$x = 3$" for the physics question, "$x = 5$" for the finance question), the affiliated y-labels on the ordinate, the intersection between the graphs, and at the axis and graph labels. When comparing both pair of items (physics and finance questions), the areas listed before received similar attention (horizontal comparison of the graphs). The qualitative slope question was also the easiest question for all students, indicating a moderating role of question difficulty on transfer abilities. A working hypothesis for upcoming work thus reads that students have more trouble to transfer mathematical procedures to unfamiliar domains if they have troubles to solve the initial question in their familiar domain. In other words, the "context gap" increases for difficult items.

Conclusion

The aim of this study was to compare physics and non-physics students regarding their understanding of graph slope and the area under the graph in the contexts of physics and finance. In doing so, two data sets from German and Croatian students have been aggregated. The thorough eye-tracking analysis sheds more light on differences when changing the context in a question using isomorphic pairs. The analysis of visual attention shows that in cases of apparent successful transfer, the main focus was on features that were relevant for solving the problem. When transfer seemed to fail, students directed their attention from relevant to irrelevant regions of the graph. This pattern suggests that transfer competence could potentially be supported by visual highlights, guiding the students' attention toward relevant areas.

Apart from that, our results broadly confirm previous findings on student understanding of graphs, i.e. graph slope is an easier concept than the area under the graph for physics and non-physics students. Area questions required more time and were therefore cognitively more demanding, indicating that more emphasis should be put on the qualitative and quantitative evaluation of the area concept. Overall, our results highlight the importance of an instructional adjustment toward more emphasis in education on graph interpretation.

References

Beichner, R. J. (1994). Testing student interpretation of kinematics graphs. *American Journal of Physics, 62*(8), 750–762.

Bollen, L., De Cock, M., Zuza, K., Guisasola, J., & van Kampen, P. (2016). Generalizing a categorization of students' interpretations of linear kinematics graphs. *Physical Review Physics Education Research, 12*(1), 010108.

Bransford, J. D., & Schwartz, D. L. (1999). Rethinking transfer: A simple proposal with multiple implications. *Review of Research in Education, 24,* 61–100.

Carpenter, P. A., & Shah, P. (1998). A model of the perceptual and conceptual processes in graph comprehension. *Journal of Experimental Psychology: Applied, 4*(2), 75.

Christensen, W. M., & Thompson, J. R. (2012). Investigating graphical representations of slope and derivative without a physics context. *Physical Review Special Topics-Physics Education Research, 8*(2), 023101.

Curcio, F. R. (1987). Comprehension of mathematical relationships expressed in graphs. *Journal for Research in Mathematics Education, 18,* 382–393.

Freedman, E. G., & Shah, P. (2002, April). Toward a model of knowledge-based graph comprehension. In *International conference on theory and application of diagrams* (pp. 18–30). Berlin, Heidelberg: Springer.

Gegenfurtner, A., Lehtinen, E., & Säljö, R. (2011). Expertise differences in the comprehension of visualizations: A meta-analysis of eye-tracking research in professional domains. *Educational Psychology Review, 23*(4), 523–552.

Goldberg, J., & Helfman, J. (2011). Eye tracking for visualization evaluation: Reading values on linear versus radial graphs. *Information visualization, 10*(3), 182–195.

Hammer, D., Elby, A., Scherr, R. E., & Redish, E. F. (2005). Resources, framing, and transfer. In *Transfer of learning from a modern multidisciplinary perspective* (pp. 89–120). Greenwich: Information Age Publishing.

Hoban, R. A., Finlayson, O. E., & Nolan, B. C. (2013). Transfer in chemistry: A study of students' abilities in transferring mathematical knowledge to chemistry. *International Journal of Mathematical Education in Science and Technology, 44*(1), 14–35.

Ivanjek, L., Planinic, M., Hopf, M., & Susac, A. (2017). Student difficulties with graphs in different contexts. In *Cognitive and affective aspects in science education research* (pp. 167–178). Cham: Springer.

Ivanjek, L., Susac, A., Planinic, M., Andrasevic, A., & Milin-Sipus, Z. (2016). Student reasoning about graphs in different contexts. *Physical Review Physics Education Research, 12*(1), 010106.

Kekule, M. (2014). Students' approaches when dealing with kinematics graphs explored by eye-tracking research method. In *Proceedings of the frontiers in mathematics and science education research conference, FISER* (pp. 108–117).

Kekule, M. (2015). Students' different approaches to solving problems from kinematics in respect of good and poor performance. In *International Conference on Contemporary Issues in Education, ICCIE* (pp. 126–134).

Klein, P., Küchemann, S., Brückner, S., Zlatkin-Troitschanskaia, O., & Kuhn, J. (2019). Student understanding of graph slope and area under a curve: A replication study comparing first-year physics and economics students. *Physical Review Physics Education Research, 15*(2),

Kozhevnikov, M., Motes, M. A., & Hegarty, M. (2007). Spatial visualization in physics problem solving. *Cognitive science, 31*(4), 549–579.

Leinhardt, G., Zaslavsky, O., & Stein, M. K. (1990). Functions, graphs, and graphing: Tasks, learning, and teaching. *Review of Educational Research, 60,* 1.

Madsen, A. M., Larson, A. M., Loschky, L. C., & Rebello, N. S. (2012). Differences in visual attention between those who correctly and incorrectly answer physics problems. *Physical Review Special Topics-Physics Education Research, 8*(1), 010122.

Madsen, A., Rouinfar, A., Larson, A. M., Loschky, L. C., & Rebello, N. S. (2013). Can short duration visual cues influence students' reasoning and eye movements in physics problems? *Physical Review Special Topics-Physics Education Research, 9*(2), 020104.

McDermott, L. C., Rosenquist, M. L., & Van Zee, E. H. (1987). Student difficulties in connecting graphs and physics: Examples from kinematics. *American Journal of Physics, 55*(6), 503–513.

Pinker, S. (1990). A theory of graph comprehension. In R. Freedle (Ed.), *Artificial intelligence and the future of testing* (pp. 73–126). Hillsdale, NJ: Erlbaum.

Planinic, M., Ivanjek, L., Susac, A., & Milin-Sipus, Z. (2013). Comparison of university students' understanding of graphs in different contexts. *Physical Review Special Topics-Physics Education Research, 9*(2), 020103.

Planinic, M., Milin-Sipus, Z., Katic, H., Susac, A., & Ivanjek, L. (2012). Comparison of student understanding of line graph slope in physics and mathematics. *International Journal of Science and Mathematics Education, 10*(6), 1393–1414.

Potgieter, M., Harding, A., & Engelbrecht, J. (2008). Transfer of algebraic and graphical thinking between mathematics and chemistry. *Journal of Research in Science Teaching, 45*(2), 197–218.

Salvucci, D. D., & Goldberg, J. H. (2000, November). Identifying fixations and saccades in eye-tracking protocols. In *Proceedings of the 2000 symposium on Eye tracking research & applications* (pp. 71–78).

Strobel, B., Lindner, M. A., Saß, S., & Köller, O. (2018). Task-irrelevant data impair processing of graph reading tasks: An eye tracking study. *Learning and Instruction, 55,* 139–147.

Susac, A., Bubic, A., Kazotti, E., Planinic, M., & Palmovic, M. (2018). Student understanding of graph slope and area under a graph: A comparison of physics and nonphysics students. *Physical Review Physics Education Research, 14*(2), 020109.

Viiri, J., Kekule, M., Isoniemi, J., & Hautala, J. (2017). Eye-tracking the effects of representation on students' problem solving approaches. In *Proceedings of the FMSERA annual symposium.* Finnish Mathematics and Science Education Research Association (FMSERA).

Wemyss, T., & van Kampen, P. (2013). Categorization of first-year university students' interpretations of numerical linear distance-time graphs. *Physical Review Special Topics-Physics Education Research, 9*(1), 010107.

Zavala, G., Tejeda, S., Barniol, P., & Beichner, R. J. (2017). Modifying the test of understanding graphs in kinematics. *Physical Review Physics Education Research, 13*(2), 020111.

Pascal Klein, Ph.D. is an Assistant Professor at the Georg-August-University Göttingen, Faculty of Physics. His research focuses on the diagnosis of students learning difficulties, problem-solving strategies, and misconceptions in tertiary physics education using eye-tracking. The results are reflected in the development of intervention measures aimed at improving the learning of physics and increasing the success of studies. He received his Ph.D. from the TU Kaiserslautern in 2016 on a topic of competence development with digital media (smartphones, tablets) in experimental physics lectures. From 2018 to March 2020 he was an Assistant Professor for physics education at the TU Kaiserslautern and from 2017–2020 he was scientific director of the immersive quantified learning lab (iQL) at the German Research Center for Artificial Intelligence in Kaiserslautern. Dr. Pascal Klein has lead several research projects and empirical studies about multiple representations in physics education, evaluation of existant diagnostic tests via visual attention (TUG-K), student understanding of graphs in different contexts, vector calculus, mechanics as well as physics with smartphones. Furthermore, Dr. Pascal Klein is Junior-Fellow of the Joachim-Herz-Stiftung, division digitalization in education.

Stefan Küchemann, Ph.D. is a postdoctoral researcher in the Physics Education Research Group at the TU Kaiserslautern, Germany. His main focus is the analysis of visual strategies of students during problem-solving and while learning with multiple external representations using eye-tracking. From 2015 to 2017 he was a research scholar at the Department of Materials Science and Engineering at the University of Illinois at Urbana Champaign, USA, with a focus on out-of-equilibrium processes in amorphous metals. In 2015, he completed his Ph.D. in Physics from the Georg-August-University Göttingen, Germany, with the topic "kinetic and thermodynamic processes of metallic glasses". He authored more than 30 peer-reviewed publications in international peer-reviewed journals both in educational research and solid-state physics research.

Ana Susac, Ph.D. is an Assistant Professor at the Faculty of Electrical Engineering and Computing, University of Zagreb, where she teaches courses in physics. Her research interests focus on physics education research and educational neuroscience. She studies students' conceptual understanding of physics at the high-school and university level, scientific reasoning, and role of mathematics in physics teaching and learning. She is interested in using eye-tracking and functional neuroimaging methods in science education research.

Alpay Karabulut studies mechanical engineering at the TU Kaiserslautern and currently works as a Teaching Assistant in the physics department at the TUK. He is interested in evaluating and promoting students learning and gained experience with eye-tracking studies, data analysis, and education research.

Andreja Bubic, Ph.D. is an Associate Professor at the Faculty of Humanities and Social Sciences, University of Split in Croatia. After studying psychology at the University of Zagreb in Croatia, she received her Ph.D. at the Max Planck Institute for Human Cognitive and Brain Sciences and the University in Leipzig, Germany, where she investigated future-oriented cognitive processing. After a postdoctoral stay at the MGH/HMS Martinos Center in Boston, USA, she is now in Split where she lectures and does research in the fields of educational psychology and decision-making. Her main interests in research include learning, decision-making, judgement, and other cognitive processes as well as their implications in the educational setting.

Maja Planinic, Ph.D. is an Assistant Professor at the University of Zagreb, Faculty of Science (Department of Physics), Croatia. Her research focuses on the investigation of student understanding and reasoning in different areas of high-school and university physics (at introductory level). An important area of her expertise is the construction and evaluation of diagnostic instruments (tests) in physics education research using Rasch modelling. She has led several research projects, and is currently leading the project focused on high-school physics teaching and the effects of the inclusion of students' investigative experiments on the development of students' conceptual understanding of wave optics and their scientific reasoning. She has participated in a broad spectrum of studies and published papers on the construction of a new test on graphs, evaluation of several widely used diagnostic tests in physics education (FCI, CSEM, TUV), student understanding of graphs in different contexts, atomic line spectra, Newtonian mechanics, electromagnetism and physics measurements, as well as on student reasoning in physics (e.g. proportional, algebraic) and transfer of knowledge between mathematics and physics. Some of the studies included eye-tracking as a research method. She co-authored two chapters in international books (published by Springer). She is currently a member of the Editorial Board of the Physical Review Physics Education Research.

Marijan Palmovic, Ph.D. is a Professor at the Department for speech and language pathology, University of Zagreb. He studied philosophy and linguistics and obtained his Ph.D. in cognitive neuroscience. His research interests include topics in language development and language processing, dyslexia, and language impairments. He carries out the research within the Laboratory for Psycholinguistic Research using experimental methods such as EEG/ERP and eye-tracking. He published two books and more than 40 articles in scientific journals.

Jochen Kuhn, Ph.D. is a Professor of Physics Education at TUK and Head of the Physics Education Research Group. He received his Ph.D. (in Physics in 2002) and his Habilitation (in Physics Education in 2009) from the University of Koblenz-Landau. He held two interim professorships (at the University of Regensburg and the University of Koblenz-Landau) before he got different offers for Full Professorship (as head of the group) in Germany and Switzerland in 2011. Since 2012 he has chaired the Physics Education Research Group at TUK. Jochen Kuhn received the Innovation Award of the IIAS, U Windsor, Ontario, Canada, in 2004 was Visiting Professor at the University of Windsor (Canada) in 2008 and is a Fellow of the International Institute for Advanced Studies in Systems Research and Cybernetics. He has successfully acquired individual third-party funding of more than 3 Mio. Euro (incl. DFG, BMBF) in the last five years. His research focuses on learning with multiple representations in STEM education in general and in Physics Education in particular using common and advanced multimedia technology (e.g. smartphones, Tablet PC, AR-smartglasses). He uses paper-and-pencil questionnaires as well as physiological methods (esp. Eye-tracking) to analyse learning before, during, and after intervention.

Chapter 13
Task-Evoked Pupillary Responses in Context of Exact Science Education

Roman Rosiek and Miroslawa Sajka

Introduction

Short Characteristics of Eye Movements

N. Wade and B. W. Tatler (2005), who conducted historical research on the first mentions of eye movement studies in the literature available to them, proved that eye movements have interested researchers since the dawn of time. According to their information, the first mention of eye movements appeared in the work of Du Laurens in 1596. According to the authors, the monograph of Johannes Müller, published in 1826, was the first to describe an eye movement study. The second precursor of eye movement research, according to Wade and Tatler (2005), is believed to be Charles Bell, who, in his work from 1823, presented the results of research on active and passive eye movements in response to visual stimuli, and drew attention to proprioception (referred to as muscle sense in the work) (for: Wade & Tatler, 2005). One should not forget about the work of Czech scientist Jan Evangelist Purkinje, despite the fact that his work concerned a broader subject, mainly physiology. Nowadays, the eye movements of a healthy person are divided into convergent movements, a tremor, gentle eye tremor, drift, microsaccades, optokinetic reflex, tracking, and jumping movements or saccadic eye movements (Soluch & Tarnowski, 2013). We also distinguish fixations, often considered a separate physiological mechanism, composed of minor movements called interfixations, i.e., tremor, microsaccades, and

---

R. Rosiek
Institute of Physics, Pedagogical University of
Krakow, Kraków, Poland

M. Sajka (✉)
Institute of Mathematics, Pedagogical University of
Krakow, Kraków, Poland
e-mail: miroslawa.sajka@up.krakow.pl

© Springer Nature Switzerland AG 2021
I. Devetak and S. A. Glažar (eds.), *Applying Bio-Measurements Methodologies in Science Education Research*,
https://doi.org/10.1007/978-3-030-71535-9_13

drift. Although fixations are, in fact, also movements, they are interpreted as focusing on a particular fragment of the image (a static process lasting several hundred ms), which is justified, because small movements are used to broaden the field of precise view—it is a compensation mechanism (Soluch & Tarnowski, 2013). Tracking movements involve following the sight of the observed object changing its position while maintaining a relatively stable position of the head. Ross et al. (1999) proved that the brain centers responsible for undertaking and maintaining tracking movements block the operation of saccadic movements. Under natural conditions, object tracking is also associated with head movement, and sometimes also with corrective saccades. These movements interact with each other, so their mechanisms are quite complex (Srihasam et al., 2009). Although the processes of visual attention were associated mainly with the mechanisms of eye movement, the role of attention mechanisms in tracking movements was increasingly being underlined (Khurana & Kowler, 1987). Hutton and Tegally (2005) showed a decrease in the speed and spatial precision of tracking movements, caused by performing additional tasks which required attention. According to the authors, this type of movement does not depend only on simple nervous mechanisms, but, as in the case of jumping movements, they result from complex processes of the cognitive system. As described by Soluch and Tarnowski (2013), apart from mechanical eye movements, brain mechanisms responsible for perception stability also play an important role in perception processes. Without them, conscious perception of continuous body movement relative to the perceived objects would most likely be impossible due to the need of recognizing the image again after each movement. One of the more important mechanisms is the optokinetic reflex. This mechanism compensates for head movements in order to maintain fixation on a particular object. Saccadic suppression is another important mechanism.

This phenomenon involves blocking certain batches of neurons within certain areas of the brain that are responsible for seeing through other batches that are responsible for saccadic eye movements (Lee et al., 2007). The purpose of this mechanism is to counteract the disturbance in perception that could be caused by rapid image movement (several times per second). This means that saccadic suppression is associated with the lack of vision of the part of the image which is on the path of the saccades. Typical presaccadic suppression causes blindness within 30–40 ms preceding the start of the saccade, and post-saccadic suppression causes a lack of vision lasting 100–120 ms of image perception. It can be concluded that it is not possible to see for all the time during which the fixation occurs, which should be taken into account in precisely determining the duration of the fixation, and more precisely, active focusing (Holmqvist et al., 2011). It is worth noting, after Soluch and Tarnowski (2013), that transsaccadic integration is currently the least known process enabling the stability of perception. The visual cortex is responsible for integration at cerebral level (Findlay & Gilchrist, 2003). This mechanism is of great importance due to the fact that each eye movement fundamentally changes the image of the external world projected onto the retina, and thus the visual system must somehow re-analyze the individual objects and their features (Soluch & Tarnowski, 2013). The integration mechanism guarantees that performing eye movements does not disturb the perception of particular elements, which is currently being explored in search of the bounds of this compensation process (Prime et al., 2004, 2007). An important component of the transsaccadic integration mechanism is not only image analysis after saccadic movement, but mainly analysis of the previous image (Gajewski & Henderson, 2005). Although it has been proven that transsaccadic integration is not associated with iconic memory, many new data (Findlay & Gilchrist, 2003; Melcher, 2009) indicate the involvement of attention processes and active image processing. Studies by Melcher (2005) have shown that transsaccadic integration results from the cortical representation of space, not the retinal image itself.

A precise explanation of this mechanism requires an approach to visual perception as an active way of seeking information, and not just as a passive reflection of the surrounding world. Research on transsaccadic integration is currently one of the most-explored problems in perception psychology, as the latest technological achievements, e.g., neuroimaging, are being used for the purposes of such research.

Pupillometry

The pupil is a movable diaphragm located between the cornea and the lens of the eye. The primary function of the pupil is to regulate the amount of light that falls on the retina of our eye. The pupil of a healthy person is approximately circular, so we can use its diameter to describe changes in pupil size. Winn et al. (1994), Beatty and Lucero-Wagoner (2000), The diameter of the pupil in young, healthy people ranges from 3 to 8 mm (Sosnowski et al., 1993). Studies on pupil diameter changes, often referred to as pupillometry, Andreassi (2000), are widely used in medicine. Examination of pupil diameter changes is used, among others, to diagnose neurological syndromes such as Addison's disease and Horner's syndrome, to monitor the effects of some medications, e.g., psychotropic, or to check for anesthesia (Wilhelm, 2011). Assessment of the rate of change in the diameter of the pupil as its response to light is used in clinical settings to detect optic nerve dysfunction, diagnose the effects of certain medications and substances, drugs, and assess the functioning of the brainstem (Lowenfeld, 1999). It can be said that it is a fast, painless, and non-invasive and relatively convenient diagnostic process. Currently, pupillometry is also used in psychology, Harrison et al. (2006), psychiatry, and in the field of educational research. We can find the first application attempts described in the literature in the 1960s as a measure of cognitive load affect (Hess & Polt, 1964; Hess, 1972), and, in particular, the observation that pupil width increases due to cognitive load (Kahneman & Beatty, 1966), Libby et al. (1973), Matthews et al. (1991). In the 1980s, the use of scientific research on pupil diameter changes was extended (Beatty, 1982). Increasingly, individual differences in the area of learning and processing information were also reflected in the differences in pupil diameter changes, providing potential for the identification and diagnosis of mental disorders.

Pupillometric Hypotheses

The neural efficiency hypothesis assumes that more intelligent people, when processing information and solving problems, are more efficient, without incurring as much mental effort as less intelligent people (Davidson & Downing, 2000; Haier et al., 1992, 415–426; Hendrickson, 1982). This hypothesis is reinforced by research in the field of psychophysiology on eye pupil reactions. The dilation of the eye pupil of the person undergoing research while solving a cognitive task is a psychophysiological measure of the burden of the process of analyzing and processing data. The greater the pupil dilation, the greater the burden of information processing or mental effort (Beatty, 1982). Ahern and Beatty (1979) also showed that there is a relationship between pupil responses and the cognitive abilities of the subjects. They showed that the changes recorded in students during multiplication were negatively correlated with their cognitive abilities. This meant that students with lower scores in the Scholastic Aptitude Test (SAT) had greater pupil dilatation during multiplication. This result is consistent with the neural efficiency hypotheses. Potential in the area of application of pupillometric tests in education and pedagogical sciences can be noted in the areas of anxiety disorder diagnosis. According to Ober et al. (2009) abnormalities in pupillometric reactions can be seen in people suffering from anxiety disorders. In people who showed a high tendency for anxiety, worrying, and rumination, smaller ranges of pupil diameter changes were observed when solving problems requiring a significant cognitive load. Such people, despite deviations in their reaction in terms of pupil diameter changes, performed the tasks in an effective manner, which means that the level of requirements did not exceed their cognitive abilities.

Methodology

The aim of the study is to identify whether monitoring the changes in pupil diameter when solving graph-related tasks concerning physics and mathematics will allow for obtaining information on the cognitive load of the subjects Paas et al. (2003), during the task-solving process and on the subjective assessment of the difficulty of the tasks. Gopher. (1994), Backs and Walrath (1992), Granholm, (2004), Partala et al. (2000), The experiment was carried out in a group of 103 people. The diameter of the examined pupils was measured using a Hi-Speed SMI eyetracker and iViewX ™ software with an assumed sampling frequency of 500 Hz. Data analysis was based on proprietary data analysis procedures which were performed on data series exported to CSV files. Błasiak et al. (2013), The analysis of data regarding the analysis of pupil diameter changes and the examination itself can be carried out assuming that the lighting conditions in the room do not change during the examination for each task and for each person. Therefore, in order to ensure the invariability of the light stream falling on the retina of the subjects' eye, the measurement was carried out when looking at a static image and, what is important, in the absence of changes in the distance between the screen and the eye retina of the subjects, with unchanging lighting conditions of the room in which the research was carried out. During the research, it was ensured that the lighting intensity in the room in which the test was conducted did not change. Therefore, the entire study was conducted with fixed, artificial interior lighting, which did not change for each of the subjects. Prior to the calibration, the subjects additionally spent several minutes in the room in which the examination was carried out, so that their eyes could adapt to the lighting conditions. This procedure was intended to ensure even greater reliability in measuring pupil diameter throughout the study. A 9-point device calibration procedure was performed prior to testing. Due to individual differences in terms of pupil diameter and their response, absolute values were not compared during data analysis. For the purpose of analyses and comparisons, only the percentage values of relative changes in pupil diameter were used. The entire procedure of comparing relative values consisted of the following activities: device calibration, registration of the task-solving process performed by the subjects, including registration of absolute pupil diameter values during the entire process with a sampling rate of 500 Hz, export of the recorded data to CSV files, identification and selection of appropriate data areas illustrating the eye fixations of the examined persons in terms of the visual representations of individual tasks, and all other important elements of the study for which the analysis of pupil width changes was made. The process of calculating the values of relative pupil diameter changes consisted of calculating the average pupil diameter value for each of the subjects individually in terms of task completion time, then calculating the instantaneous values of the percentage differences from the average for individual fixations for each of the solved tasks and for each subject. Subsequently, further assumptions were made. Due to the fact that the subjects did not have a time limit to solve the tasks, they could solve them at their own pace. This was associated with a very large dispersion of the recorded response times, and thus the total inability to impose and directly compare the records of this process for individual subjects. Therefore, for the purpose of analyzing and comparing the registered data, it was assumed that the individual response of the subjects, manifested in changes in the diameter of the pupil (Beatty, 1982) as a subjective assessment of the degree of difficulty, occurs during the first seconds of learning the content of the task, immediately after displaying the task, most often while reading the content of the task. The legitimacy of this assumption is confirmed by the analysis of the first fixations of the subjects during the analysis of the visual representations. Therefore, an analysis of the changes in the diameter of the pupil during the first ten fixations of the subjects in the observed image was made, assuming that it was when the subjects made the first subjective assessment of the category of the task and the degree of its difficulty (Ahern & Beatty, 1979). Therefore, in the first research approach described in this chapter, the relative percentage values of pupil diameter changes in selected groups of people for the first ten fixations were taken into account. This is a conscious simplification in which we assume averaging, i.e., analysis of all changes in the pupil diameter of the participants for the entire group, knowing that the reactions of individual people may occur at different times, the directions of the pupil diameter

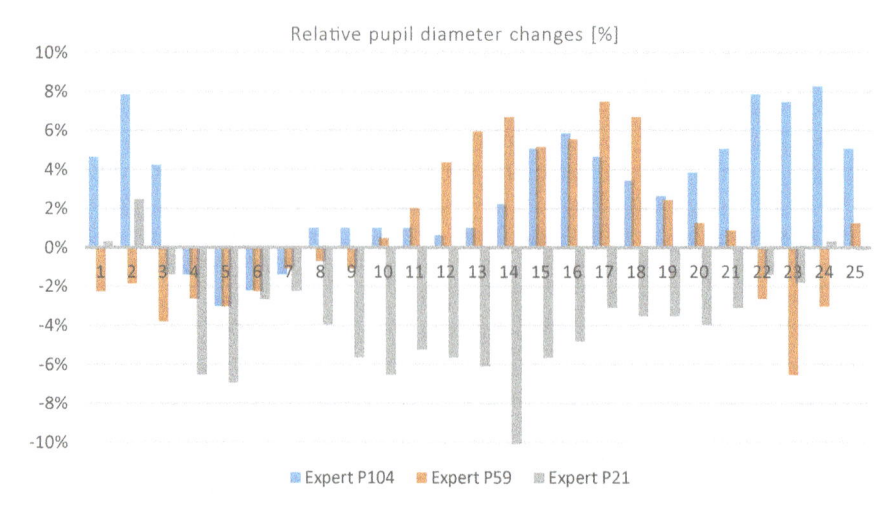

Fig. 13.1 Percentage pupil size changes of P104, P59, P21, during first 25 fixations

changes may not be consistent, and that some people can carry out tasks with very different levels of motivation (including a total lack of motivation). Assuming this method of analysis, we are aware that the calculation and visualization of average changes for selected groups of study participants is associated with the "blurring" of individual changes. We assume, however, that if it is possible to demonstrate the existence of certain tendencies for the whole group with such large simplifications, despite the process of averaging, it will mean that individual reactions combined with a precise description of their actions obtained through the interview process will provide even less ambiguous results, where the changes in relative values may be even bigger. As an example of the discrepancy of individual responses, we present the following drawings. Figs. 13.1 and 13.2 present selected examples of the pupil's reaction to the content of the task during the first fixations, including a person whose motivation, based on their interview, was very low.

Results

The following groups were distinguished from all the examined persons: experts, i.e., persons with at least a Ph.D. in physics or mathematics, Ph.D. students in physics, computer science students, biology students, high school students specializing in mathematics and physics. The mean values of relative pupil diameter changes during the first ten fixations were calculated for individual groups. The chart below presents the percentage average values of changes in relative pupil diameter, calculated for all groups of examined people (Fig. 13.3).

The second important element of the implemented tasks that affects the changes in the pupil diameter is the moment when the subject makes a decision regarding the

13 Task-Evoked Pupillary Responses ...

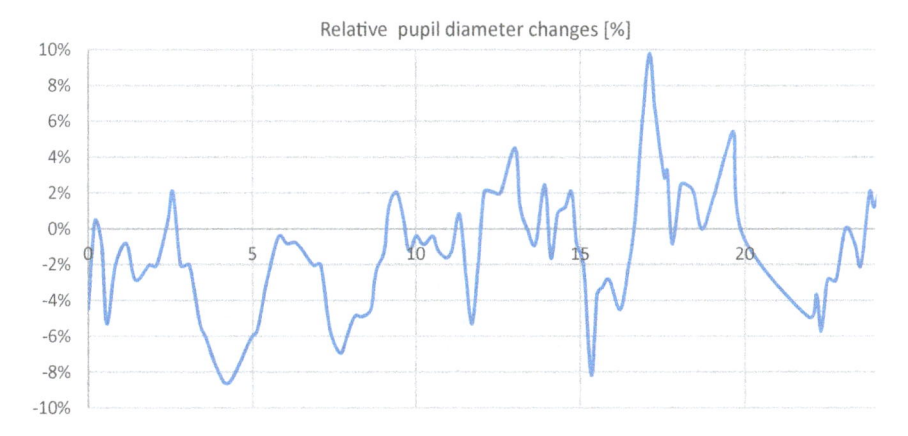

Fig. 13.2 Percentage pupil size changes of P22 during first 25 fixations

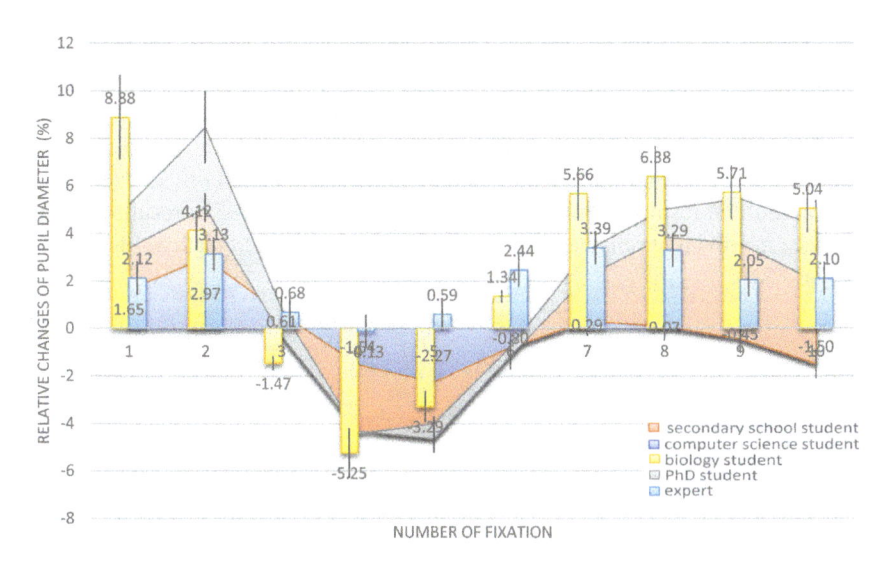

Fig. 13.3 Relative percentage changes in pupil diameter in selected groups, during first ten fixations

choice of the distractor. Due to the fact that the respondents' time was not limited for solving the tasks, they adopted various strategies for choosing the correct answer. This happened at different times during the task-solving process. In this case, a thorough analysis of the cognitive load and the degree of concentration, aside from a very detailed interview with the examined, requires the use of other measurement methods, such as EEG, including a thorough analysis of the changes in brainwave amplitudes over the course of solving a physics task. However, it is an invasive and very time-consuming method, requiring long-term preparation of the examined person, and the research method itself generates additional stress that significantly affects the size of the lesions, mainly dilation and pupil diameter. For this reason, the use of electroencephalography was abandoned in the described experiment. Such an extension will be the subject of more detailed case studies, which are being planned. Some simplification was used in this experiment. Because each of the respondents carried out the process of solving the task at a different pace, the time for solving tasks differed significantly between subjects. Due to the fact that it was not possible to directly compare these times, based on the analysis of all records of the task-solving process, it was assumed that the decision regarding the choice of answers as well as the greatest cognitive load should occur during the last 25 fixations preceding the selection of answers. For the analysis of the results, 25 fixations preceding the answer were selected for each of the subjects. For this time interval, pupil diameter values for each fixation were identified and their relative changes were calculated for each person. Then, the average values of the relative changes were calculated for each of the examined groups and all examined persons. The graphs below show the average values of the relative changes for individual groups of respondents. As for the results obtained for individual study groups, a straight equation was fitted using a linear regression method. It is represented by a dashed line in the figures. It was assumed that the directional coefficients of a straight line, including its slope, should describe the size and direction of the changes in pupil diameter. It was assumed that if the cognitive load differs in individual groups of examined people, we should obtain different amounts of relative pupil diameter changes, and hence the directional coefficients of the straight lines that illustrate this process should differ. The charts below present the results obtained, taking into account the division resulting from experience in solving problems in the field of physics and mathematics (Figs. 13.4, 13.5, and 13.6).

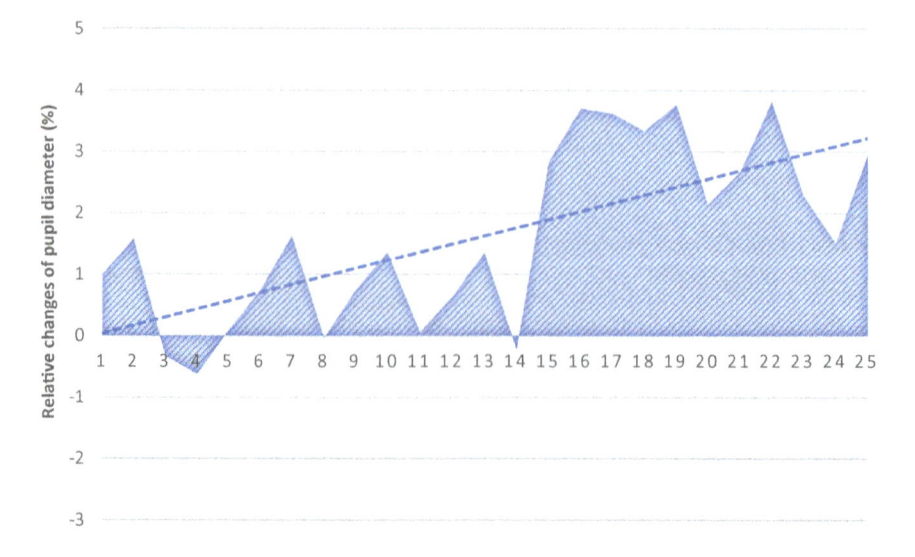

Fig. 13.4 Relative percentage changes in average pupil diameter values in expert group for last 25 fixations

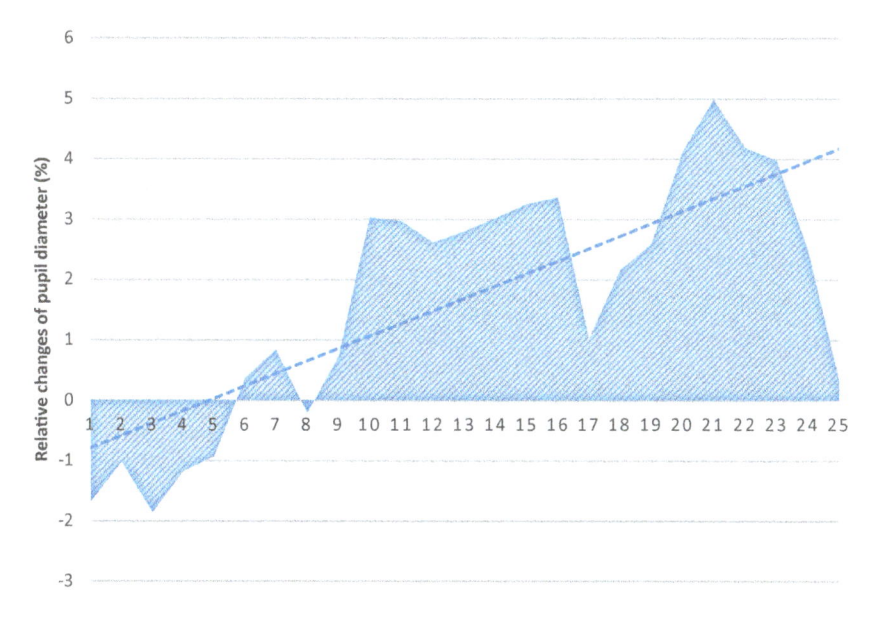

Fig. 13.5 Relative percentage changes in average pupil diameter values in Ph.D. student group for last 25 fixations

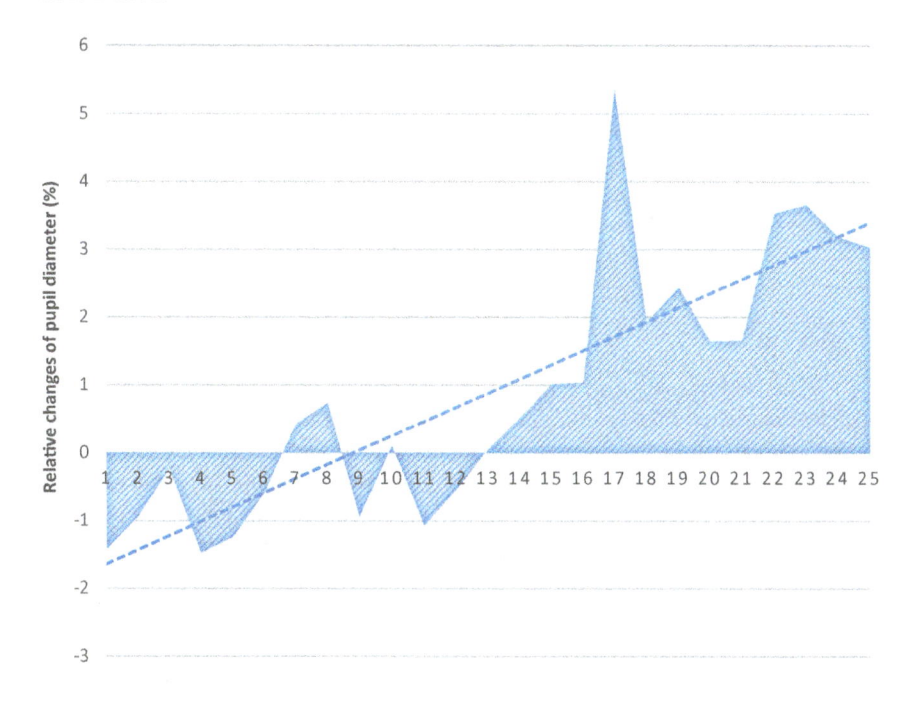

Fig. 13.6 Relative percentage changes in average pupil diameter values in secondary school student group for last 25 fixations

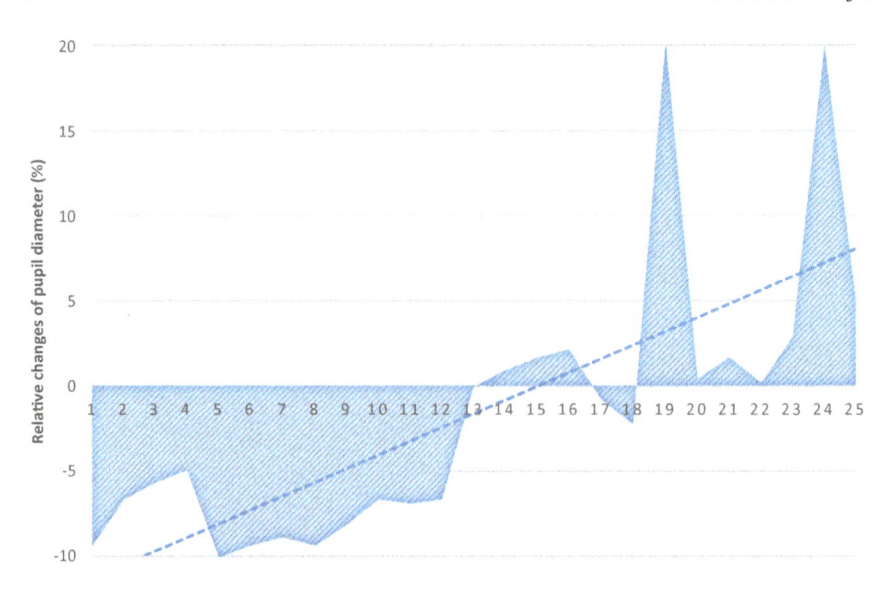

Fig. 13.7 Relative percentage changes in average pupil diameter values in biology student group for last 25 fixations

It should be noted that the following graph, due to the size of the relative changes, uses a different scale. In an interview after the study, biology students emphasized that they had already completed their education in physics, described the tasks they were solving as difficult, and that they were invited to a study which they had not been sufficiently informed about beforehand. Many of them described this situation as stressful. The highest values of relative pupil diameter changes were observed for this group of subjects (Figs. 13.7 and 13.8) (Table 13.1).

Discussion

By analyzing the relative changes in the pupil diameter during the first fixations on the task content made by the subjects, we can easily see a significant variation in the values of average relative reactions for individual groups of subjects, resulting from substantive preparation. Although the group of biology students is not fully representative, the size of the recorded reactions of pupil diameter changes in this group is the largest. High values of changes in the relative diameter of the pupils of the subjects testify to their significant cognitive effort and possible stress. The average values of relative pupil diameter changes in the group of computer science students vary to a lesser extent than in the group of high school students and to a larger extent than in the group of doctoral students. A detailed analysis of the scan path records shows that multiple subjects from IT student group chose the answer without

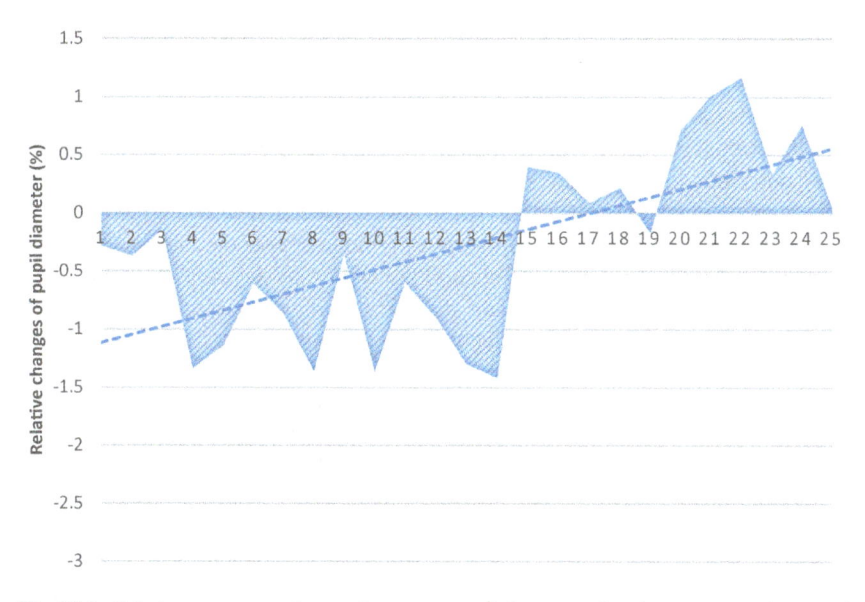

Fig. 13.8 Relative percentage changes in average pupil diameter values in computer science student group for last 25 fixations

Table 13.1 List of equations of fitted straight lines

	Formulas of linear function matched to data series
High school students	Y = 0.2 x – 1.9
Computer science students	Y = 0.1 x – 1.2
Biology students	Y = 0.8 x – 12.2
Ph.D. students	Y = 0.2 x – 1.0
Experts	Y = 0.1 x – 0.1
All	Y = 0.2 x – 2.91

a thorough analysis of the task content, sometimes randomly. This is evidenced by significantly smaller amounts of fixations in the area of the problems being solved.

Rejection of such people, based on interviews and scan path records, which did not have adequate motivation, i.e., did not read the content of the tasks, blindly guessed the solutions, did not carry out research in a reliable way, i.e., who did not make an intellectual effort when choosing answers, would allow to describe even more precise reactions of pupil diameter changes in each examined group in further studies. The linear function adjusted by the linear regression method is a function increasing for all groups of the examined persons. The value of the parameters describing the straight equation can be a quantitative description of the subjective assessment of the degree of difficulty of the problem for individuals and groups.

Conclusions

The obtained results indicate this methodology is worth using to conduct longitudinal studies in selected groups. It can be helpful, for example, for the purpose of supplementing descriptive, subjective responses regarding the assessment of the difficulty of tasks, as well as the motivation and stress levels associated with solving physics-related problems. However, it is important to remember that the largest values of relative changes in the diameter of the pupil are caused by negative emotions, e.g., stress. This should be taken into account when examining the subjective assessment of the difficulty of tasks and the intellectual effort (Madsen et al., 2012; Sosnowski et al., 1993). The results obtained during the experiment, mainly for biology students, confirm this rule. It is worth noting, however, that the study can be used as a method of determining the level of stress. The results obtained through repeated examinations of the same people, under the same lighting conditions, can be an indicator of their emotions, mainly negative, and also allow to indicate events that cause stress in students. The research results we obtained in the field of physics tasks confirm the hypothesis of neuronal efficiency. We can assume that people with a higher intelligence quotient, more experience, and broader substantive knowledge in the field of the related tasks use less resources and their psychophysiological response is smaller. This may indicate less cognitive burden, less intellectual effort when solving physics tasks. The obtained results confirm this hypothesis. However, it should be clearly emphasized that the analysis of solely the relative changes in the diameter of the pupil can only serve as an indicator, while a full interpretation of the changes must be supported by a thorough interview conducted with the examined person coupled with the analysis of other psychophysiological parameters such as heart rate, breathing, and electroencephalographic examination. By analyzing the relative changes in the diameter of the pupil during the first fixations of the subjects on the content of the task, we can easily see a significant variation in the values of average responses for individual test groups, resulting from substantive preparation. It is worth noting the need for a more detailed and thorough analysis of the results obtained for individual groups, especially the group of IT students. Further detailed analysis which involves taking note of those subjects who chose their answers without a thorough analysis of the task content among the surveyed IT students will allow to further describe the strategies and motivations when solving this task. We assume that this will become the subject of subsequent publications. In further research, it would be worthwhile to include, apart from the interview, the possibility of obtaining information on cognitive load using EEG methods.

Analysis of recorded data indicates that eyetracking methods can be very helpful in research on the didactics of STEM, e.g., in teaching subjects such as physics, IT, or mathematics. These are excellent methods supplementing our knowledge in the field and research regarding new technologies which are applicable for didactic research. It is worth using this methodology to conduct longitudinal studies in selected groups. It could be used, for example, for the purpose of supplementing the description of subjective reactions regarding the assessment of the difficulty of tasks, as well as the motivation and stress levels associated with the solving process.

References

Ahern, S., & Beatty, J. (1979). Pupillary responses during information processing vary with Scholastic Aptitude Test scores. *Science, 205,* 1289–1292.

Andreassi, J. L. (2000). Pupillary response and behavior. In N. J. Mahwah (Ed.), *Psychophysiology: Human behavior and physiological response* (pp. 218–233). Lawrence Erlbaum Association

Backs, R. W., & Walrath, L. C. (1992). Eye movement and pupillary response indices of mental workload during visual search of symbolic displays. *Applied Ergonomics, 23,* 243–254.

Beatty, J. (1982). Task-evoked pupillary responses, processing load, and the structure of processing resources. *Psychological Bulletin, 91*(2), 276–292.

Beatty, J., & Lucero-Wagoner, B. (2000). The pupillary system. In J. T. Cacioppo, L. G. Tassinary, & G. G. Berntson (Eds.), *Handbook of psychophysiology* (pp. 14–162). Cambridge, UK: Cambridge University Press.

Błasiak, W., Godlewska, M., Rosiek, R., & Wcisło, D. (2013). Eye tracking: nowe możliwości eksperymentalne w badaniach edukacyjnych, *Edukacja – Technika – Informatyka,* nr 4/2013-1.

Davidson, J. E., & Downing, C. L. (2000). Contemporary models of intelligence. In R. J. Sternberg (Ed.), *Handbook of intelligence.* Cambridge, UK: Cambridge University Press.

Findlay, J. & Gilchrist, I. (2003). Active vision: The psychology of looking and seeing. *Journal of Neuro-Ophthalmology, 26.* 10.1093/acprof:oso/9780198524793.001.0001

Gajewski, D. A., & Henderson, J. M. (2005). The role of saccade targeting in the transsaccadic integration of object types and tokens. *Journal of Experimental Psychology: Human Perception and Performance, 31*(4), 820–830.

Gopher, D. (1994). Analysis and measurement of mental load. In d'Y. Gery, P. Eelen, & P. Bertelson (Eds.), *International perspectives on psychological science: Vol. 2. The state of the art.* Hillsdale, NJ, England: Lawrence Erlbaum Associates.

Granholm, E. (2004). Introduction: Pupillometric measures of cognitive and emotional processes. *International Journal of Psychophysiology, 52*(1), 1–6.

Haier, R. J., Siegel, B., Tang, C., Abel, L., & Buchsbaum, M. S. (1992). Intelligence and changes in regional cerebral glucose metabolic rate following learning. *Intelligence, 16,* 415–426.

Harrison, N. A., Singer, T., Rotshtein, P., Dolan, R Ji, & Critchley, H. D. (2006). Pupillary contaginion: Central mechanisms engaged in sadness processing. *Social Cognitive and Affective Neuroscience, 1*(1), 5–17.

Hendrickson, A. E. (1982). The biological basis of intelligence, Part I: Theory. In H. J. Eysenck (Ed.), *A model for intelligence.* New York: Springer.

Hess, E. H. (1972). Pupillometrics: A method of studying mental, emotional, and sensory processes. In N. S. Greenfield & R. A. Sternbach (Eds.), *Handbook of psychophysiology* (pp. 491–531). New York: Holt. Rinehart & Winston.

Hess, E. H., & Polt, J. H. (1964). Pupil size in relation to mental activity during simple problem solving. *Science, 143,* 1190–1192.

Holmqvist, K., Nyström, N., Andersson, R., Dewhurst, R., Jarodzka, H., & Van de Weijer, J. (Eds.). (2011). *Eye tracking: A comprehensive guide to methods and measures.* Oxford, UK: Oxford University Press.

Hutton, S., & Tegally, D. (2005). The effects of dividing attention on smooth pursuit eye tracking: Experimental brain research. *Experimentelle Hirnforschung. Expérimentation cérébrale, 163,* 306–313. https://doi.org/10.1007/s00221-004-2171-z.

Kahneman, D., & Beatty, J. (1966). Pupil diameter and load on memory. *Science, 154*(3756), 1583–1885.

Khurana, B., & Kowler, E. (1987). Shared attentional control of smooth eye movement and perception. *Vision Research, 27*(9), 1603–1618. https://doi.org/10.1016/0042-6989(87)90168-4.

Lee, P. H., Sooksawate, T., Yanagawa, Y., Isa, K., Isa, T., & Hall, W. C. (2007). Idenity of a pathway for saccadic suppression, (w:) *Proceedings of National Academy of Science, 104*, 6824–6827.

Libby, W. L., Lacey, B. C., & Lacey, J. I. (1973). Pupillary and cardiac activity during visual attention. *Psychophysiology, 10,* 270–294.

Lowenfeld, I. E. (1999). *The pupil: Anatomy, physiology, and clinical applications.* Boston: Butterworth Heinemann.

Madsen A., Larson A., Loschky L., & Rebello N. (2012). Using scan match scores to understand differences in eye movements between correct and incorrect solvers on physics problems. In *Proceedings of the Symposium on Eye Tracking Research and Applications, EXTRA 2012*, Santa Barbara, CA, USA.

Matthews, G., Middleton, W., Gilmartin, B., & Bullimore, M. A. (1991). Pupillary diameter and cognitive load. *Journal of Psychophysiology, 5,* 265–271.

Melcher, D. (2005). Spatiotopic transfer of visual form adaptation across saccadic eye movements. In *Current Biology, 15,* 1745–1748.

Melcher, D. (2009). Selective attention and the active remapping of object features in trans–saccadic perception. In *Vision Research, 49,* 1249–1255.

Ober, J., Dylak, J., Gryncewicz, W., & Przedpelska-Ober, E. (2009). Sakkadometria – nowe możliwości oceny stanu czynnościowego ośrodkowego układu nerwowego. *Nauka, 4,* 109–135.

Paas, F., Tuovinen, J. E., Tabbers, H. K., & Van Gerven, P. W. M. (2003). Cognitive load measurement as a means to advance cognitive load theory. *Educational Psychologist, 38,* 63–71.

Partala, T., Jokiniemi, M., & Surakka. (2000). Pupillary responses to emotionally provocative stimuli. In *V. Eye Tracking Research & Application: Proceedings of the 2000symposium on Eye Tracking Research and Applications* (123–129). New York: ACM Press.

Prime, S., Niemeier, M., & Crawford, J. D. (2004). Trans-saccadic integration of the orientation and location features of linear objects. *Journal of Vision, 8,* 742.

Prime, S. L., Tsotsos, L., Keith, G. P., & Crawford, J. D. (2007). Visual memory capacity in transsacadic integration. *Experimental Brain Research, 180,* 609–628.

Ross, R. G., Olincy, A., & Radant, A. (1999). Amplitude criteria and anticipatory saccades during smooth pursuit eye movements in schizophrenia. *Psychophysiology, 36,* 464–468.

Srihasam, K., Bullock, D., & Grossberg, S. (2009). Target selection by the frontal cortex during coordinated saccadic and smooth pursuit eye movements. *Journal of Cognitive Neuroscience, 21,* 1611–1627.

Soluch, P., & Tarnowski, A. (2013). O metodologii badań eyetrackingowych. *Lingwistyka Stosowana, 7,* 115–134.

Sosnowski, T., Zimmer, K., Zaborowski, P., & Krzymowska, A. (1993). *Metody psychofizjologiczne w badaniach psychologicznych: praca zbiorowa.* Warszawa: Wydawnictwo Naukowe PWN.

Wade, N., & Tatler, B. W. (2005). *The moving tablet of the eye: The orings of modern eye movement research.* Oxford: Oxford University Press.

Wilhelm, H. (2011). Disorders of the pupil. *Handbook of Clinical Neurology, 102,* 427–466.

Winn, B., Whitaker, D., Elliott, D. B., & Phillips, N. J. (1994). Factors affecting light-adapted pupil size in normal human subjects. *Investigative Ophthalmology & Visual Science, 35,* 1132–1137.

Roman Rosiek, Ph.D. is an Associate Professor of Physics Education at the Pedagogical University, Division of Theoretical Physics and Didactics of Physics, Faculty of Exact and Natural Sciences, Krakow, Poland. His research and development focuses on issues related to the didactics of physics, in particular the education of physics and computer science teachers, understanding of concepts by students and teachers, and the use of new research and teaching methods. Since January 2014, he has been a member of the Interdisciplinary Cognitive Didactic Research Group, currently under his leadership. As part of this activity, he conducts research on the applications of eyetracking methodology, EEG and psychophysiological methods e.g.: HRV, EDA, Respiration for research in the field of cognitive science and didactics of physics. Habilitation subject: Psychophysiological Methods in Research on Didactics of Physics. The results of his scientific and research work have been published in over 50 publications. Among them there are publications indexed in the Web of Science and SCOPUS databases, chapters in monographs, a book. His works are quoted by many scientists. He delivered national and international guest lectures, plenary lectures at international conferences e.g. GIREP.

Miroslawa Sajka, Ph.D. is an Assistant Professor and the head of Chair of Mathematics Education at the Institute of Mathematics, Faculty of Exact and Natural Sciences, Pedagogical University of Krakow, Poland. She is a contributing member of the Interdisciplinary Research Group of Cognitive Didactics at Pedagogical University of Krakow from 2014, runs research on using eye-tracking in mathematics education also from physics education point of view. Her research areas: in general mathematical reasoning, specifically understanding the notion of function; teacher cognition, education and knowledge in/for teaching; comparative studies; eyetracking and other psychophysiological methods in research on mathematics and science education. Dr. Sajka authored a scientific book (monograph), over 50 scientific papers on thinking, learning and teaching mathematics and several school resources and textbooks in mathematics for secondary school students. Dr. Sajka is Vice Editor-In-Chief of a scientific journal "Annales Universitatis Paedagogicae Cracoviensis. Studia ad Didacticam Mathematicae Pertinentia" in the field of mathematics education. She undertakes co-leaderships of working groups at international conferences on mathematics education (e.g. "Comparative Studies in Mathematics Education" at CERME, "Knowledge in/for teaching at secondary level" at ICME) and she delivered national and international guest and plenary lectures at conferences.

Chapter 14
An Investigation of Visual and Manual Behaviors Involved in Interactions Between Users and Physics Simulation Interfaces

Guo-Li Chiou, Chung-Yuan Hsu, and Meng-Jung Tsai

Introduction

Improving students' in-depth understanding of physics conceptual frameworks is a central goal of physics learning. Traditional instructional approaches, however, have been repeatedly reported as having little effect on achieving this goal (e.g., Hake, 1998; Kim & Pak, 2002; Trowbridge & McDermott, 1981), and many educators have strived to advocate alternative approaches that could reach this desirable goal. Among the suggested approaches, inquiry learning has become prevalent since the early 2000s. Duschl (2008) claimed that conducting inquiry activities has the potential for developing students' understanding of the content, practice, and epistemology of science, and Abd-El-Khalick et al. (2004) proposed that inquiry should be treated as the process, as well as the outcome, of science learning. With the rapid development of digital technology, inquiry learning can now be easily implemented in computer environments, such as computer simulations (e.g., de Jong, 2006, 2011; de Jong et al., 2013; Wieman & Perkins, 2005).

Computer simulations open a productive avenue for learning physics. In general, they have two major components, a computational model and an interface (de Jong, 2011). A computational model is a computer program designed for simulating the corresponding, original natural phenomenon or physics system. An interface, on the

---

G.-L. Chiou · M.-J. Tsai (✉)
Program of Learning Sciences, School of Learning Informatics,
National Taiwan Normal University, Taipei, Taiwan
e-mail: mjtsai99@ntnu.edu.tw

G.-L. Chiou
e-mail: glchiou@ntnu.edu.tw

C.-Y. Hsu
Department of Child Care, National Pingtung
University of Science and Technology, Neipu, Pingtung, Taiwan

other hand, allows users to interact with the computational model by altering the values of the involved variables. These two components afford computer simulations with highly interactive virtual environments where multiple representations of corresponding models can be demonstrated and manipulated. These sorts of virtual environments create tremendous opportunities for conducting inquiry activities. For instance, some abstract physics concepts, such as electric current, are difficult for students to learn because of their invisibility. Using a PhET simulation (e.g., Wieman et al., 2010), students can not only observe and visualize the movements of imagined particles, electrons, as an electric current, but can also manipulate relevant variables, such as the amounts of resistance and power, to fully investigate the phenomena of electricity. Also, some experiments, such as testing airbag safety (McElhaney & Linn, 2011), can scarcely be performed by students because of their vast expense and restricted access. By using computer simulation, students have access to experiencing "professional" laboratories where they can identify and change relevant variables, test their predictions, and review and interpret the experimental results. Nevertheless, it is worth noting that models and representations displayed in computer simulations are never authentic or realistic; they are simplified and theoretical versions of complex phenomena or systems, which make them adequate media for physics learning.

Previous studies have confirmed the effect of using computer simulations on enhancing students' learning outcomes. For example, in a review study, Rutten et al. (2012) found that, compared to traditional instruction, computer simulations demonstrated a stronger effect on improving students' conceptual understanding and inquiry skills, such as making predictions. Also, Rutten et al. pointed out that representations with different formats, such as dynamic versus static, and concrete versus idealized, might have different effects on conceptual understanding. Similarly, D'Angelo et al. (2014) conducted a meta-analysis of 59 studies on computer simulations, and concluded that learning with computer simulations showed a beneficial effect on facilitating students' STEM (science, technology, engineering, and mathematics) learning in comparison to learning without them. Given that the reviewed studies were all related to STEM domains, D'Angelo et al. specifically indicated that computer simulations had a significant effect on students' inquiry and reasoning skills. Although most studies favored computer simulations over traditional instructional approaches, de Jong (2011) argued that to better understand the effectiveness of computer simulations, it is important to explore how they are used under what conditions.

As mentioned previously, interactivity is a central affordance of computer simulations that provides ample opportunities for conducting inquiry activities. Exploring how users interact with computer simulations, therefore, is an important key to unveiling the processes and outcomes of learning with simulations. Adams et al. (2008), for example, found that if students could only watch, without any interaction with, the representations displayed by a computer simulation, they tended to passively perceive what they observed as a fact without developing any new ideas or insights. In contrast, if the students were allowed to interact with the simulated representation, they might actively pose questions to the simulated models and start to conduct further investigations of the models by manipulating the relevant variables. In other words, interactivity might be a trigger for students to actively conduct inquiry activities within a computer simulation. Moreover, by conducting inquiry activities, students could retrieve their prior knowledge and integrate it with the information they had observed and received from the simulations (de Jong, 2011). They could, therefore, develop a deeper understanding of the simulated models and achieve better learning outcomes. Nonetheless, there has been little study on investigating the detailed processes of using computer simulations, and it remains unclear how the process of interacting with simulations affects students' learning achievements.

To uncover how students interact with computer simulations, some studies began to explore the detailed processes of using simulations. In particular, because the interaction between users and a simulation interface involves visual attention, specific technologies and techniques could be employed to better investigate the highly interactive processes. For example, Chiou et al. (2019) used eye-tracking techniques to study how students used computer simulations. They found that the spatial distributions and temporal sequences of visual attention could be used to account for the learning processes and outcomes of using simulations. More specifically, students who paid more attention to the target simulated phenomena were more likely to accomplish the inquiry task than those who paid scare attention to the target phenomena. In addition, regarding the sequences of visual transition, students who first fixated on the instructional information and then on the target simulated phenomena were more likely to provide a correct answer to the inquiry question. The rationales behind the usage of eye-tracking techniques are *immediacy* and *eye-mind* assumptions (Just & Carpenter, 1980). While the former assumes the direct temporal sequence between perceiving and processing external visual stimuli, the latter proposes a correspondence between what is visually perceived and what is mentally processed. These two assumptions together adequately connect the visual behaviors and mental operations while using computer simulations, and serve as a solid foundation for utilizing eye-tracking techniques for studying the interaction between users and simulation interfaces.

Visual behaviors alone, however, could not fully account for the interaction between users and computer simulations. This is because users need to manually "touch" the interface to manipulate the variables involved in the simulated models. These manual manipulations are usually done by hand and finger movements, such as clicking, dragging, and dropping the buttons of a mouse. Moreover, it is worth noting that manual manipulations could not be effectively performed without visual attention. To carefully investigate the user-interface interactions in computer simulations, it would be better to examine the concurrence between visual behaviors and manual manipulations. Nevertheless, to date, little research has been conducted to concurrently investigate both visual attention and manual manipulation as the detailed interactive process of using simulations. This present study, therefore, attempted to bridge this gap for an in-depth understanding of the process of using computer simulations for learning physics.

Research Questions

Based on the aforementioned description, this study was conducted to answer the following questions:

1. How did students interact with a simulation interface while conducting a physics inquiry activity in a computer simulation?
2. What patterns of visual and manual behaviors might lead to the successful completion of a physics inquiry activity?

Methodology

Participants

The participants in this study were 40 seventh graders (22 females and 18 males; range from 12 to 13 years old) recruited from a junior high school in southern Taiwan. They had never learned physics concepts related to refraction in a formal setting before participating in the study. These participants were divided into two groups according to their answers to the inquiry question given in the physics simulation (which will be introduced later). While those who provided a correct answer were assigned to the correct group, those who offered a wrong answer were assigned to the wrong group. As a result, there were 21 participants in the correct group (10 females and 11 males) and 10 participants in the wrong group (12 females and 7 males). The result of a t test showed that there was a significant difference in the pretest scores (this will be described later) among these two groups ($t = 2.16, p = .04$), suggesting that the two groups had different levels of prior knowledge about basic optics.

Simulation

This study adopted a computer simulation, refraction of light, developed by Chiou et al. (2019) as an inquiry environment to be investigated. The simulation was specifically developed for students to learn a physics conception, refraction. By using this simulation, students could change the media within which a light ray propagates, and could observe how the ray bends when coming into another medium. Figure 14.1 shows the interface of the simulation, including six major areas: Question, Up panel, Down panel, Light, Reflection, and Refraction. The Question displays an inquiry question that provides guidance for students to conduct an investigation within the simulation. The Up and Down panels are two control bars where students can adjust the indices of refraction of the upper and lower media, respectively. The Light area contains a flashlight by which students can change the angle of the light source, that is, the incident angle. The Reflection and Refraction areas display the reflected ray

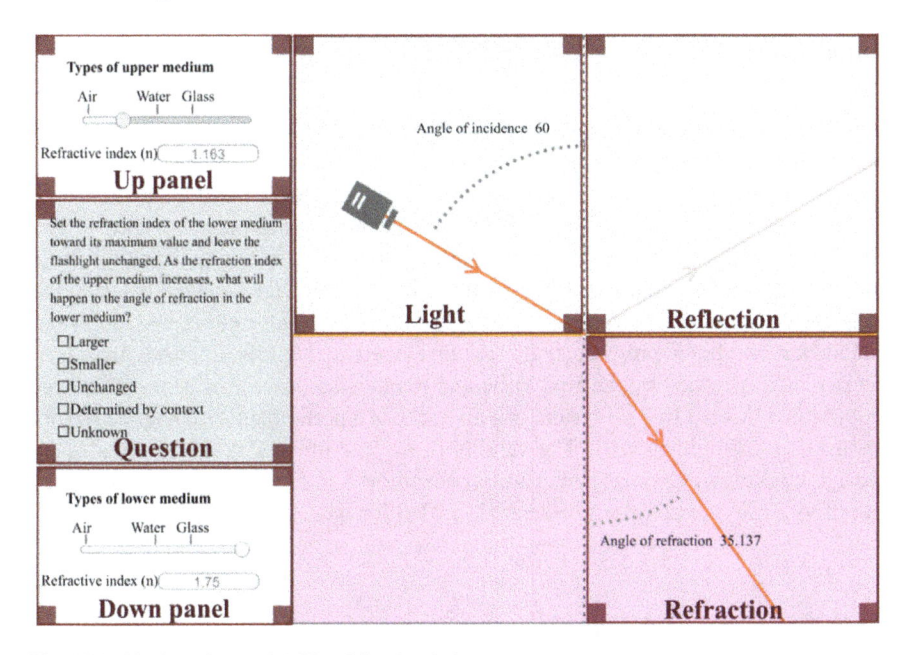

Fig. 14.1 The interface and AOIs of the simulation

and refracted ray, respectively. In brief, users can change the position of the flashlight and the materials of both the upper and lower media to observe the relationship between the incident angle and the refracted angle. Moreover, when the incident angle is greater than the critical angle, total reflection occurs and the reflected ray will be presented in the Reflection area.

Pretest

This study developed a pretest for assessing the participants' conceptual understanding of optics. This test was reviewed by two physics educators to ensure its face and content validity. In particular, because the participants had never learned the concept of refraction before participating in this study, the pretest focused only on some basic ideas of optics, such as straight propagation of light, shadow, and reflection. The pretest had 12 items in total, and the Cronbach's α is .60, indicating a proper level of internal consistency.

Apparatus

This study utilized a specifically developed eye-tracking system for recording and processing the participants' visual and manual behaviors while using the simulation. The eye-tracker used in this system was the Eye Tribe, which had a sampling rate of 30 Hz and was mounted under a 14" screen of a laptop that was used to run the computer simulation. While using the simulation, the participants rested their chin on a chin holder to prevent rapid head movement, and the distance between the participants and the laptop screen was about 65–70 cm. Our eye-tracking system included a computer program that was developed to identify fixations based on the raw data recorded by the Eye Tribe and to calculate the eye-tracking measures required for this study. The system can also record a participant's manual behaviors (mouse logs) simultaneously. The validity of the eye-movement data generated by our eye-tracking system has been carefully examined and reported with a correction rate of 99.81% (please refer to Hsu et al. [2016] for more details).

Data Collection Procedure

This study carried out the following procedure for data collection. In the beginning, the pretest was administered to each individual participant. Then, each participant went through a five-point eye-tracking calibration. If the calibration was successful, the participant started to use the computer simulation. S/he was allowed to play with the simulation to get familiar with its interface. Subsequently, the participant began to conduct an inquiry task shown on the interface: "*Please first set the refraction index of the lower medium toward its maximum value and leave the flashlight unchanged. Then try to adjust the refraction index of the upper medium and observe the path of the ray of light in the lower medium. As the refraction index of the upper medium increases, what will happen to the angle of refraction in the lower medium?*" The eye-tracking system with an Eye Tribe was employed to record the participant's visual behaviors and log data while s/he was using the simulation. The inquiry task was self-paced and the participant could spend as much time as needed to complete the task. Once the participant finished the inquiry task by submitting an answer, the eye-tracking system was immediately turned off and stopped recording. The participants spent an average of 48.99 s ($SD = 23.78$) to complete the inquiry task.

Data Analysis

This study aimed at examining the differences in the processes of using the simulations of those students who provided correct and wrong answers. Therefore, the participants were first divided into two groups based on the correctness of their answers to the inquiry question. Then, a series of between-group comparisons were made in terms of eye-movement indices, log data, lag sequential analyses, and heat maps. In the following, we will describe these four data analysis approaches.

Eye-Movement Indices

To analyze the spatial distribution of the participants' visual attention, we defined seven areas of interest (AOI) within the interface of the simulation. Six of the seven AOIs directly correspond to the major functional regions of the interface, which are the Question, Up panel, Down panel, Light, Reflection, and Refraction (please refer to the Simulation section for the meanings of these regions). The remaining AOI is Out, which refers to the region outside all of the previous six AOIs.

With respect to the seven AOIs, seven eye-movement indices were adopted for further analysis. First of all, we calculated the total time (TT) that each participant spent on each AOI. Second, the total fixation duration (TFD) of each AOI was calculated to reveal the accumulated time that each individual fixated on the corresponding region of the interface. Third, we calculated the total fixation count (TFC) of each AOI to understand the frequency with which each individual fixated on the area. Fourth, by dividing TFD by TFC, the average fixation duration (AFD) could be obtained. Fifth, to better understand the portion of time that each participant spent on each AOI, we calculated the percentage of time spent in zone (PTS) by dividing the time spent on each AOI into the total time of using the simulation. Sixth, the time to first fixation (TFF) was calculated to reveal how long it took for each participant to first allocate their visual attention to each AOI. Finally, we computed the first passing time (FPT) to represent the duration starting from the participants' first arrival to the departure of each AOI. The definitions and meanings of the seven eye-movement indices are listed in Table 14.1. Based on the values of the seven eye-movement indices, a series of non-parametric, Mann-Whitney U tests were performed to examine the differences in the distributions of the visual attention of the two groups.

Log Data Analysis

To understand how the participants manually manipulated the interface, we carefully examined the log data recorded while they were using the simulation. We first checked every adjustment each participant made with the control panels of the interface, that is, the up panel, the down panel, and the light. Each adjustment was defined as one manipulation, and the numbers of manipulations could be accumulated as total manipulation count (TMC). Moreover, each manipulation was labeled with a starting point (S) and an ending point (E), and the time period between the starting and ending point was defined as a manipulation duration (MD). With respect to each control panel, every single manipulation duration could be added up to total manipulation

Table 14.1 Eye-tracking indices for each AOI used in this study

Indices	Description
Total fixation duration (TFD)	The total time spent on fixations; this index might indicate the total visual attention devoted to an AOI in a temporal dimension
Total fixation count (TFC)	The total number of fixations within an AOI; this index might indicate the total visual attention devoted to the AOI in a frequency dimension
Average fixation duration (AFD)	The mean of every fixation duration, i.e., TFD divided by TFC; this index might indicate the depth of information processing, and is often associated with individuals' mental workload
Percentage of time spent in zone (PTS)	The time spent in an AOI compared to the total amount of time spent completing a task; this index might indicate the proportion of cognitive resources used for interacting with the information contained in the AOI
Time to first fixation (TFF)	The duration of time before the first fixation allocated in an AOI; this index might represent the salience of the information contained in the AOI
First passing time in zone (FPT)	The duration of the first time passing through an AOI, i.e., the duration from the first fixation's arrival in an AOI until the first fixation leaving the AOI; this index might indicate the length of time necessary to process specific information, and is often associated with initial information processing such as decoding

Source Cited from "Exploring how students interact with guidance in a physics simulation: evidence from eye-movement and log data analyses," by G.-L. Chiou et al. (2019), Interactive Learning Environments, p. 6

duration (TMD), which represents the total amount of time each participant spent on each control panel. By dividing the TMD into TMC, we obtained the average manipulation duration (AMD) of each control panel. In this study, we used the MC, TMD, and AMD as three major indices to examine the differences in manipulating the interface of the correct and wrong group by conducting a series of Mann-Whitney U tests.

Lag Sequential Analysis

Lag sequential analysis (LSA) is a statistical technique for examining the significance of the concurrence of any two consecutive events (please refer to Bakeman and Gottman [1997] for a detailed introduction of this technique). In this study, the LSA was applied to check whether transitions between any two visual fixations, between any two manual movements, or between visual fixation and manual movement occurred by chance or not. To achieve this aim, we first combined the eye-tracking and log data by jointly ordering each visual fixation and each manual movement with respect to its AOI in temporal sequence. Then, an eye-tracking data analysis tool, Web-based Eye-tracking Data Analyzer (WEDA; Tsai et al., 2018), was utilized to compute the frequency of transitions between any visual fixation and/or manual movement, the transitional probability of each pair of transitions, and its corresponding adjusted residuals (z scores). By checking the amount and distribution of significant transitions obtained by LSA, we could examine the difference in the patterns of interactivity between the two groups (this comparison will be described in more depth in the Results section).

Heat Map Analysis

To further examine the differences in the spatial distributions of the visual attention of the two groups, the WEDA (Tsai et al., 2018), was utilized to generate two heat maps. Based on the locations and durations of the eye fixation data of each group, WEDA calculated the normalized fixation duration allocated on each pixel of the screen. With respect to each pixel, the length of normalized fixation duration was represented by a color spectrum with one end red and the other blue; the longer the normalized fixation duration, the redder; the shorter, the bluer. By examining the insensitivity and locations of the colors shown on the heat maps, we could compare the distributions of visual attention of the two groups while they were using the simulation.

Results

RQ1: How Did Students with Different Learning Performance Distribute Their Visual Attention While Manipulating the Simulation?

Mann-Whitney U tests were conducted to examine the difference in the eye-tracking indices between the correct and wrong groups, and the results that indicate significant differences were shown in Table 14.2. According to the results, some significant differences between the two groups could be identified. For example, with respect to the TFF, the wrong group took longer than the correct group to first fixate on both the Up panel ($U = 115, p < .05$) and the Down panel ($U = 107.5, p < .05$). In addition, the wrong group appeared to spend significantly more time on the Down panel AOI than the correct group. For instance, they not only had a longer first passing time (FPT, $U = 123.5, p < .05$) but also spent more total time (TTS, $U = 120, p < .05$) and a higher percentage of time (PTS, $U = 118.5, p < .05$) on the Down panel AOI.

The correct group, in contrast, spent a higher percentage of time on the Light AOI (PTS, $U = 276.5, p < .05$).

Log Data Analysis

We conducted a series of Mann-Whitney U tests to examine the differences in the mouse control log data of the two groups. Table 14.3 shows the results that reveal statistically significant differences. As revealed in Table 14.2, the wrong group performed significantly higher frequency of manipulating both the Up panel (MC, $U = 129.00, p < .05$) and the Down Panel (MC, $U = 103.50, p < .05$). While the wrong group also appeared to spend longer manipulating both the Up panel and the Down

Table 14.2 Results of Mann-Whitney U tests on eye-tracking indices

Eye-tracking indices	Wrong group ($N = 19$)		Correct group ($N = 21$)		MWU	z
	Mean	SD	Mean	SD		
Up panel_TFF	102.44	61.46	60.25	50.43	115	−2.30*
Down panel_TFF	105.47	69.36	53.00	53.74	107.5	−2.53*
Down panel_TTS	3.48	2.18	2.31	2.76	120	−2.16*
Down panel_PTS	0.07	0.04	0.05	0.05	118.5	−2.20*
Down panel_FPT	1.17	0.90	0.62	0.60	123.5	−2.06*
Light_PTS	0.06	0.03	0.11	0.08	276.5	2.09*

*$p < .05$

Table 14.3 Results of Mann-Whitney U tests on log data indices

Log data indices	Wrong group ($N = 19$)		Correct group ($N = 21$)		MWU	z
	Mean	SD	Mean	SD		
Up panel_MC	1.95	1.61	1.05	1.36	129.00	−1.97*
Down panel_MC	2.26	1.37	1.14	0.91	103.50	−2.71*
Light_MC	0.21	0.54	0.67	1.02	247.00	1.64
Up panel_TMD	5.55	6.25	4.09	6.67	142.50	−1.55
Down panel_TMD	4.92	5.65	5.37	6.39	190.00	−0.26
Light_TMD	0.50	1.24	3.40	5.84	253.00	1.84
Up panel_AMD	2.88	4.08	3.10	6.29	168.50	−0.84
Down panel_AMD	2.10	1.86	3.47	3.81	221.00	0.59
Light_AMD	0.38	0.91	2.58	4.13	253.00	1.84

Note MC refers to manipulation count; TMD means total manipulation duration; AMD refers to average manipulation duration

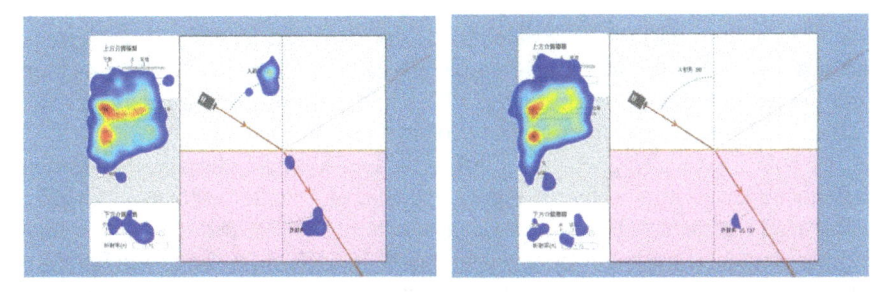

Fig. 14.2 Heat maps of fixation duration: correct group (left) and wrong group (right). Visual attentions were paid to the Lighter, Refraction, and Incident Angle areas for the correct group (left), but not for the wrong group (right)

panel than the correct group, the differences do not have statistical significance. The correct group seemed to demonstrate more manipulation of the Light panel than the wrong group, although the differences do not gain statistical significance either.

Heat Map Comparison

The heat maps that represent the distributions of visual attention of the two groups are displayed in Fig. 14.2. By comparing the two heat maps, it is apparent that the correct group paid more visual attention not only to the text of the inquiry question, but also to the degree of the incident angle, the interface where the ray starts to bend into the lower medium, and the degree of the refracted angle. It is worth noting that the incident angle, the interface, and the refracted angle are the three critical areas that provided the relevant information for correctly answering the inquiry question. The wrong group, however, only paid a little visual attention to the refracted angle, although they allocated similar amounts of visual attention to the Up panel and Down Panel AOIs. It is also worth mentioning that, while the eye-movement indices showed that the wrong group spent significantly longer (TTS) on the Down panel AOI, the intensity of their total fixation duration (TFD) on this AOI does not appear to be different from that of the correct group, as suggested by the results of the Mann-Whitney U tests.

RQ2: Did Students with Different Learning Performance Have Different Visual and Manual Behavioral Patterns?

We conducted the LSA to examine the statistical significance of the visual and manual interactive behavioral transitions demonstrated by the two groups while they were using the simulation. The results of the z scores obtained from the LSA for the two groups were shown in Table 14.4 (in each cell, the upper data is for the correct group, and the lower data is for the wrong group). Each cell indicates a behavioral transition from the corresponding behavior of its *y*-axis to the corresponding behavior of its *x*-axis. All significant transitions ($z > 1.96$) have been marked bold.

The bold z scores (higher than 1.96) were used to create the visual transition diagram of each group, as shown in Fig. 14.3. In Fig. 14.3, while the boxes with a plain background represent AOI on which visual attention was allocated, the boxes with a grey background represent a mouse manipulation. In addition, the arrows in Fig. 14.3 denote significant transitions between any visual or manual events. For example, "Up panel_S → Up panel" in Fig. 14.3a represents a significant transition from a starting click on the Up panel to the Up panel AOI, which means that the participants began to adjust the parameter on the Up panel and then changed to visually fixate on the Up panel AOI. Moreover, the number ".30" close to the arrow represents the probability of transition, which means that once the participants manually clicked the Up panel, they had a 30% probability of switching to visually fixate on the Up panel AOI. Similarly, the "Light → Light_E" illustrates a significant transition from a visual fixation on the Down panel AOI to terminate the manipulation of the Down panel, with a 64% probability.

Some common and different patterns of behavioral transitions can be identified in the two diagrams of Fig. 14.3. Regarding the commonality, both diagrams reveal temporal concurrences between visual fixation and manual manipulation. In other words, a manual manipulation either comes from or goes forward to a visual fixation on the same control panel. For example, in Fig. 14.3a, before adjusting the Up panel, the participants might first fixate on the Up panel AOI, represented by "Up panel → Up panel_S." In addition, once starting to adjust the Up panel, the participants might look back to the Up panel, "Up panel_S → Up panel," or just stop manipulating the panel, "Up panel_S → Up panel_E." Similarly, this fixation-manipulation concurrence could also be found on both the Down panel and the Light in both groups.

With respect to the differences between the two groups, the correct group seemed to demonstrate more critical behavioral transitions that are relevant for solving the inquiry question. Take Fig. 14.3a for instance; after clicking the Down panel, the correct group might transfer to visually fixate on the Refraction AOI, "Down panel _S → Refraction," which provided information about the relationship between the refracted index of the lower medium and refracted angle. Also, the transition, "Refraction → Up panel_E," suggested that the correct group would observe the Refraction AOI and then terminate the manipulation of the Up panel. This transition could provide information about the relationship between the refracted index of the upper medium and the refracted angle. In contrast, although the wrong group made some distinct behavioral transitions, these transitions appeared less important for solving the inquiry question. For example, they might end up adjusting the Light and then switch to fixate on the Refraction AOI, "Light_E → Refraction." While this transition might help the wrong group connect the incident angle with the refracted angle, this piece of information was not required for answering the inquiry question. Also, the wrong group made a visual transition from the Refraction AOI to the Reflection AOI,

14 An Investigation of Visual and Manual Behaviors ...

Table 14.4 Z scores of behavioral transitions in the correct and the wrong group

			U	D	Q	L	F	R	O	T	U_S	U_E	D_S	D_E	L_S	L_E
VISUAL		U	0	-0.38	-1.1	0.22	0.68	-0.95	-0.08	-2.28	**7.83**	1.20	0.22	-1.12	-1.33	-1.33
			0	-1.28	0.86	-0.32	-0.63	-1.72	0.13	-2.75	**5.63**	**3.95**	-2.01	-2.80	-0.89	-0.89
	D		-1.67	0	-0.30	-1.23	-0.53	-1.14	0.66	-1.56	-1.20	-1.14	**7.04**	**3.37**	-0.91	-0.91
			-2.30	0	-3.04	-0.79	-0.39	-0.01	-0.93	-1.70	-1.83	-1.8	**8.82**	**4.72**	-0.55	-0.55
	Q		1.18	0.64	0	-1.30	-0.91	-1.70	1.95	**5.48**	-1.13	-2.16	-0.43	-2.16	0.73	-2.06
			1.69	0.61	0	1.19	0.60	-0.81	1.05	**7.53**	-2.31	-2.20	-1.88	-3.87	0.85	-1.34
	L		-1.01	-1.23	-1.79	0	-0.55	1.72	-0.78	-1.59	-1.23	-0.20	-1.26	-1.26	**7.30**	**7.30**
			-0.32	-0.78	-0.14	0	-0.18	-0.49	-0.43	-0.79	0.46	1.82	-0.93	-0.92	**3.71**	**3.71**
	F		-0.74	-0.53	0.18	1.42	0	-0.51	-0.34	-0.69	-0.53	1.59	-0.55	1.42	-0.41	-0.41
			1.28	-0.38	0.74	-0.18	0	-0.24	-0.21	-0.39	-0.42	-0.41	-0.46	-0.45	-0.13	-0.13
	R		-1.59	0.83	-2.29	1.72	1.59	0	-0.72	-0.70	-0.16	**4.05**	-1.16	1.72	-0.87	1.65
			-1.72	0.02	-0.53	-0.49	**4.00**	0	1.23	-1.06	-0.15	1.89	-0.33	1.56	-0.35	-0.35
	O		1.08	0.66	0.24	-0.78	-0.34	0.76	0	1.27	-0.76	-0.72	-0.78	-0.78	-0.58	-0.58
			-0.69	-0.92	0.68	-0.43	-0.21	1.23	0	**2.63**	-1.00	1.28	-1.10	-1.08	-0.30	-0.30
	T		-1.45	-1.04	**5.34**	-1.07	-0.46	-0.99	-0.66	0	-1.04	-0.99	-1.07	-1.07	-0.79	-0.79
			-1.83	-1.11	**5.66**	-0.52	-0.26	-0.70	1.10	0	-1.21	-1.19	-1.32	-1.31	-0.37	-0.37
MANUAL	U_S		**3.31**	-1.20	-1.65	-0.31	-0.53	1.81	0.66	-1.56	0	**4.75**	-1.23	-1.23	-0.91	-0.91
			6.98	-1.80	-3.00	0.46	-0.42	1.83	-1.00	-1.82	0	**3.03**	-2.15	-2.12	-0.59	-0.59
	U_E		-0.84	-0.16	1.00	1.72	-0.51	-0.05	-0.72	0.09	-0.16	0	0.76	-1.16	-0.87	-0.87
			0.74	0.89	0.99	0.49	-0.41	-1.12	0.15	-1.13	1.17	0	-0.39	-2.09	-0.58	-0.58
	D_S		-0.32	**3.37**	-2.68	-1.26	-0.55	**3.64**	-0.78	-1.59	-1.23	-1.16	0	**6.84**	-0.93	-0.93
			-2.86	**4.73**	-3.14	-0.93	-0.46	1.51	-1.10	-2.00	-2.15	-2.12	0	**10.5**	-0.65	-0.65
	D_E		1.07	-0.31	**3.07**	-1.26	1.42	-1.16	-0.78	-0.86	-1.23	-1.16	-0.36	0	-0.93	-0.93
			-1.95	-1.33	**3.50**	-0.92	-0.45	0.63	-1.08	-1.97	0.75	-2.09	1.95	0	1.09	-0.64
	L_S		-1.27	-0.91	-0.66	**3.77**	**2.16**	1.65	-0.58	-1.19	-0.91	-0.87	-0.93	-0.93	0	**3.92**
			-0.89	-0.55	-1.22	**3.71**	-0.13	-0.35	-0.30	-0.55	-0.59	-0.58	-0.65	-0.64	0	**16.66**
	L_E		0.55	-0.91	**2.19**	1.42	-0.41	-0.87	-0.58	-0.23	-0.91	-0.87	-0.93	-0.93	-0.69	0
			-0.89	-0.55	-0.09	-0.26	-0.13	**2.65**	**3.08**	-0.55	-0.59	-0.58	1.06	-0.64	-0.18	0

Note In each cell, the upper and the lower data represent for Correct and Wrong group, respectively. A bold value means a significant transaction occurring from its corresponding behavior of the y-axis to its corresponding behavior of the x-axis. U: fixate on Up panel; D: fixate on Down panel; Q: fixate on Question; L: fixate on Light; F: fixate on Reflection; R: fixate on Refraction; O: fixate on Out; T: mouse click on Test Answer; U_S: Start adjusting the Up panel; U_E: End of adjusting the Up panel; D_S: Start adjusting the Down panel; D_E: End of adjusting the Down panel; L_S: Start adjusting the light; L_E: End of adjusting the light

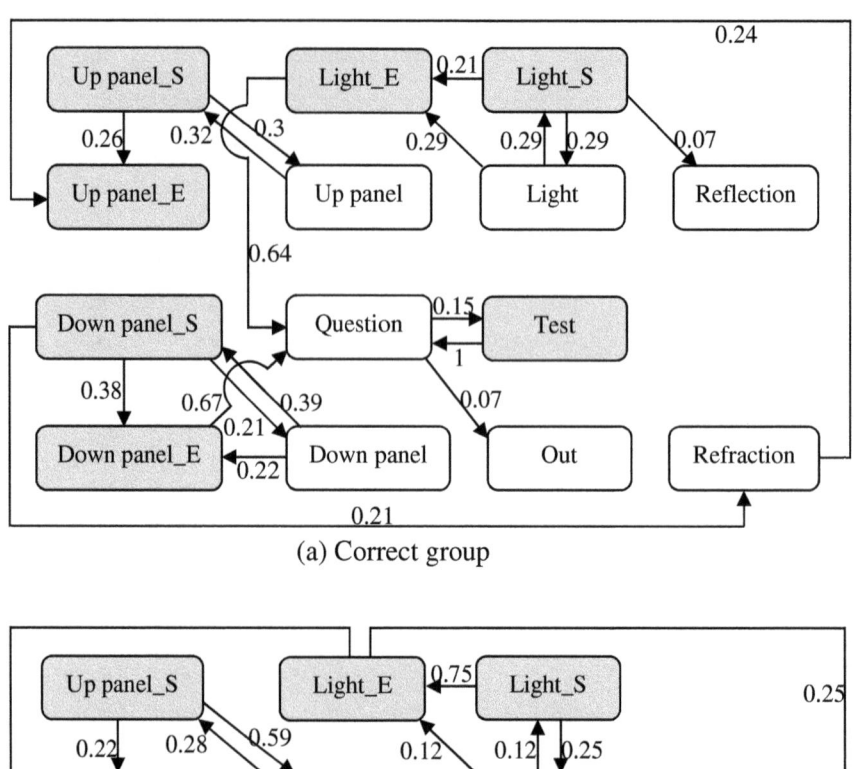

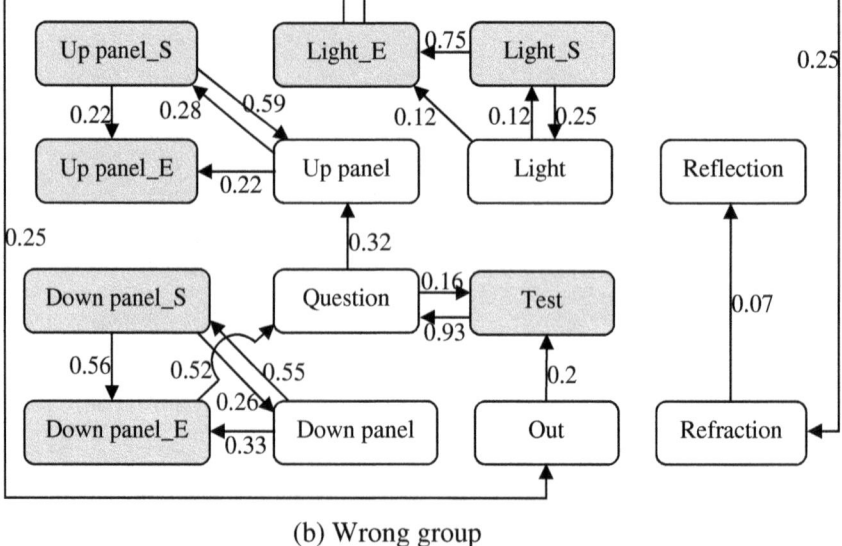

Fig. 14.3 Visual and manual transitional patterns of the two groups

"Refraction → Reflection," but, again, this transition provided scant information for solving the inquiry question.

Discussion and Conclusion

The purpose of this study was to investigate how interactions between users and simulation interface affected students' performance of inquiry activities. In particular, we combined eye-movement and log data to jointly analyze the detailed processes of user-interface interactions. This data collection and analysis approach appeared to be promising and had produced encouraging results.

Regarding the results of the eye-movement analyses, the wrong group paid significantly more visual attention to the Down panel AOI than the correct group in terms of TTS, PTS, and FPT. While adjusting the Down panel was indeed required by the inquiry question, it was never sufficient for successfully answering the question. Paying too much visual attention to the Down panel AOI, therefore, provided little help for completing the task. The results of the heat maps, nonetheless, offered alternative information about how the participants allocated their visual attention while using the simulation. Based on the heat maps, the correct group had longer visual fixations on the Light and Refraction AOIs. More specifically, they paid more visual attention to the incident angle, the point where the incident ray entered into the lower medium, and the refracted angle than the wrong group; observing these three regions was necessary for correctly answering the inquiry question. On the other hand, with respect to the results of the log data analyses, the wrong group tended to manually adjust both the Up and Down panels more often than the correct group. However, the total durations of their manipulation of the two panels were not significantly longer than those of the correct group, indicating that they might have just conducted multiple quick trials instead of carefully observing the effects of their manipulations.

Results of the LSA indicate the benefit of jointly analyzing both eye-movement and log data. In particular, it was the concurrence between the eye-movement and manual manipulation that contributed to the success of the inquiry task. The correct group, for example, made some critical transitions between the Up panel and Refraction and between the Down panel and Refraction. These two temporal transitions provided relevant information about the relationships among the refracted index of both the upper and lower media and the refracted angle, which were both necessary for answering the inquiry question. Moreover, these transitions represent the skills favored by scientific inquiry, that is, making an intervention (cause) and then observing its corresponding result (effect). Chiou et al. (2019) also identified the importance of this sort of visual transition for successful completion of an inquiry task.

The benefit of jointly analyzing both the eye-movement and log data are even more apparent in examining the behaviors of the wrong group. For example, the results of both eye-movement and log data analyses indicate that the wrong group devoted more effort to the Down panel than the correct group did. If these two sources of data were analyzed separately, it would be difficult to explain why the wrong group failed to complete the inquiry task, even though they paid more visual attention and exerted more physical effort to control the Down panel. By jointly analyzing these two sources of data, as demonstrated by the LSA results, we can quickly grasp that the wrong group failed to connect their effort on the Down panel to other relevant AOIs. Without these sorts of meaningful connections, according to de Jong (2011), their visual attention and manual manipulation might result in fragments of information and could not form an integrated understanding of the simulated models, phenomena of refraction, in this study.

Although the joint analyses of eye-movement and log data provide a promising approach to investigate the users' interaction with simulation interfaces, it remains unclear why users behave in this manner to manipulate a computer simulation. In other words, what are the factors that determine how users manipulate computer simulations? As suggested by van Joolingen et al. (2007), the result of this study indicates that the prior knowledge of the users might be an important factor that affects the learning process in a computer simulation. Based on the pretest scores, the wrong group had a significantly lower level of prior knowledge than the correct group. A lack of domain knowledge might keep the wrong group from fully understanding the concepts involved in the inquiry question and the computer simulation. Moreover, they might hold false expectations for the simulation or make incorrect predictions of the simulated phenomena. As a result, they could not form a coherent strategy for making manual intervention and visual observation, and thus failed to complete the inquiry task. Of course, there must be some other factors that affect the interactions between users and simulation interfaces, but, in this study, we could not make any postulations without further evidence.

In summary, this study highlighted the importance of investigating the interaction between users and interfaces to better understand the process of learning with computer simulations. We adopted both eye-movement and log data to jointly analyze the interactive processes of using a computer simulation. The results show that the concurrences between visual attention and manual manipulation were necessary for operating the simulation. Moreover, to successfully complete the inquiry task offered by the simulation, the participants needed to not only connect representations displayed on relevant areas of the simulation in a reasonable sequence, but also to actively integrate them into a meaningful whole. This research approach provides significant benefits over solely analyzing eye-movement or manual manipulation data. Although we identified prior knowledge as a crucial factor that might determine the behaviors of using computer simulations, more factors are needed to fully account for the individual differences in the behaviors. It is therefore suggested that future studies explore more potential factors that affect the interactions between users and simulation interfaces. In addition, more data formats, such as think-aloud verbal reports, could be employed to investigate the cognitive aspect of using computer simulations.

References

Abd-El-Khalick, F., BouJaoude, S., Duschl, R., Lederman, N. G., Mamlok-Naaman, R., Hofstein, A., Niaz, M., Treagust, D., & Tuan, H.-l. (2004). Inquiry in science education: International perspectives. *Science Education, 88*(3), 397-419. http://doi.org/10.1002/sce.10118.

Adams, W. K., Reid, S., LeMaster, R., McKagan, S. B., Perkins, K. K., Dubson, M., & Wieman, C. E. (2008). A study of educational simulations part IEngagement and learning. *Journal of Interactive Learning Research, 19*(3), 397–419.

Bakeman, R., & Gottman, J. M. (1997). *Observing interaction: An introduction to sequential analysis* (2nd edition). UK: Cambridge University Press.

Chiou, G.-L., Hsu, C.-Y., & Tsai, M.-J. (2019). Exploring how students interact with guidance in a physics simulation: Evidence from eye-movement and log data analyses. *Interactive Learning Environments.* http://doi.org/10.1080/10494820.2019.1664596

D'Angelo, C., Rutstein, D., Harris, C., Bernard, R., Borokhovski, E., & Haertel, G. (2014). *Simulations for STEM learning: Systematic review and meta-analysis.* Menlo Park: SRI International.

de Jong, T. (2006). Technological advances in inquiry learning. *Science, 312*(5773), 532–533. https://doi.org/10.1126/science.1127750.

de Jong, T. (2011). Instruction based on computer simulations. In R. E. Mayer & P. A. Alexander (Eds.), *Handbook of research on learning and instruction* (pp. 446–466). New York: Routledge.

de Jong, T., Linn, M. C., & Zacharia, Z. C. (2013). Physical and virtual laboratories in science and engineering education. *Science, 340*(6130), 305–308. https://doi.org/10.1126/science.1230579.

Duschl, R. (2008). Science education in three-part harmony: Balancing conceptual, epistemic, and social learning goals. *Review of Research in Education, 32*(1), 268–291. https://doi.org/10.3102/0091732x07309371.

Hake, R. R. (1998). Interactive-engagement versus traditional methods: A six-thousand-student survey of mechanics test data for introductory physics courses. *American Journal of Physics, 66*(1), 64–74. https://doi.org/10.1119/1.18809.

Hsu, C.-Y., Chiou, G.-L., & Tsai, M.-J. (2016). *A pilot study on developing and validating a fixation-based scaffolding learning system.* Paper presented at the Poster presented at 2016 International Conference of East-Asian Association for Science Education, Tokyo, Japan.

Just, M. A., & Carpenter, P. A. (1980). A theory of reading: From eye fixations to comprehension. *Psychological Review, 87*(4), 329–354.

Kim, E., & Pak, S.-J. (2002). Students do not overcome conceptual difficulties after solving 1000 traditional problems. *American Journal of Physics, 70*(7), 759–765. https://doi.org/10.1119/1.1484151.

McElhaney, K. W., & Linn, M. C. (2011). Investigations of a complex, realistic task: Intentional, unsystematic, and exhaustive experimenters. *Journal of Research in Science Teaching, 48*(7), 745–770. https://doi.org/10.1002/tea.20423.

Rutten, N., van Joolingen, W. R., & van der Veen, J. T. (2012). The learning effects of computer simulations in science education. *Computers and Education, 58*(1), 136–153. https://doi.org/10.1016/j.compedu.2011.07.017.

Trowbridge, D. E., & McDermott, L. C. (1981). Investigation of student understanding of the concept of acceleration in one dimension. *American Journal of Physics, 49*(3), 242–253.

Tsai, M.-J., Hsu, P.-F. & Pai, H.-T. (2018). *Lag sequential analysis in Eye-Tracking Data Analyzer (EDA) for educational researchers.* Poster presented at the 4th International Symposium on Educational Technology (ISET 2018), Osaka, Japan.

van Joolingen, W. R., De Jong, T., & Dimitrakopoulou, A. (2007). Issues in computer supported inquiry learning in science. *Journal of Computer Assisted Learning, 23*(2), 111–119. https://doi.org/10.1111/j.1365-2729.2006.00216.x.

Wieman, C. E., Adams, W. K., Loeblein, P., & Perkins, K. K. (2010). Teaching physics using PhET simulations. *The Physics Teacher, 48*(4), 225–227. https://doi.org/10.1119/1.3361987.

Wieman, C. E., & Perkins, K. (2005). Transforming physics education. *Physics Today, 58*(11), 36–41.

Guo-Li Chiou, Ph.D. is currently an Associate Professor at the Program of Learning Sciences, School of Learning Informatics, National Taiwan Normal University. His research focuses on how students learn science, particularly on mental models of physical phenomena and systems. Currently, Professor Chiou is working on how students learn physics in digital environments, with particular emphasis on simulation-based and game-based physics learning. In addition, he uses eye-tracking techniques to record and analyze students' visual behaviors while reading and solving physics problems.

Chung-Yuan Hsu, Ph.D. is currently a Professor at the Department of Child Care, National Pingtung University of Science and Technology. His research focuses on the designs of simulation-based and game-based learning environments, online inquiry learning analytics, and teacher education in educational technology. Recently, he has attempted to use eye-tracking technology to track and analyze how students interacted with the scaffolding designed in game-based learning environments.

Meng-Jung Tsai, Ph.D. is a Distinguished Professor at the Program of Learning Sciences, School of Learning Informatics, National Taiwan Normal University. Her research interests centered at information literacy education, technology-enhanced science learning and learning analytics. She has used eye-tracking technology to examine learners' visual behaviors in digital learning environments such as multimedia-based learning, game-based learning and web-based inquiry learning. Recently, she is interested in examining the role of affective factors such as interests, anxiety and self-efficacy played in human visual behaviors.

Chapter 15
Visualizing Student Navigation of Geologic Block Diagrams

Karen S. McNeal, Rachel Atkins, and Elijah T. Johnson

Introduction

Spatial thinking is defined by Uttal et al. (2013) as, "the mental processes of representing, analyzing, and drawing inferences from spatial relations…between objects…or…within objects" (p. 367). There are many spatial skills, and different disciplines and fields have varying names and definitions for them. However, the consensus is that spatial reasoning is an essential skill necessary for success in the STEM (science, technology, engineering, and mathematics) domains as these disciplines require students to visualize past or theoretical phenomena based on spatial relationships between elements in nature or a provided diagram. The importance of spatial reasoning ability in the STEM domains is solidified through a study by Wai et al. (2009) that completed an eleven-year longitudinal study and aligned their results to 50 years of preexisting psychological data to conclude that spatial ability is correlated with STEM achievement and career paths. Forty-thousand random participants were tracked for over 11 years to assess their self-organization into careers based on mathematical, verbal, and spatial ability. The data showed that students with high spatial ability excelled in physical science, math/computer science, and engineering in terminal bachelor, master, and doctorate degrees. Students with lower spatial reasoning ability tended to self-organize into education, law, and business. The occupations they pursued after college strongly resemble those same trends. Similarly, Kell and Lubinski (2013) suggest that students may self-organize into their majors and careers based on their spatial thinking ability, whereas students may

---

K. S. McNeal (✉) · E. T. Johnson
Department of Geosciences, Auburn University, Auburn, AL 36849, USA
e-mail: ksm0041@auburn.edu

R. Atkins
North Carolina State University, Raleigh, NC 27659, USA

self-select out of STEM domains due to the amount of spatial reasoning ability they possess, especially when underserved by academic institutions.

Visualizing three-dimensional (3D) structures is a challenge in science, technology, engineering, and mathematics (STEM) learning (Milner-Bolotin & Nashon, 2012). "Students in these fields are required to reason about objects or features that occur at spatial scales too large or small to be directly observed. Consequently, 3D phenomena are often illustrated using visual representations such as diagrams" (Gagnier et al., 2017, p. 884). 3D thinking is particularly important in chemistry, biology, engineering, and the geosciences. In chemistry education, students' performances on a variety of chemistry problems including problem-solving (e.g., stoichiometry) and multistep calculations, as well as balancing chemical equations were linked to their spatial thinking skill development (Staver & Jacks, 1988; Carter et al., 1987). Instructional interventions that worked to build students' spatial skills through practice lead to significant increases in student chemistry test performance (Small & Morton, 1983; Tuckey et al., 1991).

In modern biology and engineering education, 3D and 4D visualization also plays an important role in students learning both the concepts and skills needed to accurately reason about specific phenomenon (e.g., technical drawing, graphical representations of biochemical structures, or embryo development; Milner-Bolotin & Nashon, 2012). Biology and engineering classroom interventions have shown that providing students graphical software to scaffold their development of 3D and 4D skills supported student success in biochemistry classes compared to those that did not receive the software intervention (Richardson & Richardson, 2002), and in engineering, students that were exposed to such technology had gains in spatial rotation and space relations, assessed using pre-post test scores (Sorby, 2009).

The geosciences also have a high requirement for building spatial thinking skills (Sanchez & Wiley, 2014), many of which include 3D visualization. One of the spatial skills a geoscientist must employ is mental brittle transformation, which is the ability to mentally break and reconstruct objects (Resnick & Shipley, 2013). Another skill is mental rotation, which involves a person's ability to turn a 2D or 3D object about an axis (Shepard & Metzler, 1971) and may be activated by a stratigrapher examining the position of overturned strata. Yet another skill is spatial orientation, which requires an understanding of perspective and the relation of an object to a frame of reference (Ramful et al., 2017). For example, geoscience students in field camp, a capstone course required by many geology programs, may employ spatial orientation in navigating the field environment and marking the relative positions of outcrop features on a map. A more complex skill that geoscientists employ is spatial visualization which represents multiple associated tasks. Linn and Petersen (1985) define spatial visualization as "spatial ability tasks that involve complicated, multistep manipulations of spatially presented information" (p. 1484). A unique skill to the geosciences includes penetrative thinking or visualizing the subsurface or interior of an object using clues from the visible parts of the object (Alles & Riggs, 2011), like how a structural geologist generates a 2D cross-section that represents a 3D phenomenon in the real world in a drawing that represents the bedforms present at the Earth's surface and in geologic outcrops. Often these phenomena are represented in geologic block diagrams that illustrate geologic structures at scales ranging from centimeters to tens of kilometers. However, students struggle to interpret the 3D spatial relations conveyed in these diagrams (Gagnier et al., 2017). As such, we have focused on penetrative thinking skills in this study as they are key to interpreting geologic block diagrams, which are the spatial representations employed in this work.

Relevant Work in Geoscience Education Research

Many students may have natural spatial thinking ability, while others may lack this skill, which could make learning geological concepts more challenging (Ishikawa & Kastens, 2005).

Since spatial reasoning ability has been shown to be malleable (Uttal et al., 2013), many researchers have explored interventions designed to train spatial thinking skills in the geosciences. For example, Titus and Horsman (2009) performed a semester-long study training students from two different populations in spatial visualization. This study assessed the effect of spatial training with Spatial Intelligence and Learning Center (SILC)-verified assessment strategies and improved undergraduate student performance, as well as students' overall course grades. In a study conducted by Ormand et al. (2014), students in a variety of geoscience courses, from introductory to senior level major courses, and with various degrees of incoming spatial ability experienced spatial skill gains from simply being exposed to spatial concepts during the course. Gold et al. (2018) found that regular, short interventions throughout an academic semester improve students' spatial thinking skills significantly with a moderate to large effect size when compared to an instruction-as-usual control group. They also found that about 15% of the students improved their spatial skills to the point that they would be considered high enough for those commonly entering/continuing in STEM.

Spatial training using new technologies has been a growing area in the geosciences. McNeal et al. (2020) conducted a study aimed to understand the impact of using an augmented reality (AR) sandbox on students' topographic map performance. They found that students with higher spatial ability tended to perform better on the task with the AR sandbox than those with lower spatial ability, but that this performance gap was mitigated with more structured activities in the sandbox. This finding led to the idea that perhaps the AR sandbox could support students' spatial thinking skills. Johnson and McNeal (in review) have since shown that the AR sandbox has the potential to support student spatial skill development. They implemented activities with students in a lab environment to aid their development of spatial orientation, spatial rotation, and spatial visualization skills. Results indicate that the AR sandbox may have the greatest potential to assist students in developing their spatial visualization skills. Spatial visualization was the area in which students identified the most challenges and the least strategies during their problem-solving.

Spatial training with new technologies in the geosciences has also been used with geographical information systems (GIS) to explore whether GIS could impact students' spatial thinking (Lee & Bednarz, 2009; Kim & Bednarz, 2013). Lee and Bednarz (2009) grouped multiple GIS activities into spatial skills categories and administered a spatial skills test before and after the GIS course. The test showed gains in the students' spatial reasoning ability, showing that technology that has a high spatial component, such as GIS, can train spatial thinking ability.

Eye-Tracking in the Geosciences

Geoscience education researchers in classroom, field, and lab environments employ eye-tracking approaches using stationary devices and portable headsets in their studies to gain insights into student cognition. For example, a field study by Maltese et al. (2013) used eye-tracking headsets on students to both investigate its viability for observing and detailing students' experiences in the field and evaluate the variety of information that can come from eye-tracking in the field. Although there were operational and technical challenges related to using the eye-tracker in the field, the scene video they acquired elucidated how students were engaging with the geology and each other in the field.

Eye-tracking was also used by McNeal et al. (2014) to evaluate and revise an online curriculum, *EarthLabs*. College undergraduates interacted with the online modules while being eye-tracked to determine how they were engaging with the material to improve the online *EarthLabs* curriculum and user experience. Evidence from eye-tracking revealed that although students were engaging with the text portions of the modules more than the images, that engagement declined over time as the students worked through the activities in the module. Additionally, students generally found charts, graphs, and questions embedded in the text to be most useful, however, they experienced difficulty engaging with graphs depicting change over time. Learning what students are paying attention to as well as how their engagement varies over time from their eye movements in this study speaks to the usefulness of eye-tracking for user-testing.

The effectiveness of eye-tracking for user-testing is also expounded upon in a study by Maudlin et al. (2020) where male and female decision-makers and students were eye-tracked to explore gender-differences in visual attention on a decision-support website, *PINEMAP DSS*. *PINEMAP DSS* is a website service that communicates climate impacts to loblolly pine forests in the southeastern United States. Since this information is primarily for forest service professionals and decision-makers, testing the usability of the medium and getting insight into how different users interacted with the content was of high importance. The researchers found that males paid more attention to the data and map features of the website, while females paid less attention to the data itself and more time evaluating other features of the website including tabs, map legends, and text. Males also outperformed females on the questions they were given about the information on the websites. These results can be used to revise website-based tools so that content creators can effectively communicate with their intended audiences.

Atkins and McNeal (2018) also explored how users interacted with climate information by eye-tracking students looking at climate change graphs. This study compared the eye movement and attention patterns of undergraduate and graduate students to identify how knowledge, skill, and expertise affect performance on fact-extraction and extrapolation tasks. They found that undergraduates spent more time on graphical elements not pertinent to the content (i.e., axes, title, legend), while graduate students spent more time interpreting the provided data. They also found that undergraduate students with high graphical skills performed similarly to graduate students. By exploring the cognitive limitations of novice students on climate change graph understanding, scientists can improve those graphs to help communicate their findings more effectively and educators can focus their instruction on scaffolding student to be able to interpret scientific graphs.

In this study, we use eye-tracking in a similar way: to understand the visual attention of students while problem-solving. Spatial reasoning ability as a suite of cognitive skills can be improved with training (Uttal et al. 2013), so understanding the challenges students have with these skills is a first step in developing interventions designed to improve those skills. A spatial thinking skill that has been shown to be challenging for students in the geosciences is penetrative thinking, particularly when studying geologic block diagrams. Currently, eye-tracking has not been applied to understanding how students navigate block diagrams in the geosciences. We aim to investigate how students visually navigate geologic block diagrams. More specifically, we look to identify emergent patterns between students who do well (high performers) and students who do poorly (low performers) in solving geologic block diagrams. Finally, we highlight common errors made by students as identified by their visual navigation patterns while solving geologic block diagrams and provide future directions for investigating the cause of these errors.

Methods

Participants

The 58 participants in this study consisted of 45 undergraduates enrolled in an Earth Systems Science course at a large land-grant university in the southeastern United States and 13 graduate students in a graduate program at the same university. Our participants ranged in class rank from freshman to graduate students with at least 1 year of experience ages 17 to >23 with a median age of 20 years and a male to female ratio of 31:27. Participants were recruited from two sections of an introductory Earth Systems Science course taught by different instructors and received a $20 Amazon gift card as compensation after completion of the pre-test and eye-tracking study outside of class. Graduate students participated on a volunteer basis and were recruited from a departmental listserv. Human subject's research approval from the Institutional Review Board (IRB) was obtained before recruitment and the commencement of any research activities.

Experimental Design and Instrumentation

This study used two versions of the Geologic Block Cross-sectioning Test (GBCT; Orman et al., 2014). The first 16-question version was used as a pre-test to establish participants' visual penetrative thinking skills and the ability to recognize the correct vertical cross-section through a geology block diagram prior to eye-tracking. For undergraduates, the test was administered at the end of an Earth Systems Science class period, while graduate students were asked to take the pre-test immediately before the eye-tracking study. After pre-testing, participants were asked to solve five selected problems of varying difficulty from the second version of the GBCT assessment. These problems addressed geologic concepts such as dipping beds, faulted horizontal strata, dipping transverse beds, and plunging folds (Fig. 15.1). During this second assessment, participant eye movements were tracked while solving each question using a Tobii TX300 eye-tracker and participants were also asked to state their answer to the presented problem aloud.

The eye-tracker was attached to a 23-inch computer monitor, collecting at 300 Hz, and did not come in physical contact with participants. Calibration was completed

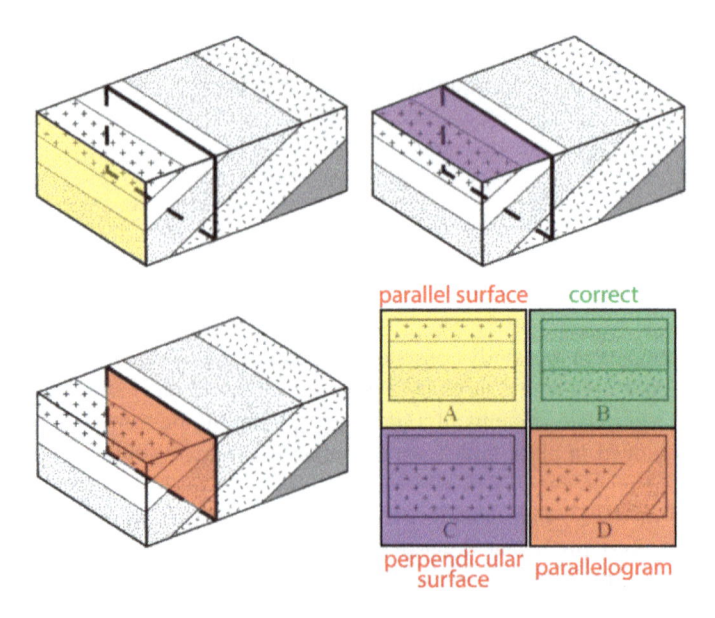

Fig. 15.1 (above) Highlights specific errors depicted in the response choices and where they may be found in the 3D block diagram. Choice B is the correct response for this question

for each participant to ensure accuracy and precision among participant trials. Participants sat ~65 cm from the monitor and gazed at the computer screen to view the provided graphs with an unobstructed view. The noncontact nature of this technology allows for the capture of natural eye movements, compared with instrumentation that is worn by the participant. The system allowed for corrective lenses to be worn without affecting results.

Ormand et al. (2014) created the 16-question Geologic Block Cross-sectioning Test which has since undergone multiple assessments to ensure validity and reliability internally and between the two versions. It was also developed to specifically address common misconceptions people have about geologic block diagrams (Kali & Orion, 1996) (see Fig. 15.1 for examples). Participants are given the same three instructions for all diagrams: "1. Study the geologic structure that is displayed in the 3D block diagram, 2. Visualize what the cross-section of that geologic structure would look like on the surface of the vertical plane intersecting the block and 3. Choose the multiple choice answer that illustrates the structure along that plane. Where more than one answer appears to be possible, choose the MOST LIKELY answer." For both the pre-test and eye-tracking test, an example with its correct answer was provided to allow participants to practice the format before being assessed.

Each question consists of a 3D box with an illustrated geologic problem inside, and a dark box highlighting the horizontal transect where the participant is asked to mentally slice through the diagram, with four possible answers to choose from (Fig. 15.1). The types of errors included in the four possible answers for each question fall under the two broad categories of penetrative and non-penetrative answers (Ormand et al., 2014). Penetrative errors reflect an attempt to visualize the inside of the structure, but do so incorrectly, whereas non-penetrative errors indicate an individual's inability to mentally penetrate the block, subsequently their answers reflect one of the visible sides of the diagram (Kali & Orion, 1996).

Data Analysis

Two aspects of eye movements that are most often studied include saccades and fixations. Saccades are the short periods of rapid eye movement between fixations that redirect participant gaze from one fixation to another (Ramat et al., 2008). These can occur up to four times in a second and participants are effectively blind while they occur (Land, 2012). Fixations are the points between saccades where the eye is nearly stationary for relatively longer periods of time (~70–100 ms). These eye movements are of particular interest as it is during these viewing times that mental processing takes place (Bojko, 2013).

A correlation analysis was done to determine the relationship between pre-test score and score on the five-question eye-tracking test. All 58 participants yielded a moderately strong correlation coefficient of 0.6581, indicating that the five selected questions appropriately represent participants' overall performance on the assessment. Therefore, we use performance on the more thorough 16-question pre-test to

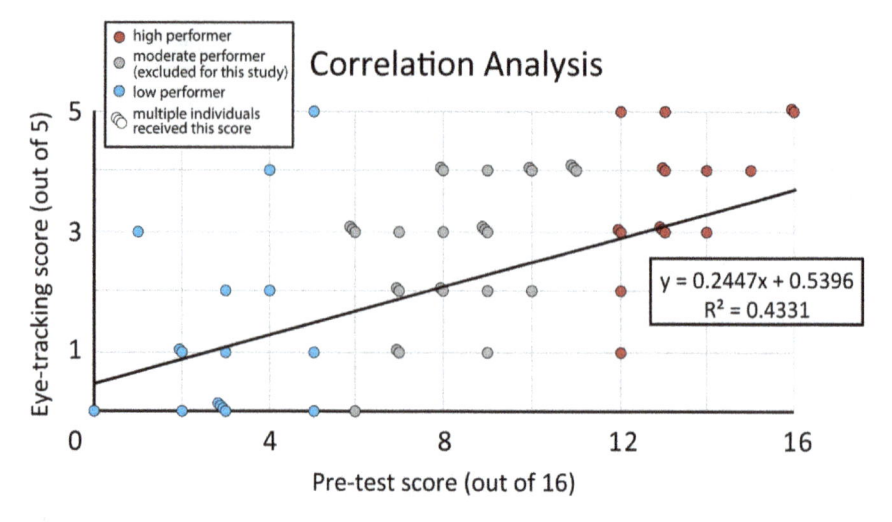

Fig. 15.2 (above) Correlation analysis between questions asked during eye-tracking (eye-tracking score) and pre-test scores reveals a moderately strong correlation, therefore we are confident that the five questions selected for the eye-tracking study yield valid results. Additionally, the pre-test was used to group participants into high performers (red points) and low performers (dark blue points) using quartiles. These groupings are used for the remainder of the analysis. Points have been jittered to indicate multiple points at each location

bin participants into **low performers** ($n = 16$, pre-test scores: 0–5 out of 16) and **high performers** ($n = 16$, pre-test scores: 12–16) by quartiles (Fig. 15.2).

Results

From the GBCT assessment, we were able to confirm that high performers on the pre-test answered the eye-tracking questions correctly more often than low performers. Further, when high performers answered questions incorrectly on the eye-tracking assessment, they all made the same error, whereas low performers made multiple different errors. Additionally, we were able to determine that the most common error made by all participants was a parallelogram error, indicating the possibility that they were solely visualizing within the parallelogram itself, and not taking into account the behavior of the layers in the rest of the diagram. This was confirmed in our eye-tracking results when comparing the visual patterns of high and low performers. For example, Fig. 15.3a shows high performers, who also answered questions correctly more often, focusing their attention on the face that indicates that the layers are dipping to inform their answer selection (see the top right corner of the bolded box for question #2). Conversely, low performers do not focus their attention in this same place, but rather distribute their gaze throughout the parallelogram, congruent with their most often incorrectly selected answer (parallelogram).

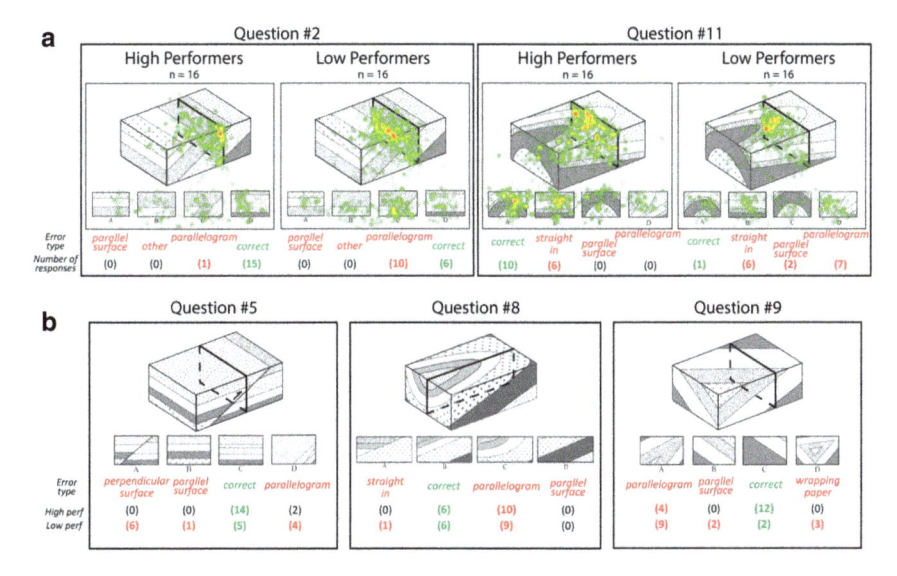

15 Visualizing Student Navigation of Geologic Block Diagrams 303

Fig. 15.3 (above) Part a shows examples of eye-tracking results from two of the five questions asked during the eye-tracking assessment. Areas of red indicate high concentrations of visual gaze and areas in green correspond to low concentrations of gaze. The type of error that each response choice depicts is indicated in RED LETTERING below each choice, GREEN LETTERING indicates correct answer, and the number of responses is in parentheses. Part b is a summary of the number of high and low performer responses for the remaining three questions without the eye-tracking heat maps for simplicity

Out of the five eye-tracking questions, only #8 and #11 had penetrative errors (labeled "straight in") as selectable options. These types of errors are unique because they indicate the participants attempt to see through the diagram and visualize the internal structure. The difference between high and low performers is highlighted in Fig. 15.3a for question #11. The only error made by high performers was a penetrative error, indicating that those who answered incorrectly were still attempting to mentally penetrate the diagram and visualize in 3D. Conversely, not only were the number of correct responses fewer for low performers, but their incorrect answers spanned all options, indicating an inability to visualize in 3D. The eye-tracking results from both high and low performers reflect these differences, showing the gaze concentrations of high performers distributed throughout the diagram, whereas low performers remain mostly in the parallelogram (Fig. 15.4).

The distribution of gaze throughout the diagram, particularly for a complicated problem such as the ones depicted in questions #9 & 11, indicates the viewers attempt at understanding the overall behavior of the geologic layers and interpreting how they may look at different angles.

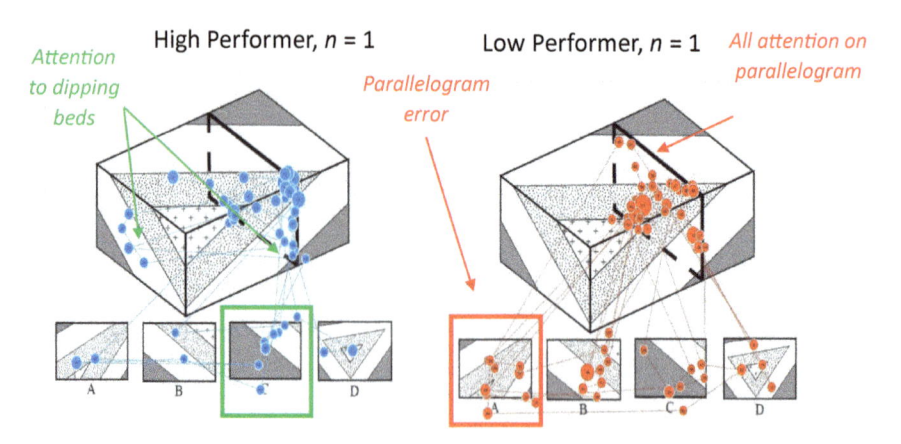

Fig. 15.4 (above) Examples of eye-tracking gaze plots from one high and one low performer from question #9. The green arrows show what the high performer is paying attention to and the selection of the correct answer indicated by the green box. The red arrows indicate the parallelogram error of the low performer and the resulting incorrect response option that was chosen as indicated by the red box

Discussion and Future Directions

Overall, eye-tracking results from all five questions indicate that high performers allocate proportionately more of their visual gaze outside of the bolded parallelogram in the provided block diagrams. We interpret this to indicate that these participants are likely attempting to visualize the whole structure (i.e., on all sides), and using this to determine their answer. By using observations from all sides of the diagram, they are able to develop a complete story, as opposed to selecting pieces of the diagram (i.e., what's depicted inside the parallelogram) to inform their answer. This exploratory work has helped to show eye-tracking as a useful tool to understand how individuals navigate spatial diagrams. This study was able to highlight the differences between high and low performers as revealed in pre-test scores and in eye-tracking results.

Some of the most informative next steps to build on findings from this work would be the addition of concurrent think alouds and/or a post-assessment interview, along with the addition of multiple spatial skill pre-assessments. The addition of a qualitative metric would provide insight into why individuals selected their answers and combined with correlations between spatial skill competencies acquired by additional assessments, could help identify specific areas to target spatial training to facilitate skill improvement. Recommendations for more rigorous spatial skill assessments include the Purdue Visualization of Rotations Test (PVRT) (Guay, 1976) for mental rotation, the Educational Testing Service (ETS) Hidden Figures test (Ekstrom et al., 1976) for disembedding skills, and Planes of Reference test (Titus & Horsman, 2009) for penetrative thinking. Furthermore, while our initial study used five questions for eye-tracking analysis, we recommend increasing the number of questions to increase statistical robustness.

Another question worthy of investigation that could be informed by eye-tracking is "do patterns exist among eye movements between diagram and answer options?" and the connection that those patterns may have with the types of cognitive processes being used to solve the problem (i.e., inductive vs. deductive reasoning). For example, if a participant investigated the answer options before viewing the diagram, they may have solved the problem by eliminating incorrect options. Conversely, first fixations at the main diagram may indicate the development of an answer before looking at the choices and may indicate a higher confidence in 3D visualization. The investigation of these spatiotemporal patterns combined with additional metrics (i.e., gender, pre-test performance) and/or assessments could answer unique questions that traditional paper testing is not able to.

We caution future researchers interested in expert/novice comparisons to perform a thorough assessment of a range of spatial skills before binning participants into expert and novice categories. Research has shown that spatial abilities are a combination of one's prior experiences that can be traced all the way back to childhood, their innate ability, and exposure to formal training (Gold et al., 2018). Not only do spatial abilities vary extensively across student populations, it has also been shown that with appropriate training, spatial skills can be improved (Lee & Bednarz, 2009; Uttal et al. 2013; Ormand et al., 2014; Gold et al., 2018). For these reasons, it is important to confirm that perceived experts (i.e., domain experts) actually exhibit high levels of spatial abilities.

Despite the needs for continuing research, this study provides insights about how to better support students' 3D problem-solving skills, especially among high and low performers. Potential classroom activities that could help mediate this performance gap is to pair high and low performers together to solve spatial problems. This would require some pre-testing of students at the beginning of a course, but such distributed expertise pairing could be helpful to students. Additionally, the replaying of eye-tracking scan patterns of high performers before completing a geologic block diagram problem may help to improve the performance of all learners. Both of these suggestions require further testing to document any potential student learning gains through intervention studies in the geosciences. However, these are activities that our research eludes to as potential actions that instructors could take to help support the learners in their classrooms develop 3D spatial problem-solving skills.

Conclusions

This exploratory eye-tracking study provided unique insights not yet obtained by the geoscience education research community about how high and low spatial performing students navigate geologic block diagrams, a 2D visualization tool used in the geological sciences to represent conditions within a 3D geologic formation. The results showed that there were differences in the visual attention that high and low performers made on the diagrams. These differences aligned with the correct/incorrect selection on the geologic block diagram assessment used in this study where high performers tended to make more fixations on all faces of the diagram where low performers tended to fixate on more specific areas of the diagram. This trend indicated that high performers seemed to be able to "see the big picture" whereas low performers could not when solving 3D visualization problems in the geosciences. Additional research is recommended to expand on and verify our exploratory results, however, the use of eye-tracking within the context of understanding how students solve spatial problems in the geosciences has provided new insights that with continued research can be used to inform how best to scaffold students as they build their spatial skills in the geosciences, a STEM field that has a high requirement for multiple spatial thinking skills to be developed among learners.

References

Alles, M., & Riggs, E. M. (2011). Developing a process model for visual penetrative ability. *Geological Society of America Special Papers, 474,* 63–80.

Atkins, R. M., & McNeal, K. S. (2018). Exploring differences among student populations during climate graph reading tasks: An eye tracking study. *Journal of Astronomy and Earth Sciences Education (JAESE), 5*(2), 85–114.

Bojko, A. (2013). *Eye tracking the user experience: A practical guide to research.* Rosenfeld Media.

Carter, C. S., Larussa, M. A., & Bodner, G. M. (1987). A study of two measures of spatial ability as predictors of success in different levels of general chemistry. *Journal of Research in Science Teaching, 24*(7), 645–657.

Ekstrom, R. B, French, J. W., Harman, H. H., & Dermen, D. (1976). *Manual for kit of factor referenced cognitive tests.* Princeton, NJ: Educational Testing Service.

Gagnier, K. M., Atit, K., Ormand, C. J., & Shipley, T. F. (2017). Comprehending 3D diagrams: Sketching to support spatial reasoning. *Topics in Cognitive Science, 9*(4), 883–901.

Gold, A. U., Pendergast, P. M., Ormand, C. J., Budd, D. A., & Mueller, K. J. (2018). Improving spatial thinking skills among undergraduate geology students through short online training exercises. *Journal International Journal of Science Education, 40*(18).

Guay, R. B. (1976). *Purdue spatial visualization test.* West Lafayette, IN: Purdue Research Foundation.

Ishikawa, T., & Kastens, K. A. (2005). Why some students have trouble with maps and other spatial representations. *Journal of Geoscience Education, 53*(2), 184–197.

Kali, Y., & Orion, N. (1996). Spatial abilities of high-school students in the perception of geologic structures. *Journal of Research in Science Teaching, 33*(4), 369–391.

Kell, H. J., & Lubinski, D. (2013). Spatial ability: A neglected talent in educational and occupational settings. *Roeper Review, 35*(4), 219–230.

Kim, M., & Bednarz, R. (2013). Development of critical spatial thinking through GIS learning. *Journal of Geography in Higher Education, 37*(3), 350–366.

Land, M. F. (2012). The operation of the visual system in relation to action. *Current Biology, 22*(18), R811–R817.

Lee, J., & Bednarz, R. (2009). Effect of GIS learning on spatial thinking. *Journal of Geography in Higher Education, 33*(2), 183–198. https://doi.org/10.1080/03098260802276714.

Linn, M. C., & Petersen, A. C. (1985). Emergence and characterization of sex differences in spatial ability: A meta-analysis. *Child Development, 56*(6), 1479–1498.

Maltese, A. V., Balliet, R. N., & Riggs, E. M. (2013). Through their eyes: Tracking the gaze of students in a geology field course. *Journal of Geoscience Education, 61*(1), 81–88.

Maudlin, L. C., McNeal, K. S., Dinon-Aldridge, H., Davis, C., Boyles, R., & Atkins, R. M. (2020). Website usability differences between males and females: An eye-tracking evaluation of a climate decision support system. *Weather, Climate, and Society, 12,* 183–192. https://doi.org/10.1175/WCAS-D-18-0127.1.

McNeal, K. S., Libarkin, J. C., Ledley, T. S., Bardar, E., Haddad, N., Ellins, K., & Dutta, S. (2014). The role of research in online curriculum development: The case of EarthLabs climate change and Earth system modules. *Journal of Geoscience Education, 62*(4), 560–577.

McNeal, K. S., Ryker, K., Whitmeyer, S., Giorgis, S., Atkins, R., LaDue, N., & Pingel, T. (2020). A multi-institutional study of inquiry-based lab activities using the Augmented Reality Sandbox: impacts on undergraduate student learning. *Journal of Geography in Higher Education, 44*(1), 85–107.

Milner-Bolotin, M., & Nashon, S. M. (2012). The essence of student visual-spatial literacy and higher order thinking skills in undergraduate biology. *Protoplasma, 249*(1), 25–30.

Ormand, C. J., Manduca, C., Shipley, T. F., Tikoff, B., Harwood, C. L., Atit, K., & Boone, A.P. (2014). Evaluating geoscience students' spatial thinking skills in a multi-institutional classroom study. *Journal of Geoscience Education, 62,* 146–154.

Ramat, S., Leigh, R. J., Zee, D. S., Shaikh, A. G., & Optican, L. M. (2008). Applying saccade models to account for oscillations. In *Progress in brain research* (Vol. 171, pp. 123–130). Elsevier.

Ramful, A., Lowrie, T., & Logan, T. (2017). Measurement of spatial ability: Construction and validation of the spatial reasoning instrument for middle school students. *Journal of Psychoeducational Assessment, 35*(7), 709–727.

Resnick, I., & Shipley, T. F. (2013). Breaking new ground in the mind: An initial study of mental brittle transformation and mental rigid rotation in science experts. *Cognitive Processing, 14*(2), 143–152.

Richardson, D. C., & Richardson, J. S. (2002). Teaching molecular three-dimensional literacy. *Biochemistry and Molecular Biology Education, 30*(1), 21–26.

Sanchez, C. A., & Wiley, J. (2014). The role of dynamic spatial ability in geoscience text comprehension. *Learning and Instruction, 31,* 33–45.

Shepard, R. N., & Metzler, D. (1971). Mental rotation of three-dimensional objects. *Science, 171,* 701–703.

Small, M. Y., & Morton, M. E. (1983). Research in college science teaching: Spatial visualization training improves performance in organic chemistry. *Journal of College Science Teaching, 13*(1), 41–43.

Sorby, S. A. (2009). Educational research in developing 3-D spatial skills for engineering students. *International Journal of Science Education, 31*(3), 459–480.

Staver, J. R., & Jacks, T. (1988). The influence of cognitive reasoning level, cognitive restructuring ability, disembedding ability, working memory capacity, and prior knowledge on students' performance on balancing equations by inspection. *Journal of Research in Science Teaching, 25*(9), 763–775.

Titus, S., & Horsman, E. (2009). Characterizing and improving spatial visualization skills. *Journal of Geoscience Education, 57,* 242–254. https://doi.org/10.5408/1.3559671.

Tuckey, H., Selvaratnam, M., & Bradley, J. (1991). Identification and rectification of student difficulties concerning three-dimensional structures, rotation, and reflection. *Journal of Chemical Education, 68*(6), 460–464.

Uttal, D. H., Meadow, N. G., Tipton, E., Hand, L. L., Alden, A. R., Warren, C., & Newcombe, N. S. (2013). The malleability of spatial skills: A meta-analysis of training studies. *Psychological Bulletin, 139*(2), 352–402.

Wai, J., Lubinski, D., & Benbow, C. P. (2009). Spatial ability for STEM domains: Aligning over 50 years of cumulative psychological knowledge solidifies its importance. *Journal of Educational Psychology, 101,* 817–835, https://doi.org/10.1037/a0016127.

Karen S. McNeal, Ph.D. is the College of Science and Mathematics Molette Endowed Professor in the Department of Geosciences at Auburn University, Auburn, Alabama, USA. Her research is in the field of geoscience education and geocognition where she applies mixed methods research (e.g., qualitative and quantitative approaches) to investigate geoscience teaching and learning challenges in formal and informal settings. Her research group is interested in addressing questions that examine how to engage individuals in geoscience related materials, the misconceptions and mental models people hold about complex Earth systems, the psychomotor responses (e.g., eye movements, skin conductance) that occur when people are exposed to geoscience related materials and experiences, and the best practices in teaching and learning about the Earth. Her team conducts research in the broad areas of: (i) Earth systems science understanding, (ii) climate change understanding and communication, (iii) spatial thinking, (iv) active learning, (v) co-production of science, and (vi) diversity and inclusion in the geosciences. She has published over 60 peer-reviewed papers and received ~$25 in external funding from a variety of federal and state funding sources. She has served as Editor for Research for the Journal of Geoscience Education and has been the President of the Geoscience Education Research Division of the National Association of Geoscience Teachers (NAGT).

Rachel Atkins is a Ph.D. candidate at North Carolina State University in Raleigh, North Carolina, USA. Her research spans multiple disciplines including tectonic and fluvial geomorphology, planetary geology and geoscience education. Her solid Earth geology research investigates how planetary surfaces evolve over human and geologic timescales as a result of anthropogenic, global contraction and river processes. Her discipline-based education research experience has centered on investigating how students learn geoscience concepts in an effort to understand how to improve teaching practices. Some of these projects include employing eye-tracking to determine how novice students visually navigate climate change graphs and using the eye-tracking technology to determine how students explore and solve 3-Dimensional problems (e.g., block diagrams) in 2-Dimensional space (e.g., a piece of paper or computer screen). These skills have been shown to be potential barriers for retention of geoscience students. Additionally, she has been involved with projects investigating the effectiveness of using Augmented Reality Sandboxes (ARS) on student learning. Some of the eye tracking projects that Rachel has been involved with have been published in the Journal of Astronomy and Earth Science Education, Journal of Geography in Higher Education and Weather, Climate and Society. She has served as a reviewer for the Journal of Geoscience Education since 2017 and received a two-year NASA Future Investigator of Earth and Space Science and Technology grant in 2019 and a Southeast Climate Adaptation Science Center Global Change Fellowship in 2015. Rachel is also an active member of several professional organizations including the National Association of Geoscience Teachers (NAGT), Geological Society of America (GSA) and the Association for Women Geoscientists (AWG).

Elijah T. Johnson is a Ph.D. student at Auburn University in Auburn, Alabama USA, a National Science Foundation Graduate Research Fellow, and a geoscience education researcher. He investigates the role and development of spatial thinking in STEM and how the geosciences can promote and improve spatial learning across disciplines. His research employs augmented-reality technology to enhance the learning environment of introductory geoscience classrooms through training topographic map skills. One of these technologies is the augmented-reality sandbox, which he is using to create student-informed curricula that effectively teach topography and improve spatial thinking skills concurrently. He is also involved in ongoing research in the field of health geography where he is using GIS to map and analyze the spatial distribution of geology, groundwater quality, demography, and chronic illness in the Southern United States. Eli is an active member of several professional organizations including the National Association of Geoscience Teachers (NAGT) and the Geological Society of America (GSA).

Index

A
Adaptive expertise, 169–171, 177–179
Area under a graph, 244, 245, 247, 254–256
Attention, 1, 3–11, 14, 17, 21, 25, 33–38, 42–46, 73, 77, 78, 84, 85, 87, 94, 99, 102, 103, 109, 121, 134–136, 145, 146, 148, 149, 157, 158, 160, 166, 172, 173, 175, 177, 185, 191–193, 197–199, 201, 203–213, 219, 220, 224, 225, 245, 251–253, 256, 257, 261–263, 279, 283–285, 287, 288, 291, 292, 298, 299, 302, 304, 305
Attention allocation, 110, 115, 116, 121, 173, 220, 224
Authentic problems, 172, 179
Authentic tasks, 2, 107

B
Biological key, 156, 159, 164, 165
Burning, 133, 138, 139, 143, 144, 148
Butterfly, 158–166

C
Case, 2, 4, 11, 23, 36, 56, 63, 81, 82, 88, 113, 116, 117, 121, 122, 131, 160, 162, 171–179, 199, 203, 206, 211, 213, 219, 224, 249, 256, 257, 262, 267, 268
Case study, 12, 13, 15, 16, 24
Chemical reaction, 78, 80, 81, 93, 94, 102, 129, 130, 133, 136, 145, 146, 148
Chemistry triplet, 129, 130, 132, 136, 148, 149

Cognitive abilities, 1, 2, 4, 7, 12, 15–17, 23–25, 36, 264
Cognitive load, 9–11, 17, 22, 24, 109, 110, 119, 121, 122, 135, 145, 148, 149, 174, 199, 201, 203, 209, 211–214, 220, 221, 228, 235, 263, 264, 267, 268, 272
Content analysis, 97, 213
Context, 6, 33–35, 38–40, 46, 56, 108, 130, 148, 159, 172, 173, 177, 178, 185, 187, 188, 211, 244, 245, 247, 249–252, 254–257, 306
Context-based, 108–117, 119, 120, 122, 123, 129, 132, 136–138, 140, 142–149, 220–223, 225–236

D
Decision making patterns, 9
Didactics, 89, 159, 208, 272
Difficulty perception, 235
Dynamic SMR, 108, 111, 112, 115–119, 121–123, 226

E
Electrodermal activity, 55, 58
Experimental video, 95, 96, 101–103
Expert, 2, 9, 15, 23, 94, 103, 109, 111, 121, 138, 148, 169–171, 173, 175, 179, 188, 189, 191, 192, 194, 196–204, 206, 209–213, 219, 220, 223, 227, 266, 268, 271, 299, 305
Explanatory key, 73–75, 77–88, 134
Eye-tracker, 45, 96, 107, 113, 159, 220, 221, 223–226, 229, 235, 282, 298, 300

© Springer Nature Switzerland AG 2021
I. Devetak and S. A. Glažar (eds.), *Applying Bio-Measurements Methodologies in Science Education Research*,
https://doi.org/10.1007/978-3-030-71535-9

Eye-tracking, 1, 8, 9, 11, 12, 24, 75, 77, 79, 94, 95, 97, 101–103, 109–111, 114, 123, 129, 134, 138, 142, 149, 157, 163, 166, 171–173, 175, 177, 179, 185, 189, 199, 208, 209, 219–222, 224, 236, 245, 246, 248, 250, 251, 257, 279, 282, 284–286, 298–306

F
Fixation density, 16–19, 24, 25, 82–85
Fixations, 9–11, 16–20, 22–25, 42, 77, 79, 82–85, 87, 109, 113, 129, 134–136, 145, 146, 159, 160, 173, 175, 179, 185, 191, 193, 194, 198–205, 207, 208, 212, 213, 219, 220, 224, 248, 251, 261, 262, 265–272, 282–285, 288, 291, 301, 305, 306
Fluency, 14–16, 41, 42, 170

G
Geosciences, 296–299, 305, 306
Graph understanding, 299

H
Heart rate, 36, 55–60, 63, 65, 67, 272
Heat map, 10, 11, 16–19, 24, 77, 82–85, 142, 143, 148, 160, 192, 193, 283, 285, 287, 291, 303
High/low performers, 23, 95, 299, 302–306

I
Intelligence, 4, 7, 12, 15, 16, 135, 272
Interest, 2, 8–11, 17, 19, 24, 34, 39, 42, 57, 77, 79, 84, 86–88, 108, 109, 117, 121, 123, 131, 132, 134, 136, 138, 139, 146, 148, 160, 165, 172, 175–177, 191, 192, 194, 209, 211, 219–221, 223–226, 228, 236, 255, 256, 261, 283, 301, 305
Interview, 8, 15, 25, 74, 75, 79, 82, 87, 88, 96, 97, 99, 174, 175, 177, 209, 220, 266, 267, 270–272, 304

K
Knowledge, 1–4, 7, 8, 12, 14–16, 22, 24, 34, 45, 46, 73, 75, 79, 87, 107, 108, 110, 118, 123, 131, 132, 134, 138, 144, 148, 155, 158, 161, 162, 165, 169–179, 189–191, 206, 208, 210, 211, 218, 219, 236, 243–245, 272, 279, 280, 292, 299

L
Lag sequential analysis (LSA), 284, 285, 287, 291
Level of understanding (LU), 97–100, 102, 103
Life sciences, 169, 171–173, 177, 179

M
Macroscopic level, 2–4, 17–20, 93, 108, 115, 117–119, 122, 123, 130, 145, 149, 218, 226–228
Mathematics education, 244
Metacognitive, 9, 34, 41, 42
Motivation, 2, 33–40, 43–45, 132, 137, 138, 141, 142, 144, 148, 199, 206, 211, 212, 266, 271–273

N
Novice, 94, 103, 109, 121, 173, 174, 219, 220, 227, 299, 305

O
Observational skills, 157–159

P
Penetrative thinking, 296, 297, 299, 300, 304
Photo, 2–4, 17–21, 73, 110–112, 115–120, 123, 134, 138, 149, 157–166, 223, 226–228, 230–233
Physics, 23, 56, 71, 111, 119, 137, 141, 187–190, 196, 197, 206, 211, 213, 218–220, 244–257, 264, 266–268, 270, 272, 277–281
Pre-service teacher, 137, 158, 161, 164, 166, 222
Problem solving, 3, 4, 7–12, 17–21, 23–25, 42, 109–111, 175, 185, 187, 213, 223, 236, 244, 245, 296, 297, 299, 305
Problem-solving process, 8, 10, 22, 24, 25, 187, 211
Problem-solving strategies, 17, 25, 140, 149, 224
Processing information, 43, 134, 263, 264
Psychophysiological analysis, 61
Pupil diameter, 109, 135, 200, 201, 203, 204, 212, 213, 221, 228, 236, 263–271
Pupillometry, 134, 136, 263

R

Redox reactions, 94–96, 98–103
Respiration rate, 56, 59, 60, 63–65, 67

S

Saccades, 9, 10, 17, 20, 21, 24, 42, 109, 113, 134, 160, 220, 224, 248, 262, 301
Self-concept, 34, 39–42
Self-determination, 38, 39, 141
Self-efficacy, 38, 42, 141
Self-regulation, 38, 45, 178
Simulation, 277–285, 287, 291, 292
Skin temperature (ST), 55, 56, 58, 60, 62, 64, 65, 67
Slope concept, 253, 255
Spatial thinking, 4, 295–299, 306
Students' achievements, 111, 132, 136, 144, 148
Students' mental abilities, 137
Sublimation, 110–114, 116, 117, 119, 120
Submicroscopic level, 2–4, 6, 17, 18, 20, 22, 25, 93, 96, 102, 108, 112, 115, 117–119, 122, 123, 132, 137, 144, 145, 149, 218, 221, 228, 229
Submicroscopic representations (SMRs), 2, 18, 72–75, 77–82, 87–89, 96, 107–110, 118, 119, 121, 122, 131, 132, 134, 218, 221, 223, 226, 228, 229, 234–236

T

Test of logical thinking (TOLT), 14–16, 132, 138, 140, 142, 222, 224, 225, 229, 233–235
Textbook, 72–78, 80–82, 88, 89, 109, 174, 229
Think-aloud, 8, 94, 96, 97, 99, 101, 225, 292, 304

V

Visual attention, 3, 9, 10, 33, 42–46, 77, 109, 121, 148, 160, 166, 173, 175, 185, 191–193, 197–199, 201, 203–207, 209–213, 219, 220, 224, 245, 251–253, 255–257, 262, 279, 283–285, 287, 288, 291, 292, 298, 299, 305

W

Working memory, 4–6, 10, 12, 13, 16, 22, 42, 108, 123, 131, 135–138, 141, 142, 144, 148